Hydraulic Behaviour of Estuaries

LIBRARIES
DRAWN FROM

Civil Engineering Hydraulics Series

General Editor: E. M. Wilson

Professor of Hydraulic Engineering, University of Salford

Hydraulic Behaviour of Estuaries

D. M. McDowell and B. A. O'Connor

Simon Engineering Laboratories, University of Manchester

First published 1977 by
THE MACMILLAN PRESS LTD
London and Basingstoke
Associated companies in New York Dublin
Melbourne Johannesburg and Madras

ISBN 0 333 12231 3

Printed in Great Britain at the
University Printing House, Cambridge

Contents

PREFACE vii

1. A GENERAL DESCRIPTION OF ESTUARINE BEHAVIOUR 1
2. HYDRODYNAMICS OF ESTUARIES 30
3. MIXING PROCESSES 48
4. SEDIMENT MOVEMENTS 83
5. THE STUDY OF TIDAL SYSTEMS: FIELD MEASUREMENTS 124
6. THE STUDY OF TIDAL SYSTEMS: MATHEMATICAL TIDAL MODELS 146
7. THE STUDY OF TIDAL SYSTEMS: WATER QUALITY MODELS 177
8. THE STUDY OF TIDAL SYSTEMS: HYDRAULIC MODELS 197
9. CONTROL OF ESTUARIES 225
10. DISCUSSION OF CASE HISTORIES 250

INDEX 279

Preface

Estuaries are meeting places—of salt water and fresh water, of fresh-water and salt-water flora and fauna, of land-borne and sea-borne sediments, of seafarers. This complexity accounts for much of their fascination for people like the authors of this book, but it makes it difficult to study estuaries and to describe them in a single introductory volume. We have confined the book to the flow of water, salts and water-borne solids; that is, to the rather narrow view of physical behaviour that excludes flora and fauna. We realise that plant and animal life have an essential role in the physical behaviour of most estuaries, in colonising mud flats, in changing the character of fine sediments and, in warm climates, in determining the topography by growth of coral and cementation of shell debris. We have concentrated on the movement of water and its immediate effects. Our purpose has been to expose the mechanics of estuarine systems and to indicate the principal methods used to control them.

There have been many descriptions of estuarine behaviour, but all, including this book, have been limited by the power of words and the poverty of our scientific formulations. In most of them, an 'if only...' approach has been used. The estuaries have been straightened and their cross-sections deformed into neat shapes while the water has been homogenised or, as a concession to reality, stratified in an attempt to make the physical processes manageable. An optimistic view has been taken of such complexities as sediment transport, use being made of equations that do not work at all well even for the unidirectional flow situations for which they were designed.

In this book, we have tried to give an honest account of how real estuaries behave and to show how they might be managed. Underlying the whole book is a description of the interaction of the many physical factors. Much of this can be read and understood without any knowledge of advanced mathematics. Inevitably some chapters are based on mathematical equations. Summaries of these have been provided so that the non-mathematician can see what can be achieved by mathematical tools.

Chapters 1, 5, 9 and 10 give an account of estuarine behaviour; of the care that must be taken in trying to measure it quantitatively; of methods of estuarine management by engineering work; and of the response of several actual estuaries to control work. These can all be read without much specialised knowledge. The other chapters are concerned with mathematical formulation of the behaviour of real estuaries and with aids to solution of real problems using physical and mathematical modelling techniques. We have tried to describe the physical basis of each technique and to demonstrate its use and limitations in dealing with real situations.

We consider that, in an introductory book such as this, it is more important to present a global view of estuarine situations than to explore techniques or specific items of estuarine behaviour in great depth. We also believe that a sound understanding of estuarine physics is a necessary prerequisite for all concerned with their management and control.

There is nothing more frightening to a beginner than an exhaustive reference list and bibliography. The selection that we have made should open the way to the tremendous amount of published work that is now available. Exclusion or inclusion of your favourite book should not be construed as our judgement of its merit!

We are indebted to many people, publishing houses and institutions whose work we have quoted and gratefully acknowledge. We have included much that has not been previously published, mostly of work done by us or with which we have been closely concerned. We would particularly like to thank the following organisations for permission to reproduce their material: Department of Civil Engineering, University of Liverpool and the Mersey Docks and Harbour Company for data concerning the Mersey Estuary; British Transport Docks Board for data concerning the Severn and Humber estuaries; the Chairman and Chief Hydraulic Engineer of Calcutta Port Trust for data concerning the River Hooghly; the Hydraulics Research Station, Wallingford and the Water Research Centre, whose work is acknowledged under the all-embracing title of Her Majesty's Stationery Office.

Manchester, 1977 D. M. McD.
B. A. O'C.

1

A general description of estuarine behaviour

The hydrological cycle begins and ends in the sea. The first stage in the cycle is evaporation of water from the surface of the sea and the last stage is the return of water to the sea through rivers. Most rivers enter the sea where there is enough tidal rise and fall to modify flow near their mouths. The part of the river system in which the river widens under the influence of tidal action is the estuary, but the behaviour of the estuary is influenced by the circulation of water and solids in the sea as well as by the whole tidal part of the river system. The estuary cannot be considered in isolation; the whole system has many inter-dependent parts, extending from the landward limit of the tidal rivers forming it to a point offshore beyond which the effect of an individual estuary on water circulation and sediment movement can no longer be discerned.

Much detailed work has been done on the behaviour of estuaries and some of this will be described in later chapters. Before any particular study can be begun, however, it is essential to understand the mechanism of the estuary; the way in which the individual parts of the system react with each other and how each affects, or is affected by, the others. Movement of water under the action of tides and river flow is closely inter-related with movement of sediment; both are affected by wave action and tidal currents in the sea outside the estuary. It is convenient to consider the various components separately, but it must never be forgotten that they cannot behave independently. For this reason, this first chapter is devoted to a general discussion of the behaviour of typical estuaries.

Estuaries are governed by tidal action at the sea face and by river flow. These are the main independent variables. The boundary shape of the estuarine system is determined by the geomorphology of the land and the properties of all alluvial materials that form the bed and banks of the channels. Usually, the overall boundary shape changes only slowly, though there may be rapid local or short term adjustments. In some cases, tectonic movements have caused changes but these are not considered here. Gradual changes take place due to accumulation and re-distribution of river-borne solids, but their importance varies very greatly in different estuarine systems. The geomorphology of an estuary basin is, essentially, a fixed boundary condition but the channels as modified by flow can be regarded as a variable boundary. Sea-borne sediment stirred up by tidal currents and wave action can enter an estuarine system from beyond any immediate zone influenced by the estuary. Where this happens, the influx of sea-borne sediment becomes another independent variable that must be reckoned with in any analysis.

1.1 TIDES AND MEAN TIDAL CURRENTS

1.1.1 Tides at coastal sites

Tides in the sea result from the gravitational pull of the moon, the sun and the planets and from local meteorological disturbances. The effect of varying gravitational pull can be predicted with quite high accuracy. The meteorological effects are random in their occurrence and, apart from some general seasonal trends, can only be predicted a short time in advance. Rise and fall of sea-level is essentially independent of conditions within an estuary, the only exception being that very large discharges of fresh water may occasionally result in a slight increase in water level for a few kilometres inland of the mouth of an estuary.

The moon orbits the Earth once in 28 days and the whole system orbits round the sun once in 365·2 days. The paths of the moon round the Earth and the Earth round the sun are both elliptical, so the gravitational force of attraction passes through a maximum and a minimum during each orbit. The axis of the Earth is inclined to the plane of its orbit round the sun, and the orbital plane of revolution of the moon around the Earth is also inclined at an angle to the Earth's axis. Consequently, the gravitational tide-producing force at a given point on the earth varies in a complex but predictable manner. The largest component of the force is due to the moon, and has a period of about 12 h 25 min. The lunar force itself reaches a maximum once in 28 days, when the moon is nearest to the Earth, or at perigee. When the moon is at apogee, furthest from the Earth, the lunar tidal force is only about $\frac{2}{3}$ of its maximum value. The total tidal force due to the combined action of sun and moon is greatest when they act together; that is, when sun and moon are as nearly in line with the Earth as possible. This occurs twice a month, when the moon is on the same side of the Earth as the sun and when they are on opposite sides, i.e. at new moon and full moon. When this happens, *spring tides* occur, having a range of movement greater than average. When sun and moon are in quadrature with the Earth, their effects give rise to smaller-than average *neap tides* which also occur twice a month.

It may not be immediately obvious that two tides will occur during each rotation of the Earth, and that spring tides will occur when the forces due to the sun and the moon appear to be in opposition to each other. The explanation follows from the law of gravity: that the gravitational force varies inversely as the square of the distance between two objects. Consider the Earth and the moon as two bodies moving around a common centre. They are kept in orbit by gravitational pull, which is just equal to the centrifugal force due to their rotation. The Earth, however, is large enough for the forces

Table 1.1

Name of component	Symbol	Period: (solar hours)	Amplitude ratio ($M_2 = 100$)	
Principal lunar	M_2	12·42	100	semi-diurnal
Principal solar	S_2	12·00	46·6	
Larger lunar elliptic	N_2	12·66	19·2	
Luni-solar semi-diurnal	K_2	11·97	12·7	
Luni-solar diurnal	K_1	23.93	58·4	diurnal
Principal lunar diurnal	O_1	25·82	41·5	
Principal solar diurnal	P_1	24·07	19·4	

* See Defant, A., *Physical Oceanography*, Vol. 2, Ch. VIII (1961).

to vary appreciably along its diameter. At its centre the forces are equal. At the point nearest the moon gravitational pull exceeds the centrifugal force. There is a resultant force tending to move water towards the moon on the side of the Earth facing the moon. On the side of the Earth remote from the moon, gravitational pull is less than the centrifugal force. The centrifugal force is dominant on that side and water tends to move towards the point furthest from the moon.

As a result of the inclination of the plane of rotation of the moon round the Earth, there are times when sun and moon are more nearly in line with the Earth than at other times. They occur twice a year, usually near the spring and autumn equinoxes, but the exact timing varies from year to year. These are the periods of greatest tidal range: during spring tides, and of least tidal activity: during the adjacent spells of neap tides. The whole progression of spring–neap tides and equinoctial variations is approximately repeated every 19 years. A full description of the tide-generating force may be found in texts on tides and oceanography, e.g., Macmillan[1], Neumann and Pierson[2], Defant[3].

The tide-generating force can be expressed as a series of harmonic components. The periods and amplitudes of some of the principal harmonic components are given in table 1.1 (after Doodson[4]). These components account for about 83% of the total tide-generating force.

Estuarial engineers are not concerned directly with the tide-generating forces themselves, but with their effects on rise and fall of sea-level close to the mouths of estuaries. The rise and fall of water in the oceans is propagated as a shallow-water wave; shallow because its length is large compared with the depth of water. This long wave has a very small amplitude in the deep oceans, usually of order 1 m. Because of the great length of the wave, however, this tidal wave is accompanied by movement of a huge mass of water. The oscillatory behaviour of this mass is conditioned by the depth and shape of the ocean basins and by the fact that motion takes place on a rotating spheroid. If a natural period of oscillation of water in an ocean basin is close to that of one of the components of the tide-producing force, oscillations of that period will build up. In many parts of the world the semi-diurnal tide becomes

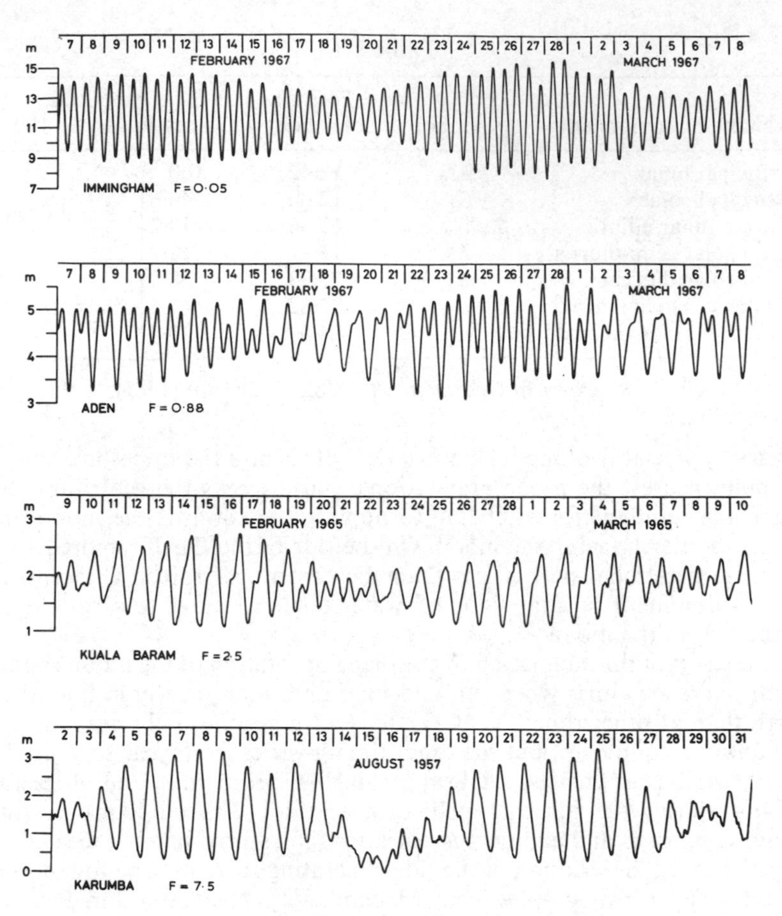

Figure 1.1. Tidal records at four ports (courtesy the Institute of Oceanographic Sciences, Bidston, UK)

dominant, while in others the diurnal component becomes amplified so much that it becomes the dominant effect, even though the diurnal components of tidal force are considerably less than the semi-diurnal components.

At tidal observatories and ports, local effects due to the shape of the ocean basins and the shallow water near the coast determine the tidal behaviour. It is not possible to calculate these local effects from the tide-generating forces. Instead, local tidal records can be analysed into components having the same periods as the harmonic constituents of the tide-generating force. Estimation of the amplitudes of these components and their phase relationships enables tidal predictions to be made, on the assumption that each component will vary with time in the same way as the corresponding component of the tide-generating force[4, 5]. Tidal predictions based on such harmonic analyses are published for all major and many minor ports. The importance of semi-diurnal and diurnal components at any locality can be estimated from the ratio

$$F = \frac{K_1+O_1}{M_2+S_2}$$

of the major local harmonic constituents. Semi-diurnal influences increase as this ratio decreases, and are dominant when it is less than unity. In the tide generating force without shallow-water effects, these four components amount to 70% of the total force, and the ratio $F = 0{\cdot}68$, showing that the tide generating force has a dominant semi-diurnal variation with a significant diurnal effect.

It is instructive to compare tides at places at which values of the ratio F differ. Defant[3] has reproduced predicted tidal curves for four ports at which F varies from 0·11 (semi-diurnal) to 18·9 (diurnal). The curves correspond to the same dates in the year 1936. Actual tides differ from predictions because of meteorological effects or variations in river flow. In the case of small amplitude tides, these differences can be of the same order as the predicted tidal range. Figure 1.1 shows actual recorded tides at four ports. In addition to showing typical tide curves for various values of F, it shows the effect of meteorological influences. The most obvious is a lowering of the mean tidal level by up to 1 m at Karumba between 14 and 19 August 1957. Meteorological records taken at the time show that there was a strong offshore wind blowing. Karumba is at the south-east corner of the Gulf of Carpentaria in northern Australia. It seems likely that the low tide level was caused by wind stress on the large, shallow surface of the Gulf. The record from Immingham also shows lowering of the same order, e.g. on 27 February 1967, but it is less noticeable because of the greater tidal range there.

1.1.2 Tides in estuaries

Tidal rise and fall of the water surface at the entrance of an estuary causes surface gradients which result in the propagation of a gravity wave into the estuary. The rate of propagation depends primarily on the depth of water and, in consequence, on the tidal range at the mouth. It is usual for waves to move at a different speed from that of the fluid through which they move; typical tidal currents flow at speeds up to 2·5 m/s with, rarely, local speeds up to 5 m/s, whereas the tidal wave travels at celerity $c_0 \approx (gH)^{\frac{1}{2}}$ relative to the water, where H is the water depth. The celerity is nearly 10 m/s through water 10 m deep. The consequences of this are vital to an understanding of the behaviour of estuaries. The tidal wave travels more slowly as the depth decreases and, consequently, the wave form becomes distorted as it travels inland. The distance travelled by the crest of the wave in a given time exceeds that travelled by the trough, so the further the wave has to travel in shallow water, the more distorted it becomes. A semi-diurnal tide in the sea has a period averaging about 12 h 25 min and the mean time of rise roughly equals the mean time of fall. Within an estuary, the time of rise from low to high water decreases at points further inland and it is quite common for the time to vary from over 6 hours at the seaward end to less than 3 hours near the upstream tidal limit. Because the whole process repeats 12 h 25 min later, the ebbing

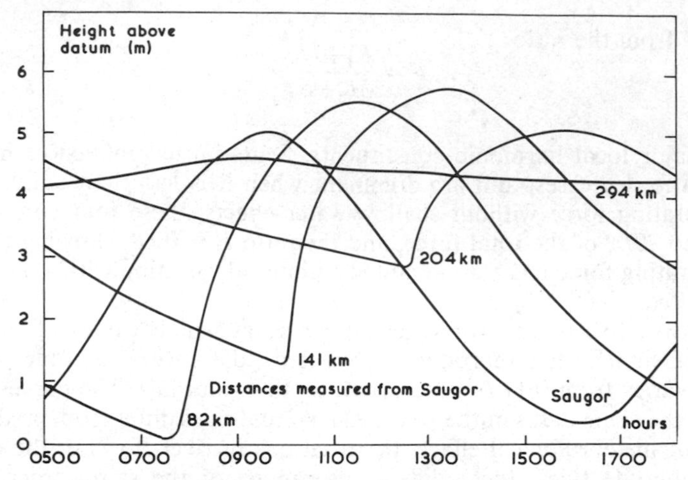

Figure 1.2. Spring tide recorded in the River Hooghly

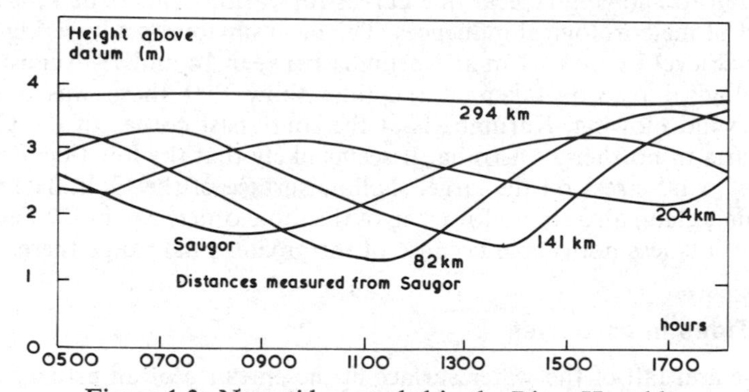

Figure 1.3. Neap tide recorded in the River Hooghly

tide must occupy the remaining time, which can be over 9½ hours near the tidal limit.

During the falling or ebb tide, surface gradients are set up at the seaward end that are initially similar in magnitude to those occurring during the rising or flood tide, but opposite in direction. As water level falls, however, the rate of wave propagation decreases and the influence of the rate of fall of the sea becomes weaker. Instead, the ebb flow within an estuary becomes a process of drainage under gravity and, where there are extensive areas of shallow water, there is a prolonged period of flow in the main deep channels, fed by surface drainage from the shoals on either side.

Tidal behaviour in an estuary can be observed by recording the rise and fall of water simultaneously at several stations spread along its length. Some typical observations are reproduced in figures 1.2–1.4. The River Hooghly in India (figures 1.2 and 1.3) has a moderate tidal range (5·5 m during spring tides) and a considerable tidal length, with the result that when high water occurs at the mouth, the previous high water has just reached the tidal limit.

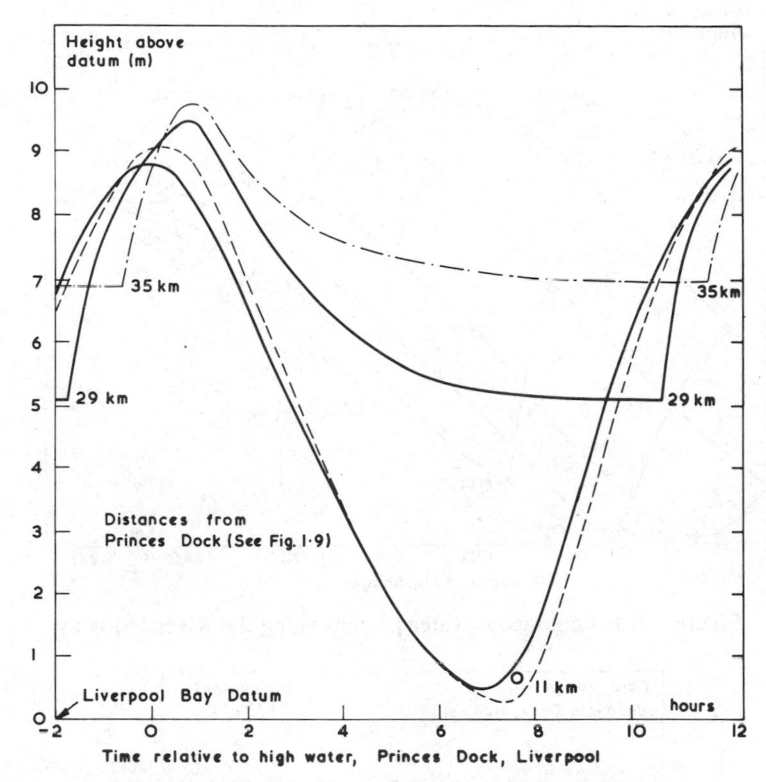

Figure 1.4. Spring tide recorded in the Mersey Estuary

The Hooghly shown in plan in figure 10.13, is tidal for a length of about 300 km. In the Amazon, which is tidal for over 850 km, seven or eight successive high tides can occur simultaneously (Defant [3]).

The Hooghly, with a tidal range of about 5·5 m during springs, and a channel depth below low water that varies considerably, but averages about 6 m over long distances, has a variation of depth between 6 and 12 m during spring tides, and this causes the considerable distortion of the tidal wave shown in figure 1.2. During neap tides, the tidal range can be about 2 m, and the low water level is over 1 m above the spring tide low water. Depths then vary between 7 m and 9 m, and as a result of this relatively small variation of depth, the tidal wave shown in figure 1.3 undergoes only slight distortion. The effect of distortion can also be seen in figure 1.5, in which the tidal observations shown in figure 1.2 are plotted to show simultaneous water surface profiles along the estuary.

A further effect of tidal propagation deserves comment. The mean tide level at the sea-face at Saugor shown in figures 1.2 and 1.3 is about the same in each case, but at stations further inland the mean level during spring tides is higher than that observed during neaps. This effect can be so great that at some landward stations, low water during neap tides can be lower than low

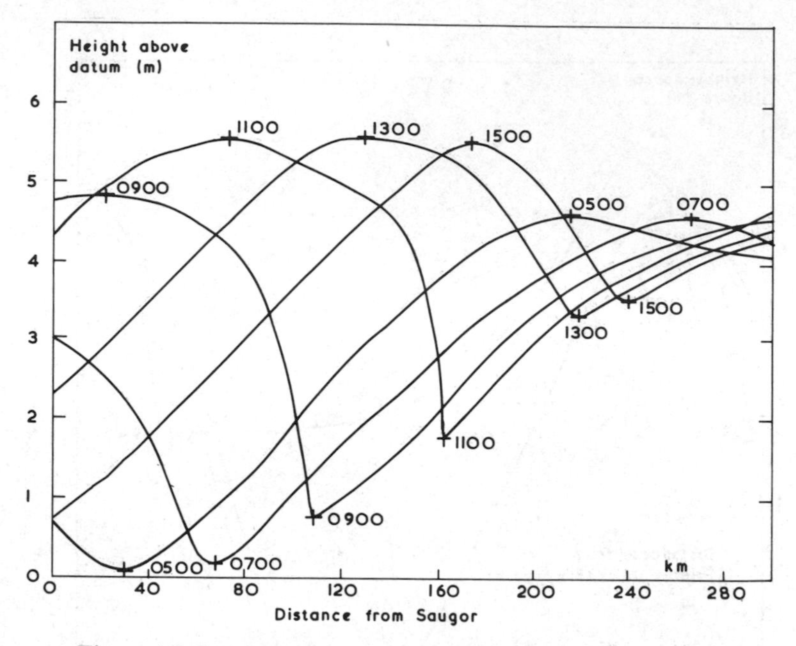

Figure 1.5. Instantaneous water profiles along the River Hooghly

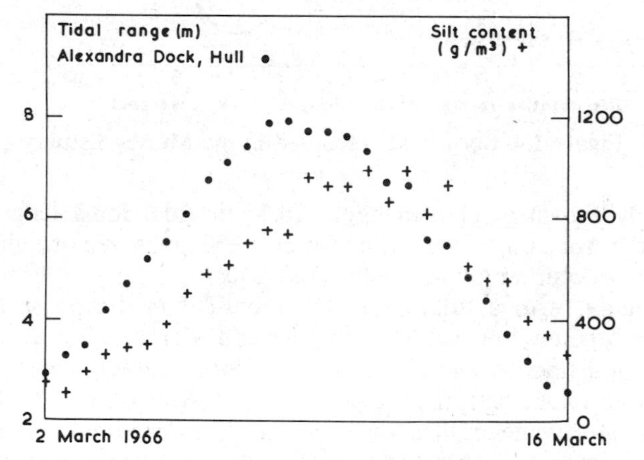

Figure 1.6. Silt content at successive high tides – Humber Estuary

water during springs. For example, at 204 km from Saugor, low water, spring tide (LWS) is 2·8 m above datum, whereas low water, neap tide (LWN) is only 2·2 m above datum. Evidently a large volume of water can accumulate in the upper part of an estuary during a spring tide cycle. This causes an increase of salinity during spring tides and a decrease during neaps. It also contributes to net landward movement of sediment during spring tides and net seaward movement during neaps, with the peak of sediment concentration occurring later than the highest spring tides. This latter effect is shown in figures 1.6 and

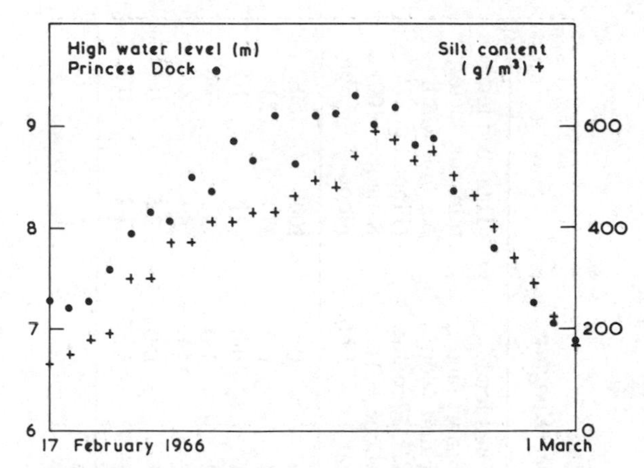

Figure 1.7. Silt content at successive high tides – Mersey Estuary

1.7 which are based on observations in the rivers Humber and Mersey in England.

The tide curves from the Mersey shown in figure 1.4 are typical of an estuary having a large tidal range compared with the depth of water and extensive areas of inter-tidal shoals, in this case between 11 km and 29 km landwards from the mouth (figure 1.9). Between 0 and 11 km landwards, the mid-channel depths decrease from about 19 m to about 4 m below datum. Between 11 km and 29 km landwards, the flow near low tide level is confined to relatively narrow channels between the shoals. The initial rapid fall of water followed by slow drainage from the shoals is shown by the curves of water surface level at 29 km and 35 km from the mouth.

Nevertheless the mean bed level, taking into account these local increases and decreases of depth, usually changes only gradually along the axis of a tidal river. If the bed and banks can freely erode, such estuaries widen exponentially towards their mouths. This erosion can produce very great widths at the seaward end. Table 1.2 shows the variation in average width of the Hooghly. Although the rate of sediment transport per unit width is usually greatest in the middle reaches of an estuary, the total rate of sediment transport may be greatest near the sea where widths become very large. Some calculated values are given in table 1.2. The figures for sediment transport factors shown in this table are based on the simplified assumption that the rate of sediment transport per unit width is proportional to the fifth power of the flow speed. They are included for comparative purposes only. These illustrate that the largest rates of sediment transport, in the absence of river flow, are landwards and that total quantities in movement are greatest near the mouth, whereas the greatest rates of transport per unit width occur in the middle reaches. The dominant rôle of flood tide sediment transport over that of the ebb is offset to some extent by the short duration of flood tide compared with the ebb in many estuaries, as illustrated in figure 1.8 which shows tide curves for sites on the River Hooghly. This may be compared with figure 6.2 which shows similar curves for the River Thames, in which the differences

Table 1.2. *Typical spring tide quantities calculated for the River Hooghly*

Distance from sea at Saugor (km)	Average width (km)	Max. rates of flow (m^3/s)		Max. cross-sectional mean speed U (m/s)		Bed load transport factor, per unit width ($\propto U^5$. Flood at Saugor = 1)		Total bed load transport factor, per unit width (Flood at Saugor = 100)	
		Flood	Ebb	Flood	Ebb	Flood	Ebb	Flood	Ebb
0	21	$2{\cdot}6 \times 10^5$	$1{\cdot}09 \times 10^5$	1·5	1·3	1·0	0·5	100	50
30	6·8	1·4	0·7	2·3	1·4	8·4	0·7	270	23
51	5·0	$6{\cdot}7 \times 10^4$	$3{\cdot}7 \times 10^4$	1·6	1·2	1·3	0·3	31	7·1
77	1·6	3·8	1·4	2·0	1·5	4·2	1·0	32	7·6
108	1·0	2·1	1·1	2·4	1·5	10·5	1·0	50	4·8
129	0·7	1·2	0·8	1·4	0·8	0·7	0·04	2·5	0·14
134	0·5	$7{\cdot}7 \times 10^3$	$7{\cdot}4 \times 10^3$	1·3	1·1	0·5	0·13	1·1	0·3

(No river flow. Last four columns are comparative figures referred to conditions at Saugor.)

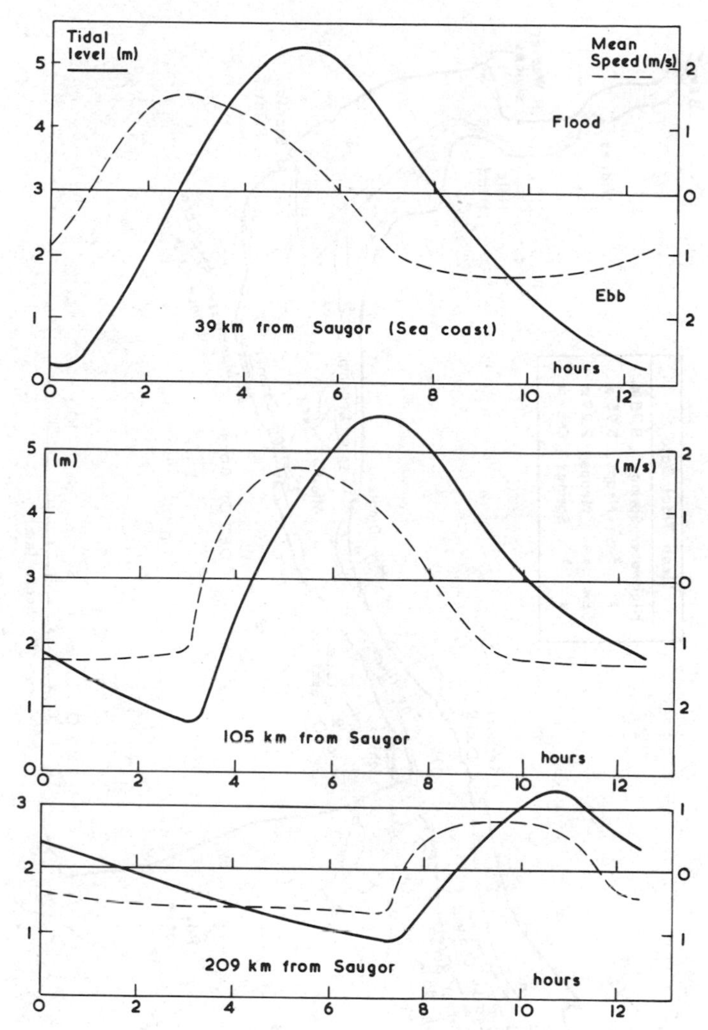

Figure 1.8. Tidal levels and computed cross-sectional mean speeds of flow in the River Hooghly

between flood and ebb durations and peak velocities are much less. The overall trend of tidal action in the absence of river flow is to transport sediment landwards.

There are many estuaries that cannot develop their channels freely because of geological formations. One example is the Mersey, which discharges to the sea through a comparatively narrow, deep channel, shown in figure 1.9. Landwards of this channel is a wide, shallow basin in which channels can develop freely in alluvium. One consequence of the constriction near the mouth is that the highest flow-speeds occur there, so the bed is scoured down to rock and gravel.

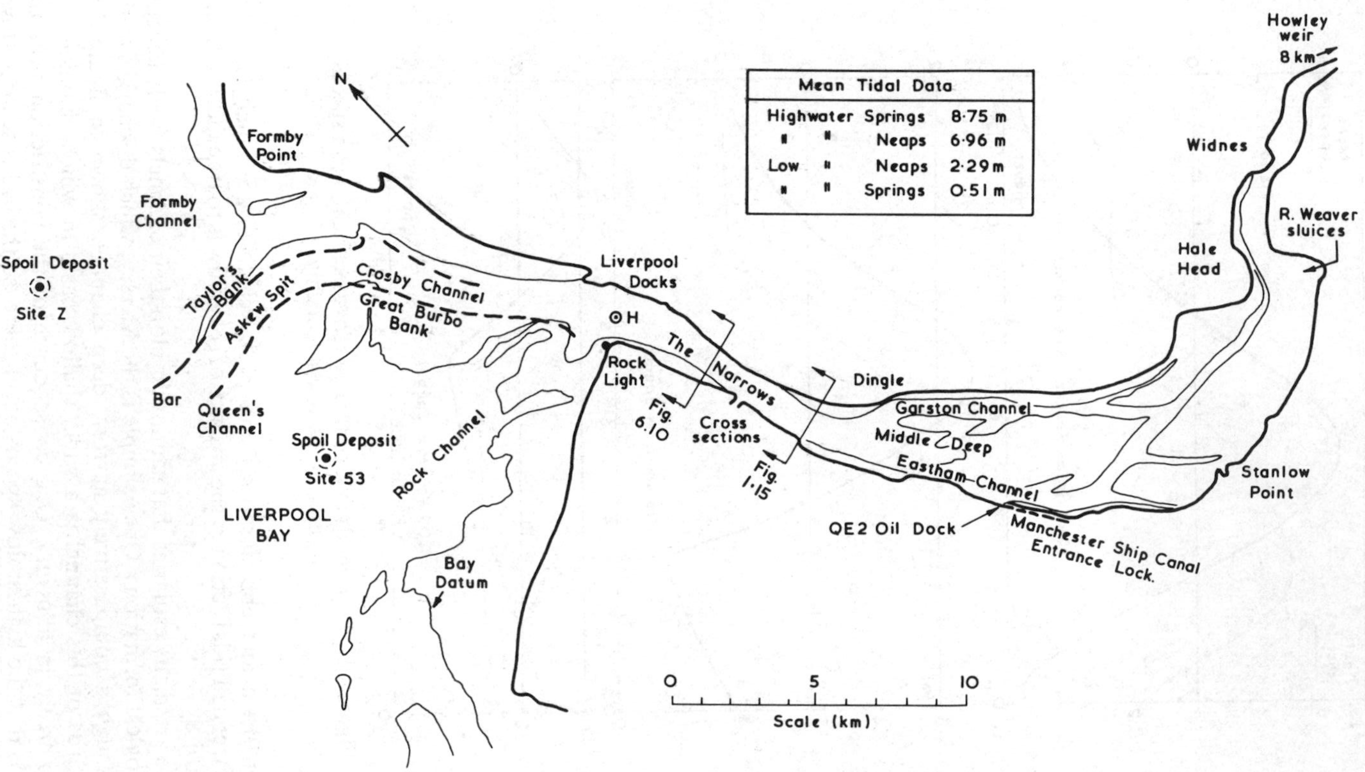

Figure 1.9. Mersey Estuary and part of Liverpool Bay

1.2 RIVER FLOW AND ITS EFFECTS

1.2.1 The flux of fresh water

So far, the description of estuarine behaviour has been concerned with the effects of tidal rise and fall. Some estuaries undergo prolonged periods with negligible fresh water flow and others have lost all contact with the rivers that formed them. But at some time in their existence all estuaries were formed by the combined action of tides and river flow.

The equilibrium of an estuary can only be maintained if the quantities of solids, fresh water and minerals in solution each remain in 'balance'. This requirement of continuity of matter is simple in principle, yet it has a most far-reaching effect on the behaviour of estuaries.

Fresh water entering an estuary must leave at the same rate, averaged over a period of several weeks, if the system is in equilibrium. Rainfall, evaporation and percolation all take part in the process but seldom contribute significantly to the balance of flow of fresh water, except during times of very low river flow. Water leaving an estuary eventually mingles with saline water until it can no longer be distinguished from it; but this process is inevitably a gradual one and may still be detected many kilometres offshore of even quite small rivers. The density of coastal waters is noticeably lower than that of the oceans; for example, in the Irish Sea, measured densities of surface waters 30 km from Liverpool are between 1022 and 1025 kg/m^3 compared with densities in the Atlantic Ocean between 1026·5 and 1027 at the latitude of the British Isles. In the case of the Amazon, densities do not approach typical ocean values for as far as 1000 km from its mouth.

The inevitable movement of fresh water away from an estuary mouth into the sea is accompanied by movement of saline water entrained with it. This saline water must be replaced if equilibrium is to be maintained. In such a situation, the exact quantity of salts entrained with fresh water, and removed from a given region in unit time, must be replaced by an equal influx of water and dissolved salts. Because there is a small increase of density with salinity at a particular temperature, the fresh water moves on the surface away from the estuary mouth whereas the saline water moves towards it near the sea bed.

1.2.2 Density gradients

Within an estuary, the effect of density gradients is considerable. It has already been shown that there is a tendency for net landward movement of sediment to occur over the middle reaches because flood tidal velocities are stronger than those of the ebb. Superimposed on this effect, the density difference between the water at the seaward end of an estuary and water entering from rivers causes net landward movement of water near the estuary bed, and a compensating seaward movement near the water surface. This can cause fine sediments to be carried landwards in suspension to a point of zero net movement, which is near the landward limit of density gradients (figure 1.10). Water is predominantly fresh upstream of this point. When river flow is high,

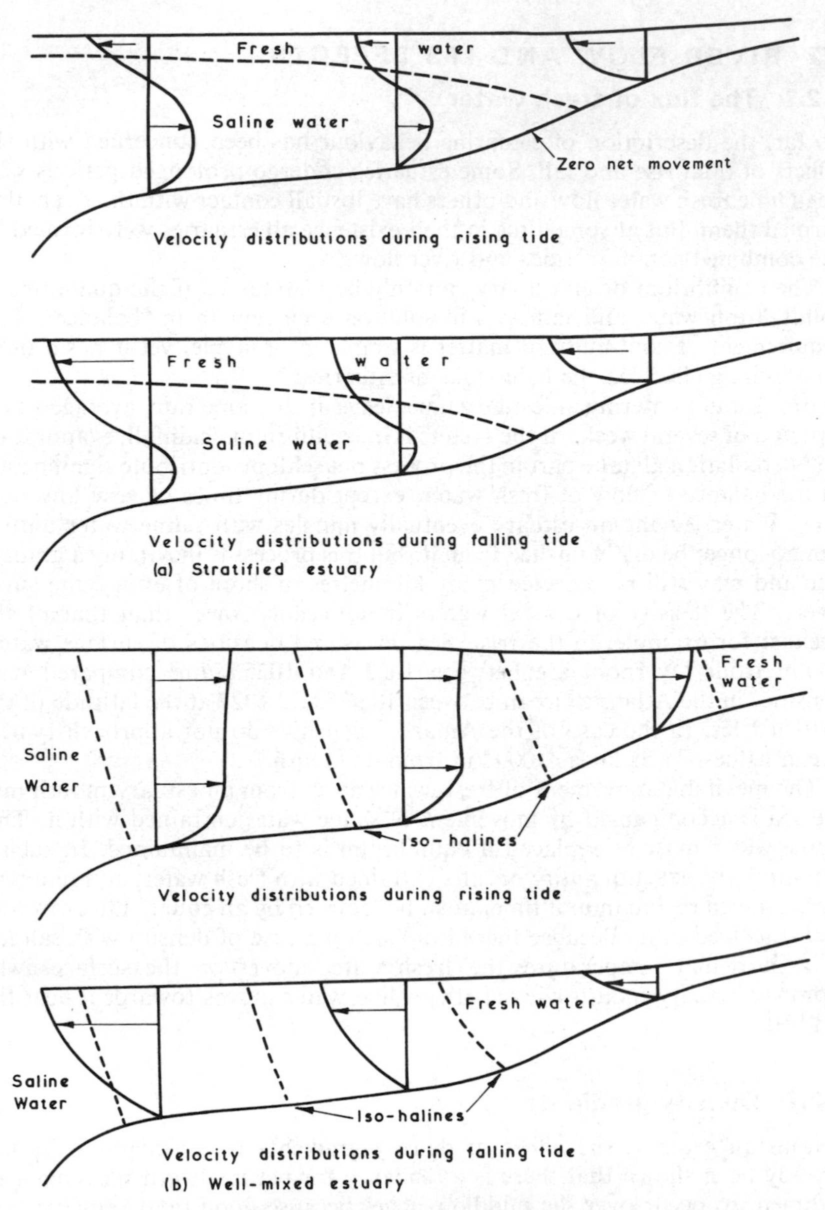

Figure 1.10. Distributions of velocity over depth in typical stratified and well-mixed estuaries

this position of zero net movement is moved seawards and conversely, when the flow is small it moves landwards.

The effect of density gradients on flow in an estuary depends on the level of turbulence that occurs. If turbulent mixing is so intense that there is only a small difference in density over the depth at any point, there must be a horizontal gradient of density ranging from 1000 kg/m^3 at the upper tidal limit to the density of sea water ($\sim$ 1026 kg/m^3) at some distance offshore. It is shown in Chapter 2 that the horizontal force due to such a density gradient increases with depth below the surface and is always directed in the direction of decreasing density. It therefore gives rise to a small landward force, zero at the surface and reaching a maximum at the bed.

The magnitude of the horizontal force due to density gradients is directly proportional to the magnitude of the horizontal density gradient. Water having a density increasing with depth, but having no horizontal density gradient will be completely stable. This state can only occur, however, in water at rest. Spread of salinity throughout the water mass can then only take place very slowly by molecular diffusion. As soon as flow takes place, there must be mixing of fresh and saline water at a rate dependent on the turbulent energy. This is discussed more fully in Chapter 3. At low rates of flow, mixing between fresh river water and sea water only occurs at a narrow interface between the two layers. Almost complete separation can then exist between the fresh and saline water. Nevertheless, at the interface between them some mixing will occur. To maintain equilibrium, this salt would have to be replaced from the only possible source and there would therefore be landward movement of saline water in the lower layers.

The situation can be summarised using the principle of continuity or conservation of matter. If an equilibrium state is to be maintained when there is steady flow of fresh water into a channel containing salt water, the net flow through any cross-section must equal the rate of inflow of fresh water. Thus, if there is a rate of flow Q_f into the channel and if, as a result, there is a landward flow of Q_s in any part of a cross-section there must be a seaward flow of $Q_f + Q_s$ in the remainder of the cross-section. Similarly with regard to salt; for equilibrium to be maintained, the net flow of salt through any cross-section must be zero. Thus, if a mass M_s per second of salt is entrained in fresh water and carried seawards, an equal mass must be moved landwards in the saline layer.

These principles show that entrainment of salt over a distance in a river must result in landward flow of salt in the lower layer, gradually decreasing in quantity as it penetrates inland, as in figure 1.11.

This effect will also occur when there is some to and fro movement due to tides. If energy losses are small and if turbulence is not so intense that the layers become rapidly intermixed, the discharges due to entrainment of salt can then be added to the tidal discharges.

Between the two extremes of well-mixed flow and stratified flow there can exist any degree of mixing. Horizontal gradients of density are fundamental components of all estuaries and their presence always results in a component of force in the direction of decreasing density, the force being greatest at the bed.

Stratified flow occurs in estuaries that have weak tidal action; that is, estuaries with small tidal range and/or steeply sloping beds, and, in consequence, small tidal storage volume. It can occur during neap tides in estuaries that would otherwise be partially mixed. In the Mississippi, for example,

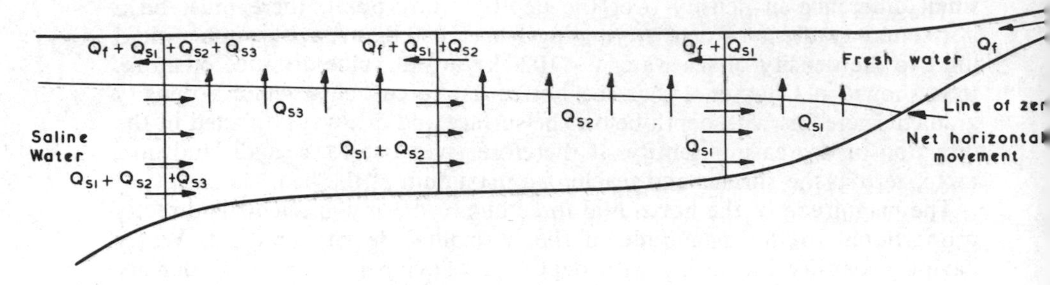

Fig. 1.11. Effect of entrainment on rates of flow in a stratified estuary

during dry weather flow, saline water has been detected at the bed 218 km upstream of the mouth of the Southwest Pass[6]. The Mississippi is tidal for up to 426 km from its mouth, where the tidal range during spring tides is about 0·6 m.

Conversely, well-mixed tides occur in estuaries having high tidal activity, such as the Mersey and Hooghly. The Mersey is tidal over a length of about 50 km and has a tidal range of up to 10 m at its mouth, whereas the Hooghly is tidal for almost 300 km and has a normal tidal range of up to 5 m at its mouth.

There have been several attempts to classify estuaries according to the degree of their stratification. Ippen[7] has suggested a stratification number G/J where G is the rate of turbulent energy dissipation per unit mass of fluid, and J is the rate of gain of potential energy per unit mass due to the increase in salinity of water as it flows through the estuary. A large value of G/J (100 or more) indicates a well-mixed estuary. G is difficult to evaluate in real estuaries and Harleman[8] has shown that an estuary number $P_tF_0^2/Q_fT$ was proportional to G/J in flume tests. The terms of the estuary number are all easily evaluated in real estuaries:

P_t is the volume of the tidal prism, defined as the volume entering the mouth of the estuary during the rising tide.

Q_f is the rate of inflow of fresh water.

T is the tidal period.

F_0 is a Froude number $U_0/\sqrt{(gh)}$ where U_0 is the maximum flood tide velocity averaged across the mouth, and h is the depth below mean water level at the mouth of the estuary.

Essentially, such systems of classification compare the stabilising effect of density increasing with depth, with the turbulent energy required to mix the layers. Mixing results in an increase of potential energy by raising the mean level of salt. This energy is obtained from the turbulence in the stream.

There is an interesting reversal of the normal situation at Dubai, where a cool river discharges into warm saline water. The river water is denser than the sea water and flows under it, entraining salt in the process, and causing a net landward movement near the surface and an augmented fresh-water flow near the bed[9].

The rate of flow of fresh water has a major effect on the location of the saline density gradient. There is a zone near the head of a tidal river in which water is always fresh, often referred to as the homogeneous fresh water zone. Tidal effects can be discerned by a slight rise of water level accompanied by a check in the rate of seaward flow of fresh water. A short distance seaward of this region, the check of speed is more pronounced during neap tides and becomes a landward flow during spring tides; but saline water may occur only during drought or extended dry seasons. The net flow of water and sediment during a tidal cycle is seawards at all depths. Further seawards, saline water occurs near the bed during spring tides and during periods of low river flow. Salinity near the bed increases steadily as the sea is approached. Throughout this zone of increasing salinity, often called the non-homogeneous estuarine zone, there is likely to be net landward movement of both water and solids close to the bed. Strong river flow can move the limit of this zone seawards, particularly during neap tides.

1.3 TRANSPORT OF SOLIDS

1.3.1 Classification of solids

Material transported by water can be divided into three broad groups:

(1) Particles fine enough to be kept in suspension in pure water by Brownian movement and mutual repulsion due to electrolytic potential. These are clay-sized particles.

(2) Particles fine enough to be easily lifted into suspension and transported, largely in suspension, by flowing water. These are silt and fine sand particles.

(3) Particles so large that they usually travel by rolling, sliding or hopping between irregularities on the bed such as ripples and dunes. These are medium and coarse sands and gravels.

There is no hard and fast division between these broad groups. The behaviour of fine clay-sized particles in suspension depends on their mineral composition, on their grain size and shape, and on the presence of ionising salts in the water. However, particles finer than 2 μm are generally clay-like in behaviour and particles larger than 60 μm are generally sand-like in behaviour, in the sense defined above. The presence of more than 10% by weight of clay-sized particles in a soil sample is sufficient to induce clay-like properties. (For a further description of sediment transport in estuaries see Postma[10].)

Clay-sized particles

Clay-sized particles in suspension in water carry an electrostatic charge that is usually negative. In water with a small electrolytic content particles repel each other, so remaining in stable suspension. As the electrolytic content of the water increases, the electrolytic potential of the particles decreases by an amount that varies with different minerals. The resulting difference in potential causes some particles to become mutually attractive, and flocculation occurs, the individual particles come together to form chains and open clusters. Flocs may have quite a large size—a hundred times the size of the individual particles is typical. Flocs have a large mass compared with the particles that form them. They are unaffected by Brownian movement and settle relatively quickly under gravity or centrifugal forces. A particle of quartz sand of diameter 5 μm has a settling velocity in sea water of about 0·02 mm/s, whereas a floc of similar sized particles having a breadth of 500 μm and containing 95% water would settle at about 4 mm/s (Postma[10]). In general, a very small salt content is sufficient to induce flocculation; 2 kg/m³ is enough to cause complete flocculation of some minerals such as kaolinites and chlorites, but the maximum rate of flocculation of others, such as montmorillonites, may occur at salinities greater than 30 kg/m³ – that is, greater than is found in most estuaries. Flocculation and its effects are discussed more fully in Chapter 4. For the present purpose, it is sufficient to note that flocculation is likely to occur when material is in suspension in water that undergoes an increase in salinity. The process is usually reversible if there is a subsequent decrease in salinity, and can result in breakdown of flocs that have settled onto the bed.

Flocculated material is easily transported by flowing water and usually forms a layer of relatively high concentration close to the deepest parts of the bed. Clay brought down by rivers in suspension is likely to flocculate as soon as it comes into contact with saline water. It then settles steadily to the bed, through the saline water. Because the movement of saline water under fresh water is predominantly landwards, the flocculated clay is also transported landwards, and is added to clay already present in this lower layer. This material moves landwards to a point near the upstream limit of saline water where the resultant of to-and-fro movement is zero close to the bed. Mud accumulates at this point of zero net movement, shown in figure 1.10 and, if left undisturbed, may consolidate slowly to form a deposit of clay.

A proportion of the fine material comprising the bed of an estuary does not settle to the bed during each slack tide. The proportion is greater during spring tides than during neaps. This material is often called 'wash load' because it is carried backwards and forwards with the tidal current. Wash load may increase as a result of disturbance of the bed by dredging or shipping movements.

The mechanism of seaward flow of sediment in fresh water and landward flow in a saline layer seems to offer no means of removal of clay-sized solids from an estuary system by natural forces. In practice there are several factors that can restore the balance and all depend on the variability of real flows. Strong river flows and prolonged freshets (river floods) can drive the point of zero net movement seawards, sometimes completely clear of the mouth of the estuary. When this happens the decrease in salinity can lead to breakdown of the flocs in the layer of mud on the bed of the estuary. Silt and clay particles may then be carried a considerable distance seawards before they encounter saline conditions and settle once more in flocs. A proportion of them will be deposited along the margins of the estuary, on shoals, in creeks and docks so that they do not return along the bed to the point of zero net movement when the latter moves back up river as fresh water flow decreases.

Tidal variability also causes the point of zero net movement to move up

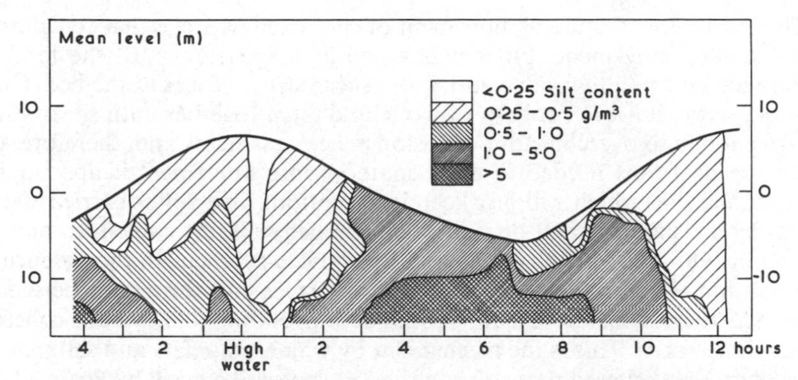

Figure 1.12. Sediment concentrations near Uskmouth, Severn Estuary

river during spring tides and down river during neaps. Freshets during neap tides have a bigger effect in displacing silt seawards than during spring tides.

In regions of large tidal range, concentration of flow in deep water channels towards low tide also results in a strong flushing action which is capable of moving unconsolidated material a considerable distance seawards. A feature of such estuaries is the presence of a belt of turbid water moving to and fro with the tides and giving rise to a layer of very high solid content close to the bed during the low water slack period and during neap tides. Concentrations in excess of 10 kg/m³ are usual in such layers and the increase in density compared with the main body of water can be sufficiently abrupt to give a clear trace on many echo-sounding charts.

Figure 1.12 shows the variation of solids content with time over the vertical at a point near Uskmouth in the Severn Estuary. Similar observations at other points showed clearly that the material in suspension was settling at slack water and was brought into suspension by strong currents. A large body of sediment was brought into the region during the ebb tide and was carried back upstream during the flood.

There is strong evidence that net landward movement of fine sediment can occur many miles offshore of the mouths of estuaries. One example is Liverpool Bay, which has been the subject of extensive hydrological studies. Seabed drifters, released at intervals over one year at stations round the Bay, returned predominantly to the estuaries of the Mersey and the Dee[11]. The implication of this is that any disturbance of the bed that would cause fresh material to be brought into suspension will lead to that material being moved predominantly landwards. Fine-grained sediments would eventually find their way back to the point of zero net movement, or to regions of weak flow such as deeply dredged areas and the margins of an estuary. The disturbance caused by dredging activity is sufficient to increase the amount of silt in motion and special precautions are often needed to limit its consequences. Spoil disposal presents special problems; the only certain way of reducing the amount of fine material available for movement is to place it permanently ashore. The problems of control of sediment in estuaries are discussed in more detail in Chapter 9 and some examples are given in Chapter 10.

There is another feature of movement of clay-sized particles that contributes to shoreward movement. Flocs will move in suspension until the level of turbulence falls to a low value and a low shear stress occurs at the bed. Once the flocs settle, however, cohesive forces hold them together until shear stress is high enough to overcome this cohesion. The shear stress and, therefore, the mean speed of flow needed to bring material into movement is appreciably greater than that which will just keep it in motion. Fine solids carried shorewards by a rising tide and deposited on the bed will not come into motion during the ebb tide until the higher shear stress is reached. In the absence of wave action which might keep material in suspension through a period of slack tide, there is therefore a differential action tending to move fine cohesive solids shorewards. This is the mechanism by which mud flats and saltings are built up in the sheltered parts of estuaries. It is described well by Postma [10].

Silt-sized particles

Silt-sized particles move predominantly in suspension at the speed of the water that is transporting them. The concentration of such particles increases towards the bed where the velocity is lower, so the mean speed of transport of solids is less than the mean speed of water. It is, however, of the same order of magnitude. Cohesive forces between silt-sized particles are negligibly small, with the result that particles of a given size come into motion as water velocity increases and cease movement on the bed as velocity decreases, at approximately the same shear stress. There is a differential action, however, because during decreasing flow speeds, some particles will still be in suspension when the shear stress falls to the critical value. For example, spherical particles of quartz of diameter 10 μm will fall at about 0·1 mm/s at 10°C; on the other hand, fine sand particles of 100 μm will settle at about 7 mm/s at the same temperature. Particles in suspension 5 m above the bed when the shear stress falls below critical will require at least 12 min to reach the bed. Some of the finer particles would never reach the bed during slack water of spring tides. Conversely, during increasing flow speed, it will take some time for fine particles to be lifted by turbulent diffusion from the bed into the upper, fast moving water. The transport of material in suspension thus lags behind the rate of flow of water. This differential action gives rise to shoreward movement of silt.

Particles coarser than silt-sized

Sand and gravel travel by rolling, sliding and bounding (saltation). Only the finest sand particles move in suspension in most estuaries, and their motion is confined to the layers near the bed. The volume rate of transport of solids along the bed is a function of shear stress at the bed and this is, to a first approximation, a function of mean speed in gradually varying flows – for example at speeds near the maximum of flood or ebb tide.

The rate of transport of sand on the bed of an estuary is a function of grain properties, water properties, and velocity distribution over the vertical. It is discussed in detail in Chapter 4. Differential action is negligibly small. There is, nevertheless, a tendency for landward bed-load movement to occur in many cases. In a well-mixed estuary, the rate of sediment transport per unit width is, to a first approximation, a function of U^n where U is the mean speed over a vertical at any time. The net transport over a tidal period is the time-integrated sum of the function of U^n. Now U never follows a symmetrical pattern during flood and ebb tide within an estuary, and the time-integrated value of U^n usually shows a landward bias under the action of tides alone. A seaward bias can only occur if there is strong river flow or, locally, if there is a concentration of ebb flow.

1.4 THE OVERALL BEHAVIOUR OF ESTUARIES

1.4.1 Three-dimensional effects

Elementary descriptions of the several factors that affect the behaviour of estuaries have been given above. Flow was separated from sediment movement and no account was taken of three-dimensional effects. No mention was made of resistance to flow. The three-dimensional nature of estuaries does have great importance, however. This arises because of the reversals of flow that occur and because the whole geometry of channels can differ between high and low tide.

Most tidal rivers in alluvium follow a meandering course in plan. Water flowing in a channel acquires the direction given to it by the boundary, and on leaving the guiding effect of a bank it tends to continue in a straight line. If the tangents to the line of deepest water of successive reaches do not coincide, the path followed by the strongest ebb current may differ from that followed by the strongest flood current. This is a common situation in tidal rivers and estuaries.

The net result is that parts of curved reaches are controlled by the ebb current while parts are controlled by the flood. Between the ebb-dominant and flood-dominant regions there is a region of neutral behaviour in which the ebb and the flood are in balance. Fine sediment can be transported through several such bends on successive flood and ebb tides, but coarse sand may only travel a short distance during each tide. Over a long period of to and fro movements, however, a considerable net drift can occur. The effect is one of cells of net movement, seawards on one side of a reach and landwards on the other, with extensive interchange of material between each reach. Figure 1.13 shows the net movement of water during flood and ebb tides for several locations in a typical well-mixed estuary. The lengths of the arrows are proportional to the mean velocity of water during rising and falling tide close to the bed. The diagram also shows cells of movement of sediment.

Towards the mouth of the estuary, where the shores widen and there is a considerable divergence between flood and ebb channels, the inertia of the ebb current has a major effect in promoting three-dimensional behaviour of the flow. Low tide occurs at the mouth before it occurs at points further inland. Ebb current still persists at the time of low water and its kinetic energy must be overcome by the rising tide before it can be brought to rest. For example, a current of 1·5 m/s has kinetic energy equivalent to a head of over 0·1 m. This seems a small difference of water level, but is typical of the difference that could occur over a distance of about 5 km in the main channel

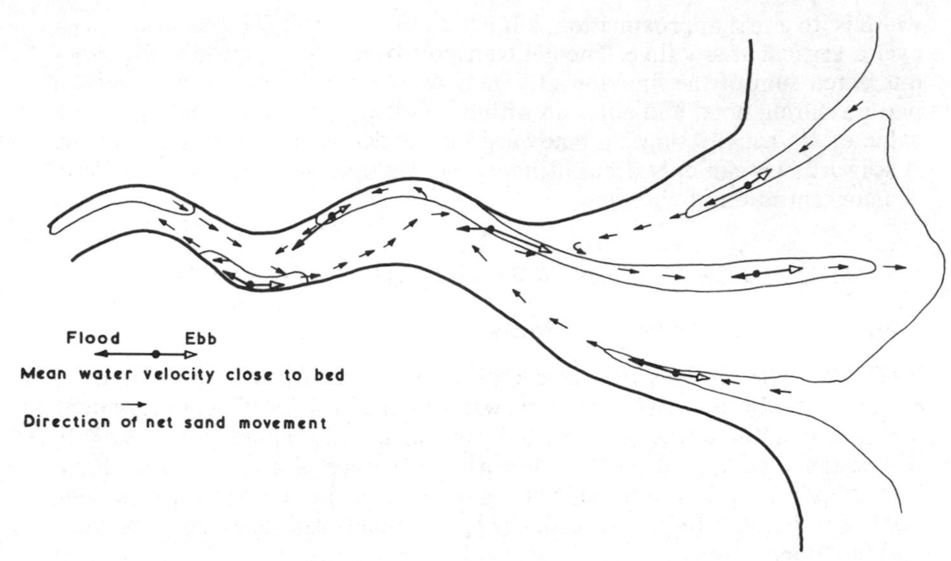

Figure 1.13. Net movement of water and solids in a typical estuary

during the ebb. The net result is that the ebb current persists in the main ebb channel after flow has reversed over the shoals alongside it. Flow over the shoals can then enter the channel from the sides, carrying sand with it, particularly if there is appreciable wave action.

The three-dimensional pattern of flow in a wide estuary may be enhanced by tidal currents in the sea outside its mouth, by wave- or wind-induced currents and by the effect of rotation of the earth. Density gradients also have a marked effect in many cases and are particularly important when the movement of sediment has to be taken into account.

Density gradients over the vertical have a marked effect on local net movement of water. Price and Kendrick[12] built a model of the Mersey Estuary and tested it, initially using fresh water. A typical result of measurement of net water movement over a full tidal cycle is shown in figure 1.14a. Net landward movement was found at one side of the channel and net seaward movement at the other, in the manner illustrated in figure 1.13. Prototype measurements had shown that there was net landward movement near the bed of the channel and net seaward movement near the surface. This was only reproduced in the model when fresh and saline waters were introduced (figure 1.14b).

The small density gradients found in near-shore regions have an effect in contributing to net landward drift of near-bed water and sediment, as demonstrated by Heaps[13] for the Liverpool Bay area of the Irish Sea.

The effect of saline penetration into estuaries due to density variations is enhanced by the three-dimensional nature of the flow. During the passage of a flood tide through a cross-section of an estuary, there is a progressive rise in water level. In all except completely stratified estuaries, this is accom-

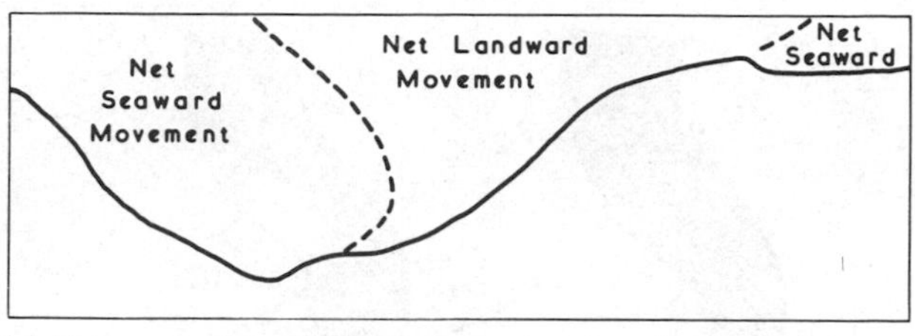

(a) Net movements with homogeneous water in the model

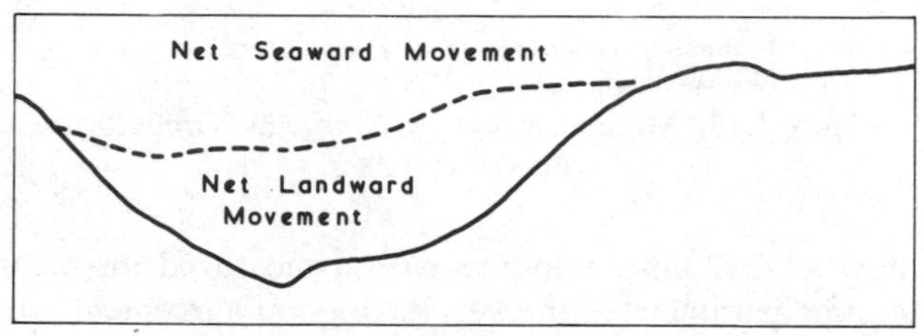

(b) Net movements with salinity variations reproduced in the model

Figure 1.14. Net movement of water during a tide cycle in a model of the Mersey Estuary (based on figure 5 of reference 12, courtesy the Institution of Civil Engineers)

panied by a rise in salinity at any point. The highest velocities occur in the deepest part of the channel and the rate of change of salinity is higher there than over shoals. There is a time lag between rise of salinity over the shoals and rise in the deeper channels and at high tide; salinity may be higher in the main channel than over the shoals. During the ebb tide, on the other hand, rapid flow in the main channel will be accompanied by a rapid fall in salinity, particularly if there is appreciable fresh water flow from rivers. The inertia of the ebb current flowing in a deep channel causes flow to persist there for a short time after low tide level. The overall effect is a time lag between salinity variations in the main channel and over shoals. Moreover, frequent changes in cross-sectional width occur in many estuaries. The flow pattern is then characterised by large-scale eddies or by regions of weak circulations of water, down-current of any sharp changes in shoreline direction. Water trapped in these areas of weak circulation is not mixed freely with the main stream and differences of salinity can also occur between them, as illustrated in figure 1.15. All these factors modify the mixing process and must be taken into account when estimating the penetration of salt into an estuary. Simple methods of study based on two-dimensional analysis and experiments in straight channels cannot be used to predict saline penetration in such cases. For example, Fischer and Dudley[14] have studied salinity intrusion into San Francisco Bay, which has a very complex plan shape. A dimensionless salinity intrusion parameter due to Rigter[15] gave the minimum intrusion length for a parti-

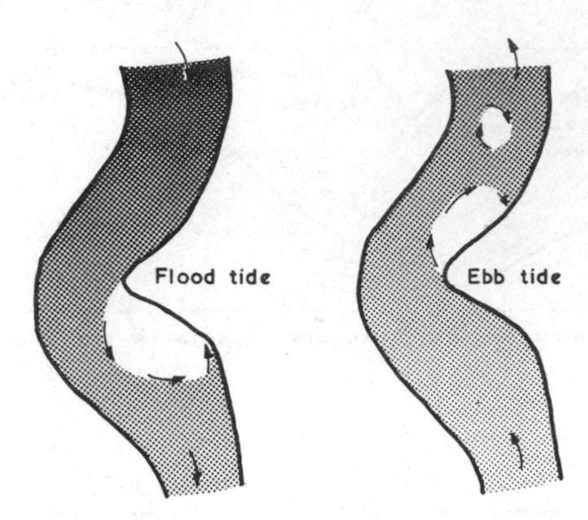

Shading indicates density gradation due to salt

Figure 1.15. Mixing caused by large-scale circulations with vertical axes

cular river flow as 4.42 km compared with a measured intrusion length of 113 km. The same parameter, however, gave good agreement with measurements in the straight, nearly parallel-sided Rotterdam Waterway and in a laboratory flume.

1.4.2 Interaction with the sea

Tidal currents near a coast are frequently out of phase with tidal levels. It cannot be assumed that low water at the mouth of an estuary will coincide, even approximately, with slack tide outside. The direction of a tidal current is particularly important. During rising tide, it may happen that the tidal current in the sea approaches an estuary from a direction that contains a source of readily available sediment. This can then be a major source of sedimentation, particularly if tidal flow is augmented by wave action. However, it cannot be assumed that sediment will approach an estuary solely from the direction of the dominant flood current during rising tide. Sediment, particularly sand, will move in a series of to-and-fro movements during successive tides. The current due to ebb and flood will determine the direction of sand movement during each tide, but the resultant movement will be the vector sum of all such movements during several tides (figure 1.16a). This effect must be taken into account when field measurements are planned.

The direction of motion as indicated by sand waves can indicate the direction of the main flood and ebb flows, but the net resultant can be at any angle to the separate flows (figure 1.16b).

Once a pattern of tidal currents has been established for a particular region it will be repeated when similar tidal and weather conditions recur. Currents caused by wind and waves or by density effects, however, can only be predicted on a statistical basis. If the statistical pattern of wind speed and direction is known for a large area of sea round an estuary it is now possible to estimate the wave climate and to predict the current pattern for fairly

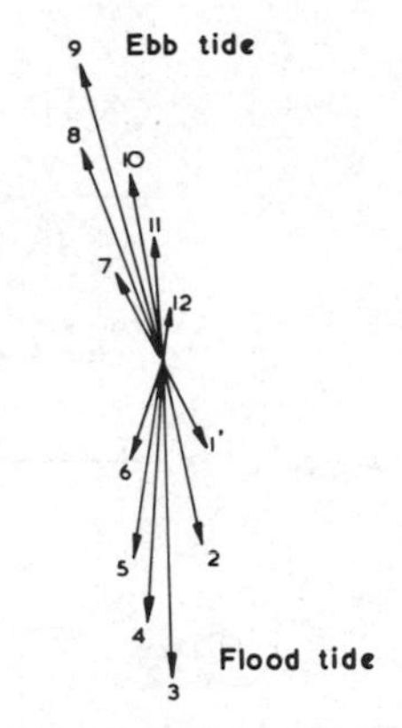

(a) Velocity Vectors at Hourly Intervals

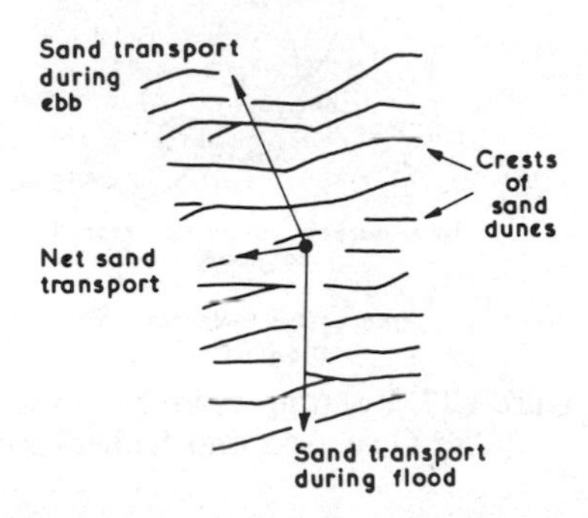

(b) Velocity Vectors and Net Sand Transport Near Shore

Figure 1.16. Velocity vectors and net sand transport

simple configurations of sea-bed profiles and shore-lines. In complex situations prediction is not yet possible without the aid of hydraulic models. Strong currents and waves in shallow water interact in a non-linear manner, so the effects cannot be studied separately without loss of accuracy.

1.4.3 Effect of rotation of the earth

Rotation of the earth is often ignored because its effects appear to be small in estuaries compared with other effects. It is, however, significant in many wide estuaries and bays. It can be represented by a force, known as the Coriolis force, with components proportional to the speed of flow and acting at right angles to it [2]. It is described in the next chapter. The Coriolis force per unit mass is usually smaller than the horizontal force per unit mass due to surface gradient, or friction, and smaller than the rate of flux of momentum. It is zero at the equator. Nevertheless it is appreciable at moderate and high latitudes, often reaching about 20% of each of the major forces between

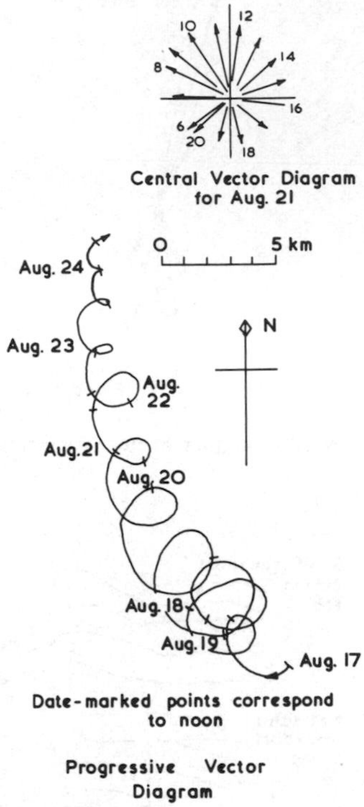

Figure 1.17. Rotating currents in the Baltic Sea
(after Gustafson and Kullenberg[17])

latitudes 50° and 60° and acting at right angles to the direction of flow. In the open sea this causes the current to rotate, clockwise in the northern hemisphere and anti-clockwise in the southern, as in figure 1.17. The effect on tidal propagation is fundamental; instead of advancing in a direction determined simply by the east–west movement of the tide-generating force, the tidal waves rotate round a series of points of zero amplitude, known as amphidromic points[3, 16]. Lines representing simultaneous high tides and points showing the appropriate spring tidal range as $2\Sigma(M_2+S_2+K_1+O_1)$ cm for water between Britain and Europe are shown in figure 1.18. Moreover it can be shown that, in a confined sea, tidal amplitude will be greater on one side than on the other, depending on the direction of main tidal current. In the seas round the British Isles, tidal ranges are much greater on the east than on the west side of the Irish Sea; on the south than on the north side of the English Channel; and on the west than on the east side of the North Sea (figure 1.18). The directions of tidal currents outside estuaries at latitudes greater than about 30° are considerably modified by Coriolis force.

In wide estuaries the effect of rotation of the earth is modified by the shoreline. The main tidal streams run nearly parallel with the axis of an estuary. Coriolis force tends to deflect the current to the right in the northern and to the left in the southern hemisphere. During ebb tide, the current will

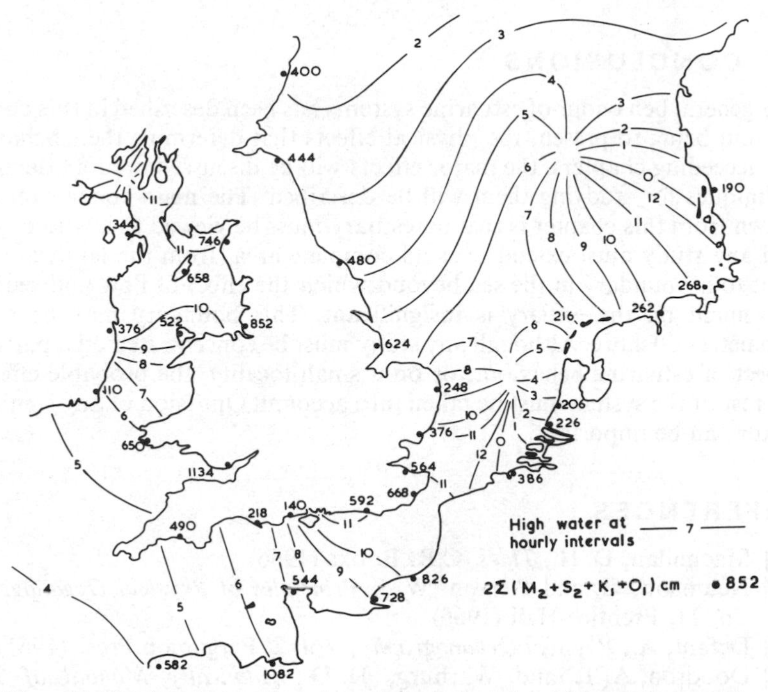

Figure 1.18. Amphidromic points and comparative tidal ranges around the British Isles (based on figures 154, 158, 161 and tables 42 and 44, of reference[3])

be deflected towards the right bank (viewed facing seawards) in the northern hemisphere, while during the flood, it will be deflected towards the left bank. Coriolis force can therefore cause net anti-clockwise circulation to be superimposed on flood and ebb flows in estuaries in the northern hemisphere.

The Coriolis force can be calculated and its effect on flow measured. It is small in estuaries, with the result that estimates of flow based on a given channel geometry with or without it often show little difference, whether these are obtained by numerical calculation or from fixed-bed hydraulic models. However, sediment transport is very sensitive to small changes in velocity with the result that channel alignment can be considerably modified by the Coriolis effect. It may be impossible to reproduce the major channel configuration in a wide tidal inlet in which channel migration can occur freely, without it. Unfortunately it is difficult to reproduce in hydraulic models. In principle, it can be represented in numerical computations, but study of the development of migrating channels would call for better knowledge of the laws of sediment transport than we have at present, while the detail required for the study of two-dimensional flow in plan would call for the use of the most powerful computers.

1.5 CONCLUSIONS

The general behaviour of estuarine systems has been described in this chapter, the aim being to present the physical effects that determine their behaviour. In succeeding chapters, the major effects will be discussed in more detail and techniques for studying them will be described. The main conclusion to be drawn from this chapter is that an estuary must be viewed as a whole system and any study must extend over its complete area, from the landward tidal limit to a boundary in the sea beyond which the effect of flow and sediment movement on the estuary is insignificant. This boundary may be several kilometres offshore. Although any study must be concentrated on a particular aspect of estuarine behaviour, or on a small locality, the probable effect on the rest of the system must be taken into account. Omission of any significant factor can be important.

REFERENCES

[1] Macmillan, D. H., *Tides*, C.R. Books (1966)
[2] Neumann, G. and Pierson, W. J. *Principles of Physical Oceanography*, Ch. 11, Prentice-Hall (1966)
[3] Defant, A., *Physical Oceanography*, Vol. **2**, Pergamon Press (1961)
[4] Doodson, A. T. and Warburg, H. D , *Admiralty Manual of Tides*, HMSO (1941)
[5] Shureman, P., *Manual of Harmonic Analysis and Prediction of Tides*, US Department of Commerce, Coast and Geodetic Survey, Special Publication No. 98, US Gov. Print. Office (1940)
[6] Smith, A. B., Southwest Pass, Mississippi River, 40 ft Ship Channel, *Permanent International Association of Navigation Congresses, Bulletin No*. 51, Brussels (1960)
[7] Ippen, A. T., *Estuary and Coastline Hydrodynamics*, Ch. 13, McGraw-Hill (1966)
[8] Harleman, D. R. F. and Abraham, G., One-dimensional analysis of salinity intrusion in the Rotterdam Waterway, *Delft Hyd. Lab., Pub. No*. 44 (1966)
[9] Loewy, E., Model test and studies for Port Rashid, Dubai, Ch. 72, *12th Coastal Engineering Conference*, vol. II, Washington (1970)
[10] Postma, H., Sediment transport and sedimentation in the estuarine environment, *Estuaries*, ed. Lauff, American Academy for the Advancement of Science (1966)
[11] Halliwell, A. R., Discussion of Paper No. 7390, The sea approaches to the Port of Liverpool, *Proc. Inst. Civil Eng*., **50**, 611–613 (1972)
[12] Price, W. A. and Kendrick, M. P., Field and model investigations into the reasons for siltation of the Mersey Estuary, *Proc. Inst. Civil Eng*., Paper No. 6669, **24**, 473 (1963)
[13] Heaps, N. S. Estimation of density currents in the Liverpool Bay of the Irish Sea, *Geophys. J. Roy. Astr. Soc*. **30**, 415–432 (1972)

[14] Fischer, H. B. and Dudley, E., Salinity intrusion mechanisms in the San Francisco Bay, California, *Proc. XVI Congress IAHR*, **1**, 124–133 (1975)
[15] Rigter, B. P., Minimum length of salt intrusion in estuaries, *Journal of Hydraulics Division, Am. Soc. Civ. Eng.*, **99**, HY9, 1475–1496 (1973)
[16] Muir Wood, A. M., *Coastal Hydraulics*, Ch. 1, Macmillan (1969)

2

Hydrodynamics of estuaries

In Chapter 1, the behaviour of an estuary was discussed in general terms. It was shown that three-dimensional effects are important and that density variations within the flow must be taken into account. The flow pattern in estuaries is unsteady, not only because of the tidal rise and fall at the mouth but also because there are large-scale eddies that vary in size and location. The complexity of the problem is such that a rigorous analysis of flow behaviour is impossible. Nevertheless some framework is needed so that the behaviour of systems can be analysed and compared, by means of field measurements, numerical analysis or by analogy with similar systems, including physical and electronic models. The equations of motion are used as the basis for development of all analytical methods in this book.

Because the major component of flow is horizontal, it is convenient to use a Cartesian system of coordinates with the x–y plane horizontal and the z-plane vertical. The datum used for charts by engineers and navigators is close to the lowest possible water level. The positive z-direction is assumed to be vertically upwards from this datum.

2.1 THE EQUATIONS OF MOTION

2.1.1 Forces and momentum

The equations of motion are concerned with forces on unit mass of a liquid and with accelerations which, because they occur in space and time, can only be described satisfactorily by the use of partial differential notation. The equations that follow are applicable to a real fluid, capable of transmitting a shear stress when in motion relative to its surroundings[1, 2]. No assumptions will be made about the nature of the shear stresses until the equations have been established.

Figure 2·1 represents an element of space, fixed relative to the Earth, within a tidal stream. The element is assumed to be so small that no sensible error will occur if stresses acting at the centre of each face are taken as the average stress acting over the whole face. Forces acting on water flowing through the element in each coordinate direction can then be equated to the rate of change of momentum in that direction. The components of forces acting on each face

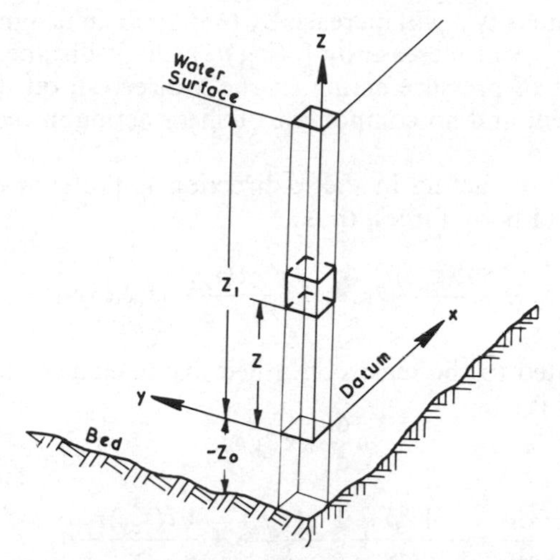

Figure 2.1. Element in space within an estuary

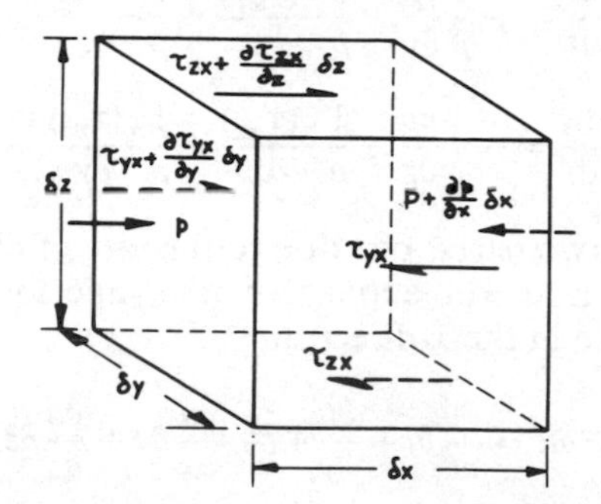

Figure 2.2. Pressure and shear stresses acting in the x-direction on element of volume of water

in the x-direction are shown in figure 2.2. In addition there are forces acting on the mass of fluid as a whole; in estuaries the main force is due to gravity, but there is also a force caused by the fact that the water is moving on a rotating earth. This force, known as the Coriolis force, can be important in wide estuaries and seas in the higher latitudes.

Let u, v, w be instantaneous components of velocity at time t in the coordinate directions x, y, z respectively.

Let ρ = fluid density

a = body force per unit mass (e.g. gravity, Coriolis force)

p = pressure intensity

τ_{yz} = shear stress acting on a plane perpendicular to the y-axis, in the x-direction.

All the forces acting on the liquid in the x-direction on the surfaces of the element are shown in figure 2.2.

The pressure intensity p will increase by $(\partial p/\partial x)\,\delta x$ in distance δx. Similarly the shear stress τ_{yz} will increase by $[\partial(\tau_{yx})/\partial y]\,\delta y$ in distance δy. There will be no component of pressure acting in the x-direction on the x–y and x–z faces of the element and no component of shear acting in the x-direction on the y–z faces.

The resultant force acting in the x-direction is the sum of the resultant pressure, shear and body forces, thus:

$$-\frac{\partial p}{\partial x}\delta x\,.\,\delta y\delta z+\frac{\partial(\tau_{yx})}{\partial y}\delta y\,.\,\delta x\delta z+\frac{\partial(\tau_{zx})}{\partial z}\delta z\,.\,\delta x\delta y+\rho a_x\delta x\delta y\delta z$$

This can be equated to the mass contained in the element multiplied by its acceleration, that is,

$$\rho\frac{\mathrm{d}u}{\mathrm{d}t}\delta x\delta y\delta z.$$

Thus
$$\frac{\mathrm{d}u}{\mathrm{d}t}=-\frac{1}{\rho}\frac{\partial p}{\partial x}+\frac{1}{\rho}\frac{\partial(\tau_{yx})}{\partial y}+\frac{1}{\rho}\frac{\partial(\tau_{zx})}{\partial z}+a_x \tag{2.1}$$

similarly
$$\frac{\mathrm{d}v}{\mathrm{d}t}=-\frac{1}{\rho}\frac{\partial p}{\partial y}+\frac{1}{\rho}\frac{\partial(\tau_{xy})}{\partial x}+\frac{1}{\rho}\frac{\partial(\tau_{zy})}{\partial z}+a_y \tag{2.2}$$

and
$$\frac{\mathrm{d}w}{\mathrm{d}t}=-\frac{1}{\rho}\frac{\partial p}{\partial z}+\frac{1}{\rho}\frac{\partial(\tau_{xz})}{\partial x}+\frac{1}{\rho}\frac{\partial(\tau_{yz})}{\partial y}+a_z \tag{2.3}$$

Note that $\mathrm{d}u/\mathrm{d}t$ has two parts; one due to the rate of change of velocity with time in the element as a whole and the other due to the rate of change of velocity with distance in the x-direction:

$$\delta u=\frac{\partial u}{\partial t}\delta t+\frac{\partial u}{\partial y}\delta x+\frac{\partial u}{\partial x}\delta y+\frac{\partial u}{\partial z}\delta z$$

$$\underset{\delta t\to 0}{\mathrm{Lt}}\frac{\mathrm{d}u}{\mathrm{d}t}=\frac{\partial u}{\partial t}+\frac{\partial u}{\partial x}\frac{\mathrm{d}x}{\mathrm{d}t}+\frac{\partial u}{\partial y}\frac{\mathrm{d}y}{\mathrm{d}t}+\frac{\partial u}{\partial z}\frac{\mathrm{d}z}{\mathrm{d}t}$$

$$=\frac{\partial u}{\partial t}+u\frac{\partial u}{\partial x}+v\frac{\partial u}{\partial y}+w\frac{\partial u}{\partial z}$$

Equations 2.1 to 2.3 are quite general and can be used for homogeneous and non-homogeneous liquids as well as for compressible or incompressible fluids. In estuaries, small variations of density due to dissolved substances or particles in suspension are often important.

If gravity is the only significant body force, $a_x = a_y = 0$ and $a_z = -g$. As previously mentioned, another body force which can be important in estuaries is the Coriolis force.

If ω is the angular velocity of Earth ($7{\cdot}27\times10^{-5}$ rad/s) and ϕ is the angle of latitude of the estuary, the effective speed of rotation about an axis normal to the surface of the Earth is $\omega\sin\phi$. The speed of rotation about axes tangential to the Earth's surface varies from $\omega\cos\phi$ for axes parallel to lines of longitude, to zero for axes parallel to lines of latitude. It can be shown [3, 4] that the effect of rotation is to cause a force per unit mass having the following components:

$2\omega v \sin\phi$ in the x-direction (a_x)

$-2\omega u \sin\phi$ in the y-direction (a_y)

$2\omega u_e \cos\phi$ in the z-direction (a_z)

where u_e is the component of velocity towards the east. The signs for the x and y components apply in the Northern hemisphere and should be reversed in the Southern.

2.1.2 Conservation of matter

The principle of conservation of matter is of fundamental importance. It requires that the total mass of any material flowing out of an element in space in a given time must equal the reduction in mass within the space in that time. The material can be a fluid, a solid dissolved in a fluid, or material in suspension in the fluid. The equations of conservation are also known as the equations of continuity. They apply to each material separately or to any combination of them. Let a material having density ρ flow through a small element of space at location (x, y, z, t) as in figure 2.3. The velocity of the material has components (u, v, w) in the three coordinate directions. The rate of change in mass in the x-direction is thus $([\partial(\rho u)/\partial x]\delta x)\,\delta y\,\delta z$. The summation of this term, with similar terms from the y- and z-directions, can then be equated to the temporal change in mass within the element to yield:

$$-\frac{\partial \rho}{\partial t}\delta x\delta y\delta z = \left[\frac{\partial(\rho u)}{\partial x} + \frac{\partial(\rho v)}{\partial y} + \frac{\partial(\rho w)}{\partial z}\right]\delta x\,\delta y\,\delta z$$

$$\therefore -\frac{\partial \rho}{\partial t} = \rho\left(\frac{\partial u}{\partial x} + \frac{\partial v}{\partial y} + \frac{\partial w}{\partial z}\right) + u\frac{\partial \rho}{\partial x} + v\frac{\partial \rho}{\partial y} + w\frac{\partial \rho}{\partial z}$$

$$\therefore -\frac{1}{\rho}\frac{\mathrm{d}\rho}{\mathrm{d}t} = \frac{\partial u}{\partial x} + \frac{\partial v}{\partial y} + \frac{\partial w}{\partial z}$$

This equation applies to all fluids, but for the special case of incompressible liquids or solids,

$$\mathrm{d}\rho/\mathrm{d}t = 0$$

and

$$\frac{\partial u}{\partial x} + \frac{\partial v}{\partial y} + \frac{\partial w}{\partial z} = 0 \tag{2.4}$$

In the case of water, changes of density with distance are insignificant over depths encountered in estuaries. The concentration of salts in solution or of solids in suspension, however, can change quite rapidly with distance. It is therefore necessary to take the rate of transport by diffusion processes into account, due to random molecular movements or due to turbulence of the fluid; or by dispersion due to interchange of material between slow-moving and fast-moving layers, e.g. as a result of gravitational settlement of solids or of particles being lifted from the stream bed. These and other related processes are described more fully in Chapter 3.

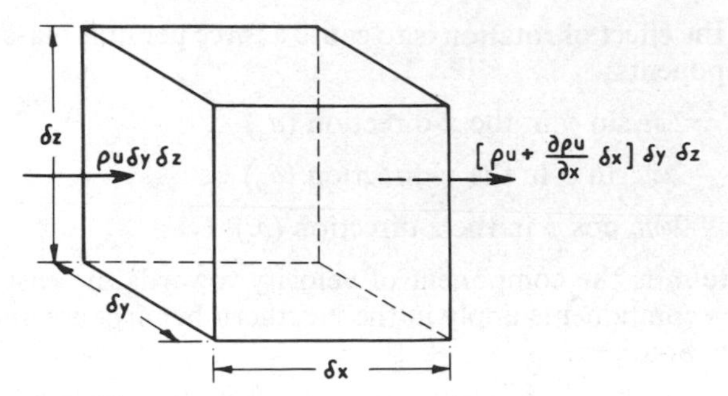

Figure 2.3. Flux through element in x-direction

2.2 APPLICATIONS OF THE EQUATIONS TO LOCALISED MEAN CONDITIONS

Equations 2.1 to 2.4 apply at a point in space and time. Real flows in estuaries are turbulent and the instantaneous conditions differ considerably from the local mean conditions. These mean conditions can be averaged over a period which is long compared with the duration of the turbulent fluctuations, but short compared with (i) the tidal period or (ii) significant changes in river hydrograph; or over volumes large compared with turbulent eddies but small compared with stream dimensions. This is shown in figure 3.1. Osborne Reynolds[5, 6] suggested that the instantaneous values in the equations could be replaced by the sum of local, mean value of quantities, (denoted by a bar, e.g. $\bar{u}$) together with fluctuating values (denoted by a prime e.g. u') so that $u = \bar{u}+u'$. The local average of the fluctuating components is zero, i.e. $\bar{u}' = 0$. He then showed that the terms in the equations could be applied to local mean conditions, provided that they included normal and shear stresses caused by the turbulent fluctuations in addition to viscous shear stress. In the x-coordinate direction, these terms would be:

$$\frac{\partial(-\rho\overline{u'u'})}{\partial x}, \quad \frac{\partial(-\rho\overline{u'v'})}{\partial y} \quad \text{and} \quad \frac{\partial(-\rho\overline{u'w'})}{\partial z}$$

where $\overline{u'v'}$ indicates the local temporal mean of the product of fluctuating values. Similar terms would apply in the other coordinate directions. They are discussed further in Chapter 3.

Equations 2.1 to 2.4 apply equally well if local mean values of velocity are used, provided that the shear stress term is modified to include the Reynolds direct and shear stresses. This form of the equations is essential for a study of estuarine problems because turbulent fluctuations affect all velocity measurements. Field measurements must be planned to take them into account; velocity measurements at a point must continue long enough to establish local mean values (see figure 3.1). If knowledge of fluid stresses is required, velocity fluctuations must also be found—an ideal that has seldom been realised for even one component of force.

It is seldom necessary to use the full three-dimensional equations of motion.

For many problems, including flow in rivers, canals, estuaries and coastal waters, when surface waves are absent, vertical components of local mean velocity are small. In consequence, vertical components of force due to changes of momentum and gradients of vertical shearing stresses are negligible compared with the force due to gravity

Equations 2.1 to 2.3 can then be simplified to:

$$\frac{\partial \bar{u}}{\partial t}+\bar{u}\frac{\partial \bar{u}}{\partial x}+\bar{v}\frac{\partial \bar{u}}{\partial y} = -\frac{1}{\rho}\frac{\partial p}{\partial x}+\frac{1}{\rho}\frac{\partial \tau_{yx}}{\partial y}+\frac{1}{\rho}\frac{\partial \tau_{zx}}{\partial z} \tag{2.5}$$

$$\frac{\partial \bar{v}}{\partial t}+\bar{u}\frac{\partial \bar{v}}{\partial x}+\bar{v}\frac{\partial \bar{v}}{\partial y} = -\frac{1}{\rho}\frac{\partial p}{\partial y}+\frac{1}{\rho}\frac{\partial \tau_{xy}}{\partial x}+\frac{1}{\rho}\frac{\partial \tau_{zy}}{\partial z} \tag{2.6}$$

$$\frac{1}{\rho}\frac{\partial p}{\partial z} = -g \tag{2.7}$$

These equations apply when localised mean parameters are used at a point in space and time within a flowing liquid. They can be solved numerically for some simple situations. Some methods of solution will be considered in Chapter 6.

2.3 EQUATIONS AVERAGED OVER THE DEPTH OF A LIQUID (figure 2.1)

Equation 2.7 can be integrated over the vertical to give

$$p_1 - p = -\int_z^{z_1} \rho g \mathrm{d}z \tag{2.8}$$

where subscript $_1$ refers to the water surface.

Local pressure intensity p can be found from this equation provided that the variation of density with depth is known. $\partial p/\partial x$ and $\partial p/\partial y$ can then be found and substituted in equations 2.5 and 2.6.

In a well-mixed estuary $\partial \rho/\partial z$ is negligible and

$$\frac{\partial p}{\partial x} = \frac{\partial}{\partial x}[\rho g(z_1 - z)] = \frac{\partial(\rho g h)}{\partial x} = g\left(\rho\frac{\partial h}{\partial x}+h\frac{\partial \rho}{\partial x}\right) \tag{2.9}$$

where $h = (z_1 - z)$, the depth below the surface.

For the case of uniform density at all points,

$$\frac{\partial \rho}{\partial x} = 0 \quad \text{and} \quad \frac{\partial p}{\partial x} = g\rho\frac{\partial h}{\partial x} = -g\rho i_x$$

where $i_x = -\partial h/\partial x = -\partial z_1/\partial x$ is the component of surface slope in the x-direction

Equation 2.5 becomes:

$$\frac{\partial \bar{u}}{\partial t}+\bar{u}\frac{\partial \bar{u}}{\partial x}+\bar{v}\frac{\partial \bar{u}}{\partial y} = g i_x+\frac{1}{\rho}\frac{\partial \tau_{yx}}{\partial y}+\frac{1}{\rho}\frac{\partial \tau_{zx}}{\partial z}$$

If the bed is at elevation z_0 above datum and $z_1 - z_0$ = stream depth H,

$$\frac{1}{H}\int_{z_0}^{z_1}\left(\frac{\partial \bar{u}}{\partial t}+\bar{u}\frac{\partial \bar{u}}{\partial x}+\bar{v}\frac{\partial \bar{u}}{\partial y}\right)\mathrm{d}z = \frac{g}{H}\int_{z_0}^{z_1} i_x \mathrm{d}z + \frac{1}{\rho H}\int_{z_0}^{z_1}\left(\frac{\partial \tau_{yx}}{\partial y}+\frac{\partial \tau_{zx}}{\partial z}\right)\mathrm{d}z$$

$$= gi_x + \frac{1}{\rho H}\int_{z_0}^{z_1}\frac{\partial \tau_{yx}}{\partial y}\mathrm{d}z + \tau_{z_1 x} - \tau_{z_0 x}$$

$\bar{u}(\partial \bar{u}/\partial x)$ and $\bar{v}(\partial \bar{u}/\partial y)$ are non-linear terms and can only be integrated over the vertical if their distributions are known. Fortunately they are usually much smaller in magnitude than $\partial \bar{u}/\partial t$ in tidal streams and it is sufficient to assume that

$$\frac{1}{H}\int_{z_0}^{z_1}\left(\bar{u}\frac{\partial \bar{u}}{\partial x}+\bar{v}\frac{\partial \bar{u}}{\partial y}\right)\mathrm{d}z = \beta_1' U\frac{\partial U}{\partial x}+\beta_2' V\frac{\partial U}{\partial y} \tag{2.10}$$

Values of the coefficients β' have been found for typical variations of velocity over the vertical. β' seldom exceeds 1·05 in real flows (Henderson [7]). Because of the smallness of the terms, however, it is common practice either to omit the term completely or to take $\beta' = 1$.

Most real tidal streams are wider, by an order of magnitude, than they are deep, and shear stresses on vertical planes are small except near steep banks and vertical walls. τ_{yx} and τ_{xy} are thus usually small and can be omitted from equations 2.5 and 2.6.

The shear stress at the surface $\tau_{z_1 x}$ is significant during periods of strong winds, but is small in estuaries compared with the shear stress $\tau_{z_0 x}$ at the bed. It needs to be taken into account in the open sea.

The equations of motion for an homogeneous liquid averaged over the vertical become

$$\frac{\partial U}{\partial t}+U\frac{\partial U}{\partial x}+V\frac{\partial U}{\partial y} = gi_x - \frac{\tau_{z0x}}{\rho H} \tag{2.11}$$

$$\frac{\partial V}{\partial t}+U\frac{\partial V}{\partial x}+V\frac{\partial V}{\partial y} = gi_y - \frac{\tau_{z0y}}{\rho H} \tag{2.12}$$

Note that, in equations 2.11 and 2.12 in which velocities are averaged over a vertical, the components of surface slope are local values and will differ from slopes averaged over a cross-section as measured by tide gauges. Local surface slopes are very difficult to measure at points remote from a fixed platform. The shear stress in these equations is the local shear stress at the bed and can be inferred from velocity gradients at the site. These equations can only apply to situations in which horizontal shear forces on vertical planes are negligible or can be lumped together to give an effective shear stress at the bed. This will be discussed further in Chapter 5 (on field measurements).

Equations 2.11 and 2.12 are made dimensionless by dividing all the terms by gravitational force per unit mass, g. The resulting terms are then equivalent to gradients:

$$\frac{1}{g}\frac{\partial U}{\partial t}+\frac{U}{g}\frac{\partial U}{\partial x}+\frac{V}{g}\frac{\partial U}{\partial y} - i_x + \frac{\tau_{z0x}}{\rho g H} = 0 \tag{2.13}$$

$$\frac{1}{g}\frac{\partial V}{\partial t}+\frac{U}{g}\frac{\partial V}{\partial x}+\frac{V}{g}\frac{\partial V}{\partial y} - i_y + \frac{\tau_{z0y}}{\rho g H} = 0 \tag{2.14}$$

The principle of continuity can also be applied to a point in plan by integrating equation 2.4 over the depth of the stream:

$$\int_{z_0}^{z_1} \left(\frac{\partial u}{\partial x} + \frac{\partial v}{\partial y} + \frac{\partial w}{\partial z} \right) \mathrm{d}z = 0$$

Putting
$$U = \frac{1}{H} \int_{z_0}^{z_1} u \mathrm{d}z \text{ and } V = \frac{1}{H} \int_{z_0}^{z_1} v \mathrm{d}z$$

and noting that $(w)_{z_1} = \mathrm{d}z_1/\mathrm{d}t$ and $(w)_{z_0} = 0$

$$\frac{\partial(HU)}{\partial x} + \frac{\partial(HV)}{\partial y} + \frac{\mathrm{d}z_1}{\mathrm{d}t} = 0 \tag{2.15}$$

2.4 EQUATIONS AVERAGED OVER A CROSS-SECTION

Simplification can be taken one stage further. It is often only necessary to determine the effect of major channel changes on mean water levels and flows. In this case, there is no direct interest in local velocities and it is usual to regard the estuary as having a straight channel and to introduce additional friction to allow for losses due to bends, changes in cross-section and three-dimensional circulations. This is directly equivalent to the procedure adopted with rivers, in which steady flow is described in terms of the Manning or Chezy equations, involving mean slope, mean speed of flow and a friction coefficient dependent on channel size and sinuosity as well as on the roughness of the channel bed.

If there are considerable changes of depth over the cross-section between the main flow channels and shoals, the difference between values of $U\partial U/\partial x$ based on integration over the cross-section and values based on mean velocities will be appreciable. It may be necessary to retain the coefficient β' described in equation 2.10.

If the s-coordinate is assumed to run along the axis of the river, and velocity components at right angles to it in plan are ignored, equation 2.13 becomes

$$\frac{1}{g}\frac{\partial U}{\partial t} + \beta' \frac{U}{g}\frac{\partial U}{\partial s} - i_s - i_f = 0 \tag{2.16}$$

and
$$i_f = -\frac{U|U|}{C^2 R}$$

$$\text{where } R = \text{hydraulic radius} = \frac{\text{sectional area}}{\text{wetted perimeter}}.$$

Alternatively
$$i_f = -\frac{n^2 U|U|}{R^{\frac{4}{3}}}.$$

It should be noted that the Chezy coefficient C or Manning coefficient n, will not necessarily be equal to values based on unidirectional flows in rivers equivalent in size, sinuosity and roughness. The reasons for this are that flow reversals in estuaries cause different bed forms from those found in unidirectional rivers; and accumulation of fine silt and mud at the bed of

Table 2.1. *Typical values of Manning's friction coefficient* n *based on field measurements in the River Hooghly, and values based on Ganguillet and Kutter's formula for a large river having similar cross-sections*

Location in River Hooghly	Hydraulic mean depth R m	Manning's n (measurements in River Hooghly)	Manning's n (values for a large river)
Saugor – at mouth	8·3	0·013	0·019
148 km upstream	6·1	0·019	0·020
200 km upstream	2·8	0·026	0·021

estuaries near the point of zero net movement causes the friction term to be modified. Nevertheless losses due to sinuosity are likely to be similar in rivers and estuaries and it has been found, in practice, that the friction equations often give acceptable results using coefficients little different from those for equivalent river channels. Some typical values of friction coefficient based on field measurements in the River Hooghly are given in table 2.1. Values based on river channels having equivalent size and sinuosity are also shown.

The surface slope in equation (2.16) is the component of surface slope in the direction of the axis of the flow channel, averaged over the cross-section. If gauges situated at the banks of the channel are used to measure this slope, care must be taken to ensure that they are not affected by local conditions at the site, or by curvature of the tidal stream. It might be necessary, in some cases, to instal a pair of gauges on opposite banks and to use a value of surface level based on both readings, as described in Chapter 5.

The coefficient β' in equation 2.16 may be as high as 1·4 when the cross-section of the channel is irregular and consists of deep channels and extensive shallow areas (Henderson [7]).

The terms of the equations of motion 2.13, 2.14 and 2.16 are each gradients of energy per unit mass of liquid. The equations illustrate the balance between potential and kinetic energy within the flow. Integration over a chosen distance leads to a statement of the change in potential and kinetic energy and the dissipation or energy by friction over that distance.

The equation of continuity for axial flow in a channel can be written in a simple and very useful form. Let the channel be divided into short reaches having length Δx between cross-sections. A relationship can be drawn up between the volume of water needed to produce a given change of level in a given time and the flow of water into and out of the reach in the same time.

At a particular instant of time, t, let the mean surface width over the reach be B, such that the plan area of the water surface is $B\Delta x$ (figure 2.4). If Δx is so small that the whole water surface can be assumed to rise or fall at the same rate, the volume of water needed to change the surface level in time Δt is $B\Delta x(\partial z_1/\partial t)\,\Delta t$ while the volume of water flowing out of the reach is

$$\frac{\partial Q}{\partial x}\,\Delta x\,\Delta t$$

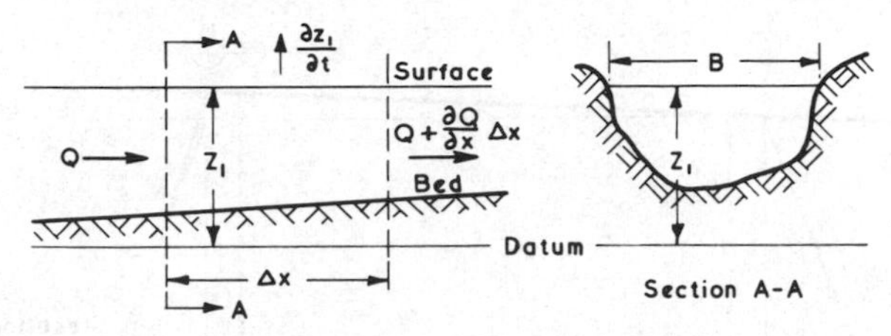

Figure 2.4. Principle of conservation applied to flow through a reach of an estuary

The condition of continuity of flow requires that

$$\frac{\partial Q}{\partial x}\Delta x\,\Delta t = -B\,\Delta x\frac{\partial z_1}{\partial t}\Delta t$$

or

$$\frac{\partial Q}{\partial x}+B\frac{\partial z_1}{\partial t} = 0 \qquad (2.17)$$

Thus if the discharge decreases in the x direction, there will be a proportional rise in water level and vice versa.

2.5 NON-HOMOGENEOUS LIQUID

When density variations are likely to affect the flow, the simplified equations 2.13, 2.14 or 2.16 are inadequate. If density variations over the vertical are significant it is necessary to use equation 2.8 to find local pressure gradients. This leads to

$$\frac{\partial p}{\partial x} = \frac{\partial}{\partial x}\left[\int_z^{z_1}\rho g\,\mathrm{d}z\right]$$

which can be solved numerically if adequate field measurements are available.

Two simpler cases can be considered. One is the case of well-mixed flow, in which density gradients in a horizontal direction can be assumed to be constant over the depth at a given cross-section. In this case, equation 2.9 gives $\partial p/\partial x = g[\rho(\partial h/\partial x)+h(\partial\rho/\partial x)]$. When $(\partial\rho/\partial x)$ is constant over a vertical this can be averaged to give:

$$\frac{1}{H}\int_{z_0}^{z_1}\frac{\partial p}{\partial x}\mathrm{d}z = -\rho g i_x+\frac{gH}{2}\frac{\partial\rho}{\partial x}$$

Equations 2.5 and 2.6 averaged over the vertical then become

$$\frac{1}{g}\frac{\partial U}{\partial t}+\frac{U}{g}\frac{\partial U}{\partial x}+\frac{V}{g}\frac{\partial U}{\partial y}-i_x+\frac{H}{2\rho}\frac{\partial\rho}{\partial x}+\frac{\tau_{z_0x}}{\rho g H} = 0 \qquad (2.18)$$

$$\frac{1}{g}\frac{\partial V}{\partial t}+\frac{U}{g}\frac{\partial V}{\partial x}+\frac{V}{g}\frac{\partial V}{\partial y}-i_y+\frac{H}{2\rho}\frac{\partial\rho}{\partial y}+\frac{\tau_{z_0y}}{\rho g H} = 0 \qquad (2.19)$$

and averaged over the cross-section,

$$\frac{1}{g}\frac{\partial U}{\partial t}+\beta'\frac{U}{g}\frac{\partial U}{\partial s}-i_s+\frac{H}{2\rho}\frac{\partial\rho}{\partial s}-i_f = 0 \qquad (2.20)$$

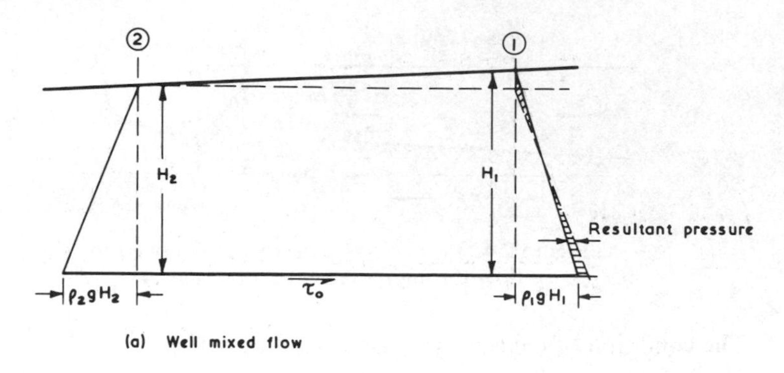

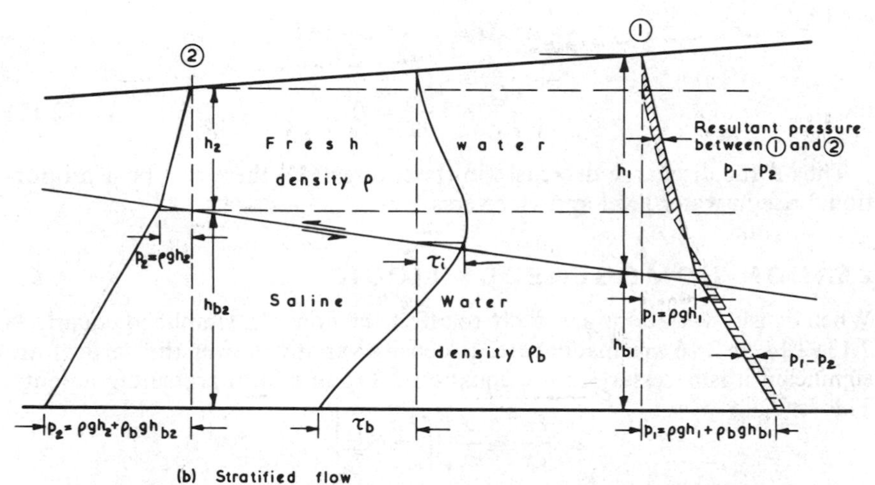

Figure 2.5. Effect of density on forces on a reach of an estuary. The diagrams represent instantaneous conditions during a tidal cycle

Equations 2.18 to 2.20 will be used in later chapters for the solution of some problems in estuaries. The other case is that of stratified flow, in which density can be regarded as constant in each layer. Equations 2.13, 2.14 or 2.16 can then be applied to each layer provided that the shear stress at the interface can be defined[8].

In well-mixed flow, the driving force is the horizontal variation of pressure caused by the horizontal variation of density and the surface slope of the water, as in figure 2.5(a). In stratified flow, the horizontal variation of density is negligible within each layer and the driving force arises from the surface and interfacial slopes, as in figure 2.5(b).

Each layer in a stratified flow behaves as a gravity wave. Vertical motion of the interface can be quite large because of the small difference in density between the layers. It can take up a steep radial slope during flow round a bend and this can give rise to lateral oscillations of the interface downstream of a bend. The effect of rotation of the Earth can also cause the interface to take up a pronounced slope lateral to the current. There will be a critical velocity in each layer with respect to the other, analogous to the critical velocity in open-channel flow with homogeneous water when Froude number $U/(gH)^{\frac{1}{2}} = 1$. This critical relative velocity is $U_1 = \pm(gh_1 \Delta\rho/\rho)^{\frac{1}{2}}$ where U_1 and h_1 are velocity and depth in one of the layers. $\Delta\rho$ is the difference in density between the layers.

A discussion of the behaviour of interfaces was given by Schijf and Schönfeld in 1953[9]. A considerable amount of work has been done since then[10], with attention directed particularly towards diffusion, shear stress and stability of the interface. Diffusion is discussed further in section 3.5.

2.6 PHYSICAL BEHAVIOUR ACCORDING TO THE EQUATIONS OF MOTION

Much of the physics of estuarial flows can be understood by considering how each of the components of the equations of motion vary. For simplicity, it will be assumed that the main flow is in the x-direction so that terms involving v and horizontal shear will be ignored.

Consider equation 2.5 with the pressure gradient expressed in terms of liquid density as in equation 2.9.

$$\frac{1}{g}\frac{\partial \bar{u}}{\partial t} + \frac{\bar{u}}{g}\frac{\partial \bar{u}}{\partial x} = -\left(\frac{\partial h}{\partial x} + \frac{h}{\rho}\frac{\partial \rho}{\partial x}\right) + \frac{1}{\rho g}\frac{\partial \tau_{zx}}{\partial z} \tag{2.21}$$

This equation applies at a point at height z above datum so that $h = z_1 - z$.

Each term in the equation varies with depth. Thus velocity is zero at the bed, and at maximum flow reaches a maximum near the surface, so $(1/g)\,\partial\bar{u}/\partial t$ and $(\bar{u}/g)\,\partial\bar{u}/\partial x$ must do the same. Shear stress, on the other hand, is zero at the surface and in steady, or gradually varied flows, reaches a maximum at the bed. The gradient of shear stress, however, is constant at all depths in steady, uniform flows. It can vary in tidal flows but will not consistently reach a maximum at any particular level. $\partial h/\partial x$ is equal to the surface slope and is thus constant at all levels at any given time. In steady flows of homogeneous liquids, $(\bar{u}/g)\,\partial\bar{u}/\partial x$, $\partial h/\partial x$ and $(1/\rho g)\,\partial\tau_{zx}/\partial z$ are in balance at any value of z. In unsteady flows, on the other hand, $\partial h/\partial x$ varies with time and, in consequence, flow acceleration $\partial\bar{u}/\partial t$ takes place.

In the greater part of most estuaries $(\bar{u}/g)\,\partial\bar{u}/\partial x$ is small compared with $\partial h/\partial x$ or $(1/g)\,\partial\bar{u}/\partial t$. The factor causing acceleration of the water is the varying surface gradient and its effect is the same at all depths. In the absence of friction, the pressure gradient $\partial h/\partial x$ would simply cause equal acceleration $(1/g)\,\partial\bar{u}/\partial t$ at all depths. Friction modifies the flow, and velocity is, in consequence, zero at the bed. If friction forces are large, as in a shallow stream over a rough boundary, with relatively high current speeds, the instantaneous velocity profiles will approximate to steady flow profiles[11], but in other cases, the existence of a force equal at all depths but varying with time will cause an increase in velocity gradient near the bed in accelerating flows and a decrease in velocity gradient during decelerating flows.

The effect is small in real estuaries and may be concealed by density gradients and secondary flows. It has been measured in laboratory

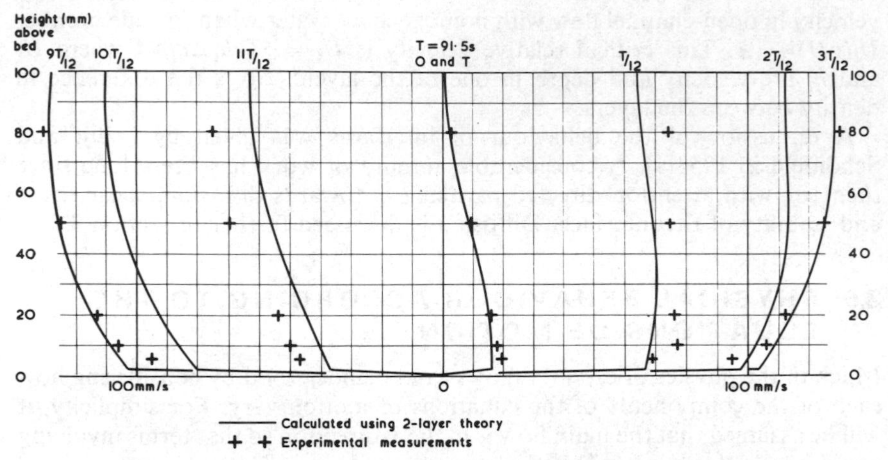

Figure 2.6. Velocity distributions calculated for turbulent flow using Agnew's two-layer theory for flow period $T = 91{\cdot}5$ s and depth 100 mm

channels[11, 12, 13] and can be analysed theoretically for simple laminar flows. The analysis of turbulent flows is more difficult, but an approximate solution was obtained by Agnew[14] who divided the flow arbitrarily into layers, to each of which he assigned a different coefficient of viscosity (figure 2.6). In gradually varied flows with relatively high friction, maximum shear stress at the bed coincides with maximum speed of flow averaged over the vertical. In more rapidly varied flow with frictional forces appreciably smaller than pressure gradients, maximum shear stress at the bed occurs before maximum speed of flow. A similar effect is noticed in plan; in this case, the friction at the sides of estuaries is small and reversal of flow at the margins of channels before reversal at the centre is a common and well-known effect[14].

The term due to density in equation 2.21: $(h/\rho)\ \partial\rho/\partial x$, is directly proportional to depth below the surface. Density increases in the seaward direction, so the term is directed upstream. Over a considerable length of many estuaries, $\partial\rho/\partial x$ is nearly constant at any instant[15] (see figure 2.7). The fresh water flow that gives rise to the density gradient must produce a net seaward drift. Evidently this net seaward drift will be greatest at the surface and least at the bed. If density effects are relatively large at a point in an estuary the net drift can be up-river near the bed[16], as shown in figure 1.14 and figure 3.6. Heaps[17] has shown that the small density gradients existing in coastal waters are sufficient to cause a net landward movement at the sea bed in Liverpool Bay, England.

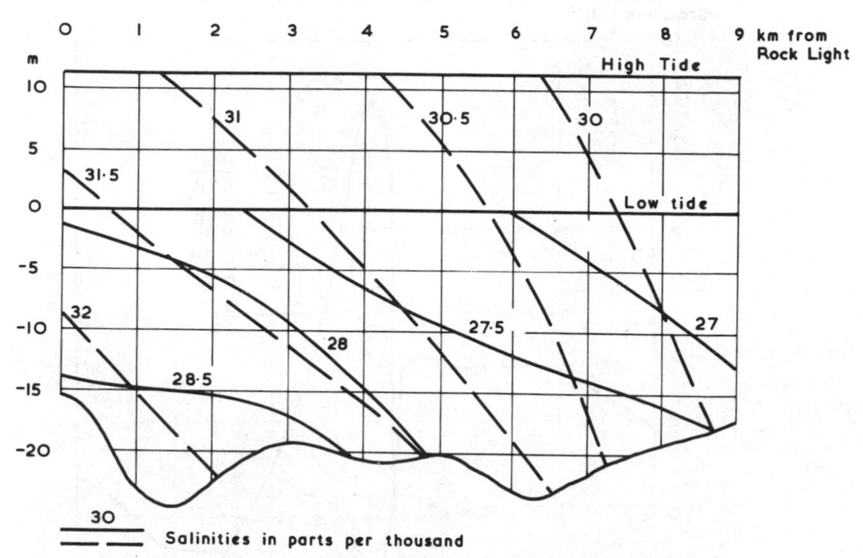

Figure 2.7. Iso-halines at high and low water in the Mersey Estuary (see figure 1.9). (Based on measurements by Hughes[15], courtesy the Royal Astronomical Society)

2.7 ORDERS OF MAGNITUDE OF TERMS IN EQUATIONS OF MOTION

The use of the equations in the analysis of estuarine flows will be covered in later chapters, but at this stage it is useful to have some idea of the magnitudes of the different terms that are likely to be encountered. The terms vary with time and do not reach their maxima simultaneously. Figure 2.8 illustrates how they behaved at one reach of the River Hooghly for one particular spring tide. At this point in the river spring tides cause a tidal bore to develop, resulting in very high local values of the surface gradient and acceleration. It is worth noting that the convective acceleration, $(U/g)\,\partial U/\partial s$, was small even under these rather extreme conditions.

Values of the terms of equation (2.20) in the Thames, Mersey (UK) and Hooghly estuaries are given in table 2.2. The values quoted here are typical maximum values occurring during normal spring tidal cycles in middle reaches of each estuary.

The Coriolis term has a negligible magnitude in the x-direction because the speed of flow V at right angles to the river axis is very small. In the y-direction, however, it is appreciable. At latitude 55°:

$$\frac{2\omega U \sin\phi}{g} = 1{\cdot}2U \times 10^{-5}\ \text{m/s}$$

which is close enough to the components of the equation in the s-direction for there to be a significant effect in wide channels.

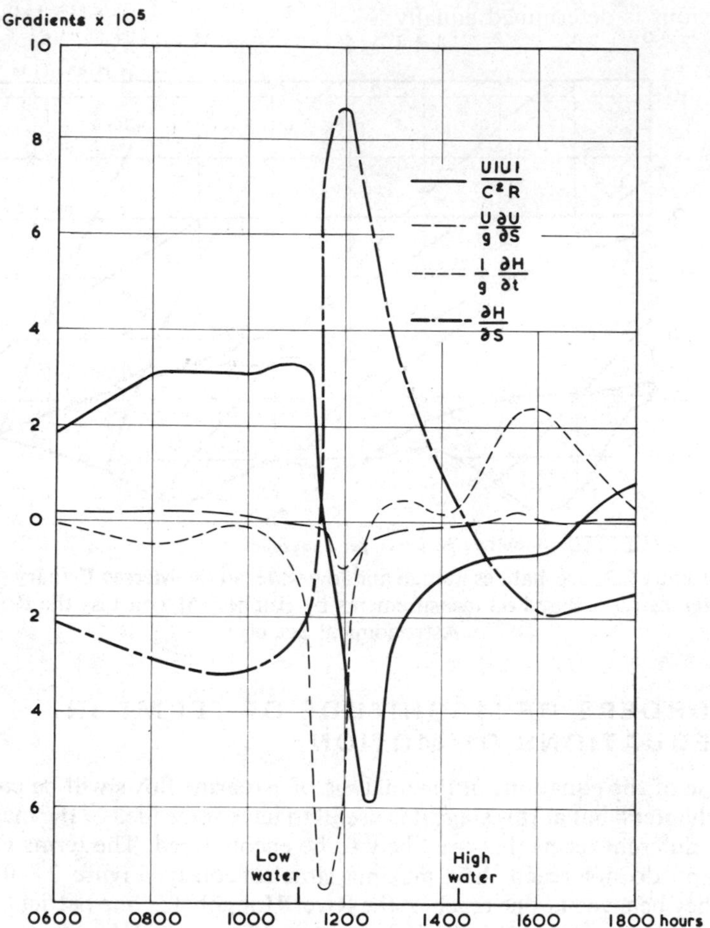

Figure 2.8. Terms of equation of motion in the River Hooghly near Calcutta. The small density term has been omitted

Table 2.2

	$\frac{\partial z_1}{\partial s} = -i_s$	$\frac{1}{g}\frac{\partial U}{\partial t}$	$\frac{U^2}{C^2 H} = -i_f$	$\frac{U}{g}\frac{\partial U}{\partial s}$	$\frac{H}{2\rho}\frac{\partial \rho}{\partial s}$
Thames	3×10^{-5}	2×10^{-5}	4×10^{-5}	$1{\cdot}5 \times 10^{-6}$	1×10^{-6}
Mersey	7×10^{-5}	4×10^{-5}	6×10^{-5}	$1{\cdot}3 \times 10^{-5}$	$2{\cdot}9 \times 10^{-6}$
Hooghly	$1{\cdot}4 \times 10^{-4}$	8×10^{-5}	6×10^{-5}	4×10^{-6}	$1{\cdot}5 \times 10^{-6}$

It would be useful to be able to classify estuaries into types, but a glance at these orders of magnitude shows how difficult this can be. $\partial z_1/\partial s$ depends on the rate of rise and fall at the mouth and on the slope of the bed. This term shows the biggest variation between different estuary systems, but the overall behaviour is determined equally by the rate of dissipation of energy, while the behaviour over depth of the stream is determined by density gradients, which in turn depend on the magnitude of fresh water flow. A further difficulty is that the relative magnitudes of the terms vary at different points along an estuary. In particular, salinity gradients may be large where rapid changes of bed level occur, as at Eastham on the River Mersey, and small elsewhere (see figure 1.9).

2.8 SUMMARY

The equations of motion and conservation of matter apply at a point in space and at an instant in time. Their direct practical application is limited because of the turbulent nature of the flow and because of the complexity of boundary conditions.

When averaged over a time interval long enough to smooth out the effects of turbulence, they can be used as a basis for study of estuarine behaviour. In particular, they demonstrate the order of magnitude of forces due to friction and pressure gradients, and the resultant accelerations, which give rise to small phase differences of velocity over the water depth. The effect of rotation of the Earth can also be demonstrated.

It is shown that density gradients give rise to a force, directed landwards, that is greater at the stream bed than at the water surface. When this force is superimposed on the oscillating tidal force, it can give rise to a differential net water movement over the stream depth. Depending on the relative strength of river flow and tidal flow, this may give rise to a stronger net seaward flow at the surface than at the bed or, when tidal flow is relatively weak, a net landward flow at the bed. This has far-reaching consequences on transport of sediment, salt and pollutants in estuaries.

It is also shown that rotation of the Earth produces a significant effect in wide estuaries at latitudes away from the equator.

When averaged over the depth or over a cross-section of an estuary, the equations 2.13 to 2.16 are simple enough to form the basis of numerical solutions of estuary problems and of the laws of scaling of physical models, described in Chapters 6 and 8.

2.9 NOTATION

a	body force on unit mass
B	width of water surface at time t
C	Chezy's coefficient
g	gravitational force on unit mass
h	depth below water surface, $z_1 - z$
H	total depth of water, $z_1 - z_0$
i_j	surface slope in j-coordinate direction
i_f	total dimensionless energy loss per unit length
n	Manning's coefficient
p	pressure intensity
Q	volume of water flowing per second

R	hydraulic radius (cross-sectional area/wetted perimeter); $R \approx$ cross-sectional mean depth in most estuaries
s	suffix indicates axial direction in one-dimensional representations of estuaries
t	time
T	tidal period
u, v, w	local components of velocity at time t, in x, y, z coordinate directions respectively
$\bar{u}, \bar{v}, \bar{w}$	temporal mean values of u, v, w
u', v', w'	instantaneous departure of u, v, w from $\bar{u}, \bar{v}, \bar{w}$
U	cross-sectional mean velocity in one-dimensional equation
U, V	temporal mean velocity components averaged over depth in two-dimensional equations
x, y, z	co-ordinate directions; x and y horizontal, z positive upwards; suffixes indicate components in x, y, z-directions
z_1, z, z_0	height of water surface, point, bed above datum, respectively; suffixes indicate conditions at surface, point and bed respectively
β'	momentum coefficient
ϕ	angle of latitude on Earth
ρ	fluid density
ω	angular velocity
τ	shear stress
τ_{ij}	component of shear stress normal to i-axis parallel to j-axis

REFERENCES

[1] Daily, J. W. and Harleman, D. R. F., *Fluid dynamics*, Ch. 5, Addison-Wesley (1966)

[2] Lamb, H., *Hydrodynamics*, 6th Edition, Ch. XI, Cambridge University Press (1932)

[3] Daily, J. W. and Harleman, D. R. F., *Fluid dynamics*, Ch. 2, Addison-Wesley (1966)

[4] Muir Wood, A. M., *Coastal Hydraulics* Ch. 1, Macmillan (1969)

[5] Osborne Reynolds, *Phil. Trans. Roy. Soc.*, **174**, 935 (1895)

[6] Osborne Reynolds, *Scientific Papers*, Vol. 2, p. 535, Cambridge University Press (1901)

[7] Henderson, F. M., *Open channel flow*, Ch. 1, Sect. 1.9, Macmillan (1966)

[8] Ippen, A. T., (Ed) *Estuary and coastline hydrodynamics*, Ch. 12 and 13, McGraw-Hill (1966)

[9] Schijf, J. B. and Schönfeld, J. C., Theoretical considerations on the motion of salt and fresh water, *Vth Congress IAHR*, Minnesota, pp. 321–333 (1953)

[10] *International Symposium on Stratified Flows, IAHR*, Novosibirsk (1972)

[11] McDowell, D. M., The vertical distribution of water velocities in tidal streams. Paper A1, *VIth Congress, IAHR*, The Hague (1955)

[12] McDowell, D. M., Some effects of friction on oscillating flows in laboratory channels, Paper 3.4, *XIth Congress, IAHR*, Leningrad (1965)

[13] Yalin, M. S. and Russell, R. C. H., Shear stresses due to long waves, *J. Hyd. Res.* **4**, 2 (1966)

[14] Agnew, R., Estuarine currents and tidal streams, *Proc. 8th Coastal Engineering Conference*, (1960), Ch. 28, 510–535, and Ph.D. thesis Queen's University of Belfast (1959)
[15] Hughes, P., Tidal mixing in the Narrows of the Mersey estuary, *Geophys. J. R. astr. Soc.*, **1**, 4, Dec. (1958).
[16] Price, W. A. and Kendrick, M. P., Field and model investigation into the reasons for siltation in the Mersey Estuary, *Proc. Inst. Civil Eng.*, **24**, 473 (1963)
[17] Heaps, N., Estimation of density currents in the Liverpool Bay area of the Irish Sea, *Geophys. J. R. astr. Soc.*, **30**, 415–432 (1972).

3

Mixing processes

A sound understanding of flow and mixing processes in tidal waters is essential if engineers are to ensure that their works produce little adverse effect upon the aquatic environment. Typical engineering schemes which can cause a change in tidal mixing processes and thereby in the quality of tidal waters are:

(i) disposal of liquid and solid pollutants;
(ii) large scale extraction of water for power station cooling systems;
(iii) river diversion schemes;
(iv) construction of river regulating structures, including dams;
(v) large-scale land reclamation projects;
(vi) construction of tidal barrages and navigation channel training works;
(vii) dredging of navigation channels to accommodate larger ships.

Failure to fully understand the basic mechanisms which control tidal mixing processes has produced many instances of inadequately planned engineering schemes. For example, the rapid expansion of industrialised coastal communities has generally been followed by excessive discharges of untreated domestic and industrial effluents, with the result that estuarine and coastal waters have become badly polluted in many parts of the world.

The object of the present chapter is to examine the factors which contribute to tidal mixing processes and to consider the equations which engineers can use in order to assess the consequences of their proposed schemes. The chapter starts by considering some details of the turbulent flow structure in tidal systems and provides a basis for understanding how mixing processes are modified by engineering works. Next, mixing processes are considered in mathematical form so that the component parts are clearly seen. Various engineering approximations are then introduced into the equations so that they can be used to evaluate the impact of engineering schemes on the aquatic environment. Detailed solution methods for the equations are, however, deferred until Chapter 7. The latter part of the chapter deals with a discussion on the variability of diffusion/dispersion coefficients in various flow conditions and the means whereby they may be determined.

3.1 FLOW STRUCTURE

Tidal flows are unsteady, non-uniform turbulent motions. Unsteady implies that the velocity at a point in the flow varies with time; the prime cause being the periodic water-surface level changes at the estuary mouth. Non-uniform means that velocities vary spatially; while turbulent means that the instantaneous velocity varies randomly with respect to both space and time about some mean value. It is possible to produce a visual record of the flow structure, by placing a heated wire probe or a small current meter in the flow. An example is shown in figure 3.1 for a fixed point in space, which shows the component of speed in the horizontal direction (x). This figure illustrates an immediate difficulty of defining turbulence in a tidal environment. It is usual in steady flow situations to consider the flow velocity to be composed of a mean ($\bar{u}$) and a fluctuating component (u'), which is zero when integrated over a short time interval (t_1). This is not so in a tidal situation and the determination of ($\bar{u}$) represents a considerable theoretical problem. It can be determined in some oscillatory systems, such as tidal models with identical successive tides, by taking an ensemble-average of values at the same relative time on several cycles. Even this is impossible in real tides because of diurnal inequality, the spring–neap cycle and the vagaries of winds and fresh water flows. Analysis of the type used in Chapter 2 and subsequently can only be true if $\bar{u}$ is considered to be an ensemble-average for identical times or is considered to be independent of time over the averaging period (t_1). The latter is a reasonable approximation in many river and estuarine flows if t_1 is taken to be the order of 1–10 minutes, but would not be so in an oceanic environment where large size eddying motion occurs.

The velocity record of figure 3.1 can be considered the result of the mean flow ($\bar{u}$) convecting a large number of vortices (or eddies) past the velocity meter. The eddies have a wide range of size and orientation and have been produced initially in a variety of ways. These include flow instability within the so-called laminar sublayer at the estuary boundaries[32], flow separation from sediment particles and bed irregularities as well as from dock entrances and channel sides, interfacial disturbances between opposing tidal streams, wind and wave action and ship motions.

Once eddying motion has been produced, it is distorted by local velocity gradients ($\partial u/\partial x$, etc.) and by the temporal acceleration of the flow. Each eddying motion can be considered as an isolated vortex tube, which is rotating at a constant angular velocity, and is transported bodily through the surrounding velocity field. The effect of gradients in the velocity field is to produce an increase in tube length and a decrease in tube cross-sectional area. The consequence of reducing the cross-sectional area of the tube is to make the vortex rotate faster and this leads in turn to increased local velocity gradients at right angles to the original direction of spatial motion. Other eddies are, therefore, also reduced in size. This interactive three-dimensional process continues throughout the flow field until the large eddies are reduced to such a size that viscous shear stresses damp out the eddying motion; kinetic energy being converted into heat continuously. A particular turbulent motion will, therefore, contain a whole unique spectrum of eddies, ranging

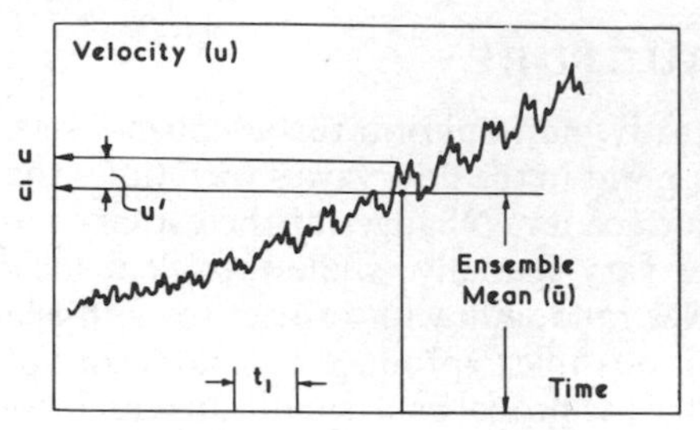

Figure 3.1. Variation of horizontal water velocity with time

in size from perhaps a millimetre up to the physical limit of the flow system.

The eddy reduction process is clearly driven by the turbulent-mean velocity gradients within the flow. Large turbulent-mean gradients will cause a rapid reduction of eddy size and will lead to the presence in the flow of a large number of rapidly rotating eddies. Conversely, small velocity gradients will lead to the presence of slowly rotating eddies. It is not surprising, therefore, to find large velocity flunctuations (u', v', w') near solid boundaries and small velocity fluctuations near the water surface. The standard deviation of velocity fluctuations ($\sigma_u = \sqrt{(\bar{u}'^2)}$, σ_v, σ_w) provides a simple description of the degree of turbulence of a particular flow. Alternatively, the *intensity* of turbulence σ_u/U_*, where $U_* = \sqrt{(\tau_{z0}/\rho)}$ and τ_{z0} is the bed shear stress, is used as a characteristic parameter in steady flows, and has typical values between one and three.

The intensity of turbulence clearly varies over a river flow cross-section but its value is also controlled by longitudinal changes in channel geometry. Flow in a converging channel will experience spatial acceleration and a reduction in turbulent-mean velocity gradient since near-boundary fluid responds readily to imposed forces. Spatial flow acceleration will, therefore, lead to a reduction in turbulent intensity while flow deceleration, as occurs in a diverging channel, will produce an increase in turbulent intensity. Temporal acceleration or deceleration of a particular flow has a similar effect to that produced by the corresponding spatial effect.

It is now possible to see how the construction of engineering works can affect the turbulent structure of a particular tidal flow. Reclamation schemes cause changes in the spatial and temporal accelerations of the flow and introduce solid boundaries in areas where previously a liquid surface existed. Open piled structures increase local turbulence levels in the wake of individual piles, while the construction of a solid pier perpendicular to the flow causes the formation of large slow moving lee-eddies.

The mixing of suspended or dissolved substances in a turbulent flow is directly affected by changes in both turbulent and mean flow rates. This can be seen by considering the turbulent-average mass transport rate per unit area (tr) in the x direction, i.e.

$$\text{tr} = \frac{1}{t_1}\int_t^{t+t_1} \rho u c \, \mathrm{d}t \tag{3.1}$$

where c is the instantaneous concentration of any quantity present in the flow defined as the mass of the quantity per unit mass of fluid, and ρ is the fluid density defined as the mass of fluid per unit volume. If both u and c are expressed in terms of a mean ($\bar{u}$, $\bar{c}$) and a fluctuating quantity (u', c') equation (3.1) can be written as:

$$\text{tr} = \frac{1}{t_1}\int_t^{t+t_1} \rho(\bar{u}\,\bar{c}+u'c')\,\mathrm{d}t = \bar{u}\,\bar{c}+\overline{u'c'} \tag{3.2}$$

The overbar indicates a time average over the interval t_1, and the integral of the product of a mean and a fluctuating quantity is taken as zero. Equation 3.2 shows that the rate of transport of a substance depends upon both mean and fluctuating quantities; consequently engineering works which modify the structure of a flow will also produce a change in the transport rate of substances contained in that flow.

Additional changes in mass transport rates are also produced by the presence of dissolved and suspended matter in the flow, irrespective of the construction of any engineering works. The analogous equation to equation 3.2 for the vertical direction (z) indicates that turbulent velocity components are capable of transporting material away from the solid boundaries of a flow system. If the density of the fluid near the solid boundaries is greater than that near the water surface, i.e. a vertical density gradient exists, energy must be expended in order to mix the denser and less-dense fluids together. This additional energy must be extracted from the total energy of the moving flow. Because it is easier to overcome the momentum of the velocity fluctuations, the energy loss appears as a reduction in the kinetic energy of the turbulent eddies, so that velocity fluctuations are reduced in the presence of vertical density gradients such as occur in the ocean due to differential solar heating, and in tidal estuaries due to the mixing of salt and river water.

Vertical density gradients can be modified by the direct addition of large quantities of low or heavy-density liquid pollutants to the aquatic environment, and by the presence in the water of large amounts of suspended sediment. Any change in flow structure due to the presence of sediment is, however, dependent upon the size, density and numbers of sediment particles present in the flow. Steady-flow laboratory experiments[1] show that large volume concentrations (30 %) of neutrally-buoyant (same density as the flow) sand-sized (100 μm) particles produce an increase in longitudinal turbulent intensity but no change in flow resistance. However, table 3.1 shows that large volume concentrations of gravel-sized quartz particles produce a reduction in both longitudinal and vertical turbulent intensities and, by implication, a reduction in flow resistance.

The difference between the neutrally-buoyant and glass particle tests is due to the net change in the energy distribution within the flow. Large volume concentrations reduce the volume of liquid within which turbulent energy is dissipated and thus turbulent intensities are increased above clear water values. Large sized sediment particles are transported with a large vertical density gradient (see Chapter 4), turbulent energy losses are high and turbulent intensities are reduced. Silt- and clay-sized particles would thus be expected to have little effect upon the structure of a tidal flow since such particles are

Table 3.1. *Effect of large concentrations of gravel-sized* (2–4 mm) *quartz particles on turbulent intensities in a* 560 mm *pipeline.* (*Based on Pechenkin*[2])

Volume concentration (%)	Relative* longitudinal turbulent intensity	Relative* vertical turbulent intensity
10	0·97	0·89
20	0·89	0·60
30	0·77	0·35
40	0·59	0·16
50	0·39	0·025

* Relative to a clear flow. Flow Reynolds number $\approx 1{\cdot}5 \times 10^5$.

transported with a small vertical density gradient and volume concentration values are generally less than 10 % in most tidal situations.

Clearly, vertical density gradients can have a considerable effect on the turbulent structure of a flow. They may also affect the cross-sectional distribution of the turbulent-mean flow since a change in turbulent velocity will produce a corresponding change in the transport of momentum within the flow cross-section. A reduction in the vertical turbulent velocity fluctuations will thus produce a steeper gradient of the turbulent-mean velocity profile over the flow depth which will, in turn, tend to increase turbulent intensities and off-set the reduction produced by the vertical density gradient. Of course, any change in the mean and turbulent structure of the flow will also produce a change in the magnitude of vertical density gradients. It is important, therefore, to consider the effect of engineering works on both the structure of a tidal flow and on vertical density gradients.

There are three methods available to engineers in order to assess the effects of their schemes on the aquatic environment. These are to:

(i) construct a small-scale hydraulic model of the tidal system;

(ii) describe the tidal mixing processes by mathematical equations and solve them either numerically or analytically, i.e. construct a mathematical model;

(iii) use past and present measurements from the actual tidal system together with any necessary mathematical techniques in order to predict future conditions.

All three techniques are considered in more detail in later chapters (see Chapters 5, 6 ,7 and 8), but the basic equations used in method (ii) are presented and discussed in the following section.

3.2 GENERAL TRANSPORT EQUATIONS

Mixing processes in a tidal environment can be described by use of the mass continuity principle given in Chapter 2.

Consider an elemental control volume of size ∂x, ∂y, ∂z and orthogonal coordinate axes (x, y, z). The rate of change of mass per unit volume at the centre of the control volume can be written down by following a similar method to that shown in Chapter 2 (p. 33). The result is the equation

$$\frac{\partial}{\partial t}(\rho)+\frac{\partial}{\partial x}(\rho u)+\frac{\partial}{\partial y}(\rho v)+\frac{\partial}{\partial z}(\rho w) = -\frac{\partial}{\partial x}(F_x)-\frac{\partial}{\partial y}(F_y)-\frac{\partial}{\partial z}(F_z) \quad (3.3)$$

where ρ is the density (mass of solution/volume of solution) at the centre of the control volume, and F_x, F_y, F_z represent the rates of transport of mass through unit area due to random molecular motion in the x-, y-, z-directions respectively.

The molecular mass flux is assumed to be described by Fick's Law, which relates it to the concentration gradient in the direction of motion. For example, in the x direction:

$$F_x = -D\frac{\partial}{\partial x}(\rho) \quad (3.4)$$

where D is called the coefficient of molecular diffusion and is a constant for a particular substance at a given temperature. It has the dimensions (L^2/T), i.e. of a velocity (V_1) times a length (L_1), which can be considered as characteristic of the mean velocity of the molecules and the mean free path through which they travel before colliding with other molecules. Since mean free paths are very small in fluids, it is not surprising to find that D is small (10^{-3} mm^2/s) and that the spreading action due to molecular motion is also small.

Similar equations to 3.3 can be used to describe the spreading of dissolved or suspended matter in a tidal environment. For example, the estuarine distribution of salinity (s), defined as the mass of salt per unit mass of sea water, can be described by the equation:

$$\frac{\partial}{\partial t}(\rho s)+\frac{\partial}{\partial x}(\rho us)+\frac{\partial}{\partial y}(\rho vs)+\frac{\partial}{\partial z}(\rho ws) = 0 \quad (3.5)$$

in which molecular transports and the effects of pressure and temperature are assumed to be negligible.

Equations 3.3 and 3.5 combine to give an equation describing the spatial and temporal change of salinity

$$\frac{\partial}{\partial t}(s)+\frac{\partial}{\partial x}(us)+\frac{\partial}{\partial y}(vs)+\frac{\partial}{\partial z}(ws) = 0 \quad (3.6)$$

where it is assumed that molecular effects are negligibly small and the mean density is only due to the presence of salt in the water. Equation 3.5 is, however, a general equation and could be used to describe the movement of any conservative (i.e. no loss or gain of mass with time) dissolved substance in the flow. The salinity (s) is simply replaced by the substance concentration

(c) defined in terms of the mass of the substance divided by the mass of the solution. Notice that in this case, equation 3.6 follows directly from 3.5 for constant density systems.

Equation 3.5 may also be used to describe the spreading of suspended particles in the flow provided that they do not interfere with the mean or turbulent motion. This implies low volume concentrations (see table 3.1) and small particle sizes. In this case the only modification to equation 3.3 is the inclusion of the vertical motion of the sediment particles under gravitational attraction. This is considered to be described by the terminal fall velocity (w_f) of the particles in still water. A term $-w_f(\partial c/\partial z)$ would then be added to the left-hand side of equation 3.5 assuming that z is measured vertically upwards from the estuary bed.

If the mixing of a non-conservative substance (i.e. one involving the generation or decay of mass with time) is to be described by a similar equation to 3.6 then an appropriate loss-of-mass term must be included in the equation. For example, if the total number of coliform bacteria present in a given volume of sea water reduces with time in an exponential manner, this is equivalent to adding the term $(-K_1 c)$, where K_1 is a decay rate coefficient, to the right-hand side of equation 3.5. K_1 may be negative for certain chemical reactions, in which case the term $(K_1 c)$ represents an addition of mass to the control volume.

Application of the first law of thermodynamics to the control volume produces a similar equation to 3.1 and describes the variation of temperature (θ) over space and with time:

$$\frac{\partial}{\partial t}(\theta)+\frac{\partial}{\partial x}(u\theta)+\frac{\partial}{\partial y}(v\theta)+\frac{\partial}{\partial z}(w\theta) = \alpha\left[\frac{\partial^2\theta}{\partial x^2}+\frac{\partial^2\theta}{\partial y^2}+\frac{\partial^2\theta}{\partial z^2}\right]+\dot{q} \quad (3.7)$$

where α is a molecular thermal diffusion coefficient equal to $\kappa/\rho C_p$ in which κ is the thermal conductivity of the fluid and C_p is its specific heat at constant pressure; $\dot{q}$ represents the rate of addition of heat to the control volume, including the effect of viscosity, although in most practical problems viscous gains can be neglected.

The effect of turbulence is now introduced into equation 3.6 by writing quantities in terms of a mean and a fluctuating component, ($s = \bar{s}+s'$) bearing in mind the theoretical difficulty of time-averaging a tidal flow, as discussed earlier. Insertion of these quantities into equation 3.6 and time-averaging over the period t_1 results in the equation:

$$\frac{\partial(\bar{s})}{\partial t}+\frac{\partial}{\partial x}(\bar{u}\bar{s})+\frac{\partial}{\partial y}(\bar{v}\bar{s})+\frac{\partial}{\partial z}(\bar{w}\bar{s}) = -\frac{\partial}{\partial x}(\overline{u's'})-\frac{\partial}{\partial y}(\overline{v's'})-\frac{\partial}{\partial z}(\overline{w's'}) \quad (3.8)$$

in which an overbar indicates a turbulent-mean quantity, i.e.

$$\frac{1}{t_1}\int_t^{t+t_1}(\,)\mathrm{d}t.$$

It should again be remembered in deriving equation 3.8 that the temporal

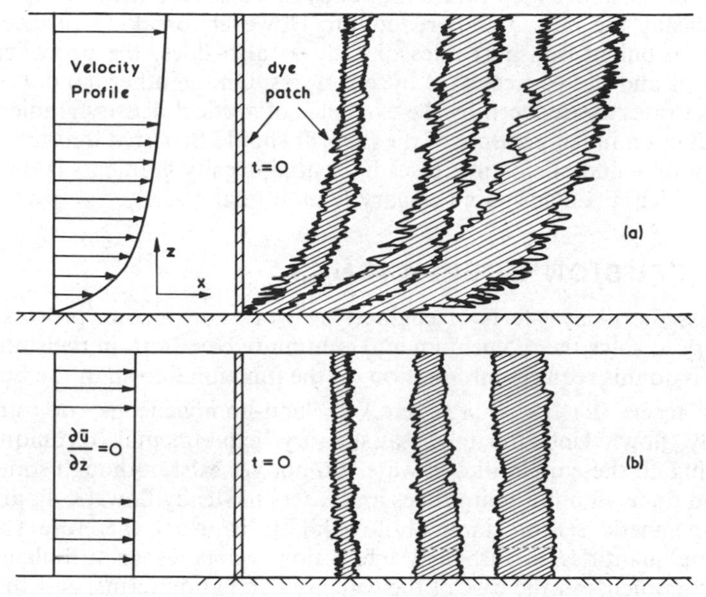

Figure 3.2. Longitudinal movement of a vertical dye patch

mean of a fluctuating quantity, or the product of a mean quantity and a fluctuating quantity, is zero.

Equation 3.8 implies simply that the change of salinity with time at a point in the flow is determined by spatial differences in transport rates associated with both the turbulent-mean flow ($\bar{u}$, $\bar{v}$, $\bar{w}$) and the velocity fluctuations (u', v', w'). The term *advection* is often used to describe transport of material by the mean flow field to distinguish it from *convection* which is associated with thermally-induced motion.

The physical interpretation of equation 3.8 is that salt water is advected back and fourth along an estuary by the turbulent mean flow field while at the same time turbulent-mean velocity gradients ($\partial\bar{u}/\partial\bar{z}$, etc.) accelerate the longitudinal and lateral spreading of the salt water through the estuary and the turbulent eddies mix the salt and fresh water together. The longitudinal spreading action of velocity gradients is illustrated in figure 3.2 by observing the motion of a concentrated vertical column of dye at successive time intervals.

Longitudinal and lateral advective spreading of the salt water will also be encouraged by velocity gradients resulting from the movement of water masses by wave action, wind-induced shear forces at the water surface, Coriolis force, local centrifugal forces produced by channel curvature, and by horizontal density gradients produced by spatial differences in the concentration of dissolved or suspended matter. It should be noted, however, that the extra spreading effects of wind and wave action and of Coriolis induced currents are small in narrow rivers and estuaries, particularly in comparison with the effects of channel curvature and horizontal density gradients.

Many of the above factors also indirectly increase turbulent transport rates by increasing the rate of eddy-reduction. However, breaking surface waves increase turbulent transport rates directly as also does, the propeller wash from ships and the vortices shed by structures. On the other hand, turbulent transport rates are reduced by the existence of vertical density gradients and by a reduction in temperature of the flow. It should be noted that the modification of flow and mixing processes by spatial density gradients is the major feature which distinguishes an estuary from a tidal river.

3.3 DIFFUSION COEFFICIENTS

To assess the likely effects of engineering works on mixing processes it is necessary to solve the momentum and continuity equations in their turbulent form. To do this requires information on the functional form of the turbulent transfer terms ($\overline{u'v'}$, $\overline{w'v'}$, $\overline{u's'}$, etc.) in non-homogeneous, non-uniform, unsteady flow. Unfortunately, satisfactory experimental techniques for measuring all these quantities in water do not yet exist, although some work has been done with hot film probes and lasers in steady flows[3, 4], and with electro-magnetic sensors in tidal flows[5]. It is usual therefore to relate turbulent quantities to turbulent mean flow velocities or turbulent mean velocity gradients in the case of the velocity fluctuation terms, and to turbulent mean concentration gradients, in the case of the mass transport terms.

This requires the introduction of a series of coefficients known generally as *eddy diffusion* or *diffusivity* coefficients. They were first introduced by Boussinesq[6] to relate turbulent shear stresses to turbulent mean velocity gradients, in an analogous fashion to the expression for viscous shear stress; the proportionality constant being, in this latter case, the coefficient of dynamic viscosity (μ). Using Boussinesq's argument, the shear stress τ_{zx} (see Chapter 2 for notation) can be expressed as

$$\tau_{zx} = -\rho\overline{u'w'} = -\rho\epsilon\left(\frac{\partial\bar{u}}{\partial z} + \frac{\partial\bar{w}}{\partial x}\right) \tag{3.9}$$

in which the overbar indicates a temporal average with respect to the turbulence and ϵ is a coefficient of eddy viscosity with dimensions (L^2/T).

In order to describe the movement of dissolved and suspended matter etc. in the flow it is usual to invoke a further analogy due to Reynolds[7], that the transfer of all quantities in the flow occurs in the same manner as momentum. The transport of material in the z direction is then described by the expression

$$-\overline{c'w'} = \epsilon\frac{\partial\bar{c}}{\partial z} \tag{3.10}$$

in which $\bar{c}$ and c' are turbulent mean and deviation concentrations respectively.

Equation 3.10 may be written down directly by analogy with Fick's law of molecular diffusion. It is apparent, however, from equation 3.9 that ϵ is not constant over the flow field as is μ or D. In steady, uniform flow $\partial\bar{w}/\partial z = 0$ and the gradient of shear stress over pipe radius or stream depth is constant. According to equation 3.9 a constant value of ϵ gives rise to a parabolic distribution of velocity as in laminar flow. Any other distribution would require ϵ to vary with distance from the boundary. In unsteady, non-uniform flow ϵ would vary in both space and time.

The validity of the Reynolds analogy is also questionable. Laboratory experiments[8] show the analogy to be true for neutrally buoyant dyes and for low concentrations of fine sediment particles ($< 100\ \mu$m). There is also evidence from laboratory[9] and estuarine[10] experiments that the transfer of salt and momentum differ. It is argued that pressure fluctuations can transport momentum but not salt. The analogy is thus applicable to those quantities which do not interfere with the transfer process. Sediment, salt and heat do not appear to fulfil this requirement. Because of this the Reynolds analogy is sometimes used with a proportionality coefficient. For example, the transfer of sediment in a vertical direction is often described by equation 3.10 multiplied by the constant β which decreases from unity, with increasing particle size, implying that larger sediment particles have higher inertia and so do not respond to all the velocity fluctuations within the flow.

To avoid some of the difficulties with equation 3.9, Hinze suggested[11] that the eddy diffusion coefficient has three components in each co-ordinate direction. The three diffusive transport terms may then be written as

$$-\overline{u'c'} = \epsilon_{xx}\,\partial\bar{c}/\partial x + \epsilon_{xy}\,\partial\bar{c}/\partial y + \epsilon_{xz}\,\partial\bar{c}/\partial z \quad (3.11\,a)$$

$$-\overline{v'c'} = \epsilon_{yx}\,\partial\bar{c}/\partial x + \epsilon_{yy}\,\partial\bar{c}/\partial y + \epsilon_{yz}\,\partial\bar{c}/\partial z \quad (3.11\,b)$$

$$-\overline{w'c'} = \epsilon_{zx}\,\partial\bar{c}/\partial x + \epsilon_{zy}\,\partial\bar{c}/\partial y + \epsilon_{zz}\,\partial\bar{c}/\partial z \quad (3.11\,c)$$

Equation 3.11 thus implies that axes exist in the flow field along which there is only one transport component. These axes are referred to as the principal axes of the diffusing substance. If, for example, they coincide with the convenient set of axes (x, y, z) then the transport term in the vertical direction would have the simple form

$$-\overline{w'c'} = \epsilon_{zz}\,\partial\bar{c}/\partial z \quad (3.12)$$

which is the same as equation 3.10.

Unfortunately the orientation of the principal axes cannot be calculated directly but it is unlikely that they coincide exactly with the convenient axes (x, y, z). Certainly the orientation of the larger, near-bed eddies does not coincide with these reference axes[11]. Because of this difficulty many engineers adopt the simpler expression 3.10 although studies of atmospheric turbulence have been made using equation 3.11.

The dilemma arises from the desire to express the diffusive transport as a gradient-type phenomenon. It is more likely to depend on local concentration gradients associated with the smaller high intensity eddies and on an advective-type process associated with the larger eddies. The dilemma is resolved somewhat in estuaries and rivers since the dominant diffusive term is often in the vertical direction, while lateral and longitudinal concentration gradients are often small compared with those in the vertical direction. Equation 3.11 then approximates to equation 3.10. The equation commonly used by engineers is therefore based on equation 3.10 and similar equations for diffusive transport in the x, y directions and has the form:

$$\frac{\partial \bar{c}}{\partial t}+\bar{u}\frac{\partial \bar{c}}{\partial x}+\bar{v}\frac{\partial \bar{c}}{\partial y}+\bar{w}\frac{\partial \bar{c}}{\partial z} = \frac{\partial}{\partial x}\left(\epsilon_x \frac{\partial \bar{c}}{\partial x}\right)+\frac{\partial}{\partial y}\left(\epsilon_y \frac{\partial \bar{c}}{\partial y}\right)+\frac{\partial}{\partial z}\left(\epsilon_z \frac{\partial \bar{c}}{\partial z}\right) \quad (3.13)$$

which has been simplified by incorporating any proportionality factors such as β into the diffusion coefficients and by using the equation of water continuity 2.4, to eliminate some of the advective terms.

The solution of equation 3.13 indicates the concentration of any conservative quantity provided the flow and diffusion parameters are known at all points and times throughout the flow system. Unfortunately, the flow and diffusion parameters are unique to each flow system and cannot be determined at present by analytical or numerical techniques. In theory, the introduction of a trace-substance into the flow enables the diffusion coefficients to be determined, provided sufficient flow and concentration observations are made with suitable instruments. The practical difficulties involved in such tests are, however, prohibitive in large three-dimensional situations and an alternative approach is clearly required to tackle large-scale environmental flows.

3.4 FURTHER ENGINEERING APPROXIMATIONS

Equation 3.13 can be simplified by neglecting or reducing the relative importance of the time-dependent term or by confining attention to one- or two-dimensional situations. Such approximations depend upon the nature of the engineering problem under consideration and the particular environmental and physical parameters making up the flow system. Consideration must, therefore, be given to the following:

(1) the degree of vertical stratification present in the flow and its change with tide and freshwater conditions;

(2) the geometrical characteristics of the flow system;

(3) the time- and space-dependent characteristics of the engineering problem itself.

For example, consider the problem of assessing the contribution of a discharge from a particular outfall on pollutant levels in a narrow, winding, partially-mixed estuary such as the Tees (UK). If the outfall discharge varies significantly in quantity and quality during the tidal cycle or is situated near the estuary mouth so that a significant amount of pollutant can escape into the ocean environment during the tidal cycle, it is necessary to examine the variation of concentration levels with time. The first term in equation 3.13 is, therefore, extremely important. The third and sixth terms in the equation are likely to be less significant since bends in the estuary will promote rapid lateral mixing and the estuary is also narrow. The fourth and seventh terms will also be important since large vertical density gradients are likely to exist in the estuary which will inhibit vertical mixing and maintain high vertical pollutant gradients. The fifth term, indicating horizontal turbulent transport, is also likely to be small compared with vertical turbulent transport rates since many estuaries are shallow compared to their length, and ϵ_x and ϵ_z are of the same order of magnitude.

The small order terms in equation 3.13 could be neglected but to do so is not really very helpful since there may be a considerable lateral variation in $\bar{u}$. Equation 3.13 is, therefore, simplified by lateral integration. All the flow and concentration quantities are first expressed in terms of a lateral-mean value for example, (U_L) and a deviation value (U'_L) respectively, i.e.

$$\bar{u} = U_L + U'_L; \quad \bar{v} = V_L + V'_L; \quad \bar{w} = W_L + W'_L; \quad \bar{c} = C_L + C'_L; \tag{3.14}$$

where, for example,
$$U_L = f_1(x, z, t) = [\bar{u}] \tag{3.15}$$

and
$$U'_L = f_2(x, y, z, t) \tag{3.16}$$

and the square brackets [ ] indicate a lateral average, i.e.

$$\frac{1}{\bar{B}}\int_a^b (\,)\mathrm{d}y$$

$\bar{B}(= b-a)$ represents the average lateral width of the estuary, and (b, a) indicate the lateral position of the estuary boundaries at elevation (z) relative to the estuary co-ordinate axes. $f_1(\,)$ and $f_2(\,)$ imply general functions of the bracketed variables. The result of the integration [12] and the use of Leibnitz's rule and lateral boundary conditions of zero advective and diffusive transport through the flow boundaries is the equation:

$$\begin{aligned}\partial(\bar{B}C_L)/\partial t + \partial(\bar{B}U_L C_L)/\partial x + \partial(\bar{B}W_L C_L)/\partial z \\ = -\partial(\bar{B}[U'_L C'_L])/\partial x - \partial(\bar{B}[W'_L C'_L])/\partial z + \partial(\bar{B}[\epsilon_z \partial C_L/\partial z])/\partial z\end{aligned} \tag{3.17}$$

on the assumption that $C'_L \ll C_L$.

The vertical transport due to velocity deviations from the lateral-mean value (the 5th term) may be neglected in comparison with the vertical diffusive term (the 6th term) provided

$$nW'_L/(\epsilon_z \partial/\partial z\{\ln (C_L)\}) \ll 1 \tag{3.18}$$

where
$$n = C'_L/C_L \tag{3.19}$$

Equation 3.18 appears to be satisfied in the Tees Estuary. Typical parameter values [13] are:

$$n = 1/20; \quad \epsilon_z = 0{\cdot}5\ \mathrm{m^2/h}; \quad W'_L = W_L/40; \quad W_L = \pi R_0/T;$$

$$R_0(\text{Tidal range}) = 2{\cdot}5\ \mathrm{m}; \quad T(\text{Tidal period}) = 12{\cdot}42\ \mathrm{h};$$

$$\partial C_L/\partial z \approx 2\ \mathrm{ppt/m}; \quad C_L = 17‰ \quad \text{(parts per thousand).}$$

These show the lateral term to be less than 1·5 % of the vertical term, based on the distribution of salinity in the estuary.

Equation 3.17 may be further simplified by relating the laterally averaged velocity and concentration term (the 4th one) to the laterally averaged longitudinal concentration gradient $\partial C_L/\partial x$, i.e.

$$[U'_L C'_L] = -E_{xL}\partial C_L/\partial x \tag{3.20}$$

where E_{xL} is called a *dispersion* coefficient in order to distinguish it from the eddy diffusion coefficients introduced earlier (see equations 3.9 and 3.10). Equation 3.17 may thus be written as:

$$\partial(\bar{B}C_L)/\partial t + \partial(\bar{B}U_L C_L)/\partial x + \partial(\bar{B}W_L C_L)/\partial z$$
$$= \partial(\bar{B}E_{xL}\partial C_L/\partial x)/\partial x + \partial(\bar{B}[\epsilon_z]\,\partial C_L/\partial z)/\partial z \qquad (3.21)$$

provided the effect of secondary flows on vertical velocities is neglected, there is no transport of mass through the flow boundaries, and the substance is conservative.

Application of equation 3.21 to estuaries, such as the Tees, shows that the first two terms are of major importance at times of maximum tidal flow. The dispersion and diffusion terms are marginally more important than the advective terms near the time of slack water when the first term is almost zero. The dispersive term (the 4th one) will assume more importance if large velocity and concentration changes occur laterally due, for example, to the presence of extensive tidal flats or channel embayments (see also Chapter 1).

Many estuaries become very wide in their seaward reaches or discharge into wide, but shallow, semi-enclosed coastal bays. The use of equation 3.21 in such situations is clearly undesirable since a laterally-averaged value may bear little resemblance to other values in the cross-section. Equation 3.13 may still be simplified in wide, shallow systems, i.e. those in which $\bar{B} \gg H$ (the flow depth). Concentration changes are likely to be much smaller over the flow depth than over the flow width in wide, shallow systems, particularly in coastal zones where vertical density gradients are minimal and additional mixing is often provided by wind and wave action. Equation 3.13 may thus be simplified by integrating it over the flow depth. Again, quantities are expressed in terms of a depth-mean value (U_d) and a deviation quantity (U_d'), i.e.

$$\bar{u} = U_d + U_d'; \quad \bar{v} = V_d + V_d'; \quad \bar{w} = W_d + W_d'; \quad \bar{c} = C_d + C_d'; \qquad (3.22)$$

where, for example $$U_d = f_3(x, y, t) = \{\bar{u}\} \qquad (3.23a)$$

$$u_d' = f_4(x, y, z, t) \qquad (3.23b)$$

The chain brackets { } represent a depth average, i.e.

$$\frac{1}{H}\int_0^H (\ )\,dz.$$

Vertical integration, together with the use of Leibnitz's rule, and boundary conditions of zero diffusive and advective transport, produce the equation [12]:

$$\partial(HC_d)/\partial t + \partial(HU_d C_d)/\partial x + \partial(HV_d C_d)/\partial y$$
$$= -\partial(H\{U_d' C_d'\})/\partial x - \partial(H\{V_d' C_d'\})/\partial y + \partial(H\{\epsilon_x \partial\bar{c}/\partial x\})/\partial x$$
$$+ \partial(H\{\epsilon_y \partial\bar{c}/\partial y\})/\partial y \qquad (3.24)$$

The deviation terms are again written in dispersion form and equation 3.24 then becomes

$$\partial(HC_d)/\partial t + \partial(HU_d C_d)/\partial x + \partial(HV_d C_d)/\partial y$$
$$= \partial(HE_{xd}\partial C_d/\partial x)/\partial x + \partial(HE_{yd}\partial C_d/\partial y)/\partial y \qquad (3.25)$$

with
$$\{U'_d C'_d\} = -E_{xd}\partial C_d/\partial x \qquad (3.26)$$

$$\{V'_d C'_d\} = -E_{yd}\partial C_d/\partial y \qquad (3.27)$$

The diffusion terms have been neglected in equation 3.25 on the grounds that $E_{xd} \gg \epsilon_x$ and $E_{yd} \gg \epsilon_y$ (see section 3.5). Alternatively, the diffusion terms can be incorporated in equations 3.26 and 3.27 so that the dispersion coefficients include diffusive effects. Equations 3.26 and 3.27 could also be replaced by more complicated forms analogous to equation 3.11 (without the z components). This more complicated equation provides a better description of the dispersion mechanism since horizontal velocities can vary in direction over the flow depth due to Coriolis force but it is hardly justified on practical grounds particularly if the x axis is chosen to be in the dominant flow direction so that E_{yd} is small.

Many engineering problems are concerned with narrow tidal rivers where vertical density gradients are negligibly small, or with narrow estuaries involving high tidal energy so that lateral concentration fluctuations and vertical density gradients are very small. The major change in concentration of a particular quantity occurs in the longitudinal direction (x) with only minor changes over the flow cross-section. Equation 3.13 may thus be simplified by integration over the flow cross-section, assuming that no transport occurs through the flow boundaries. The result[12] is

$$\partial(AC)/\partial t + \partial(AUC)/\partial x = -\partial(A\langle U'C'\rangle)/\partial x - \partial(A\langle\epsilon_x \partial\bar{c}/\partial x\rangle)/\partial x \qquad (3.28)$$

where U, C, U', C' are cross-sectional-average and deviation quantities respectively. For example,

$$\bar{u} = U + U' \qquad (3.29)$$

$$U = f_5(x, t) = \langle\bar{u}\rangle \qquad (3.30)$$

$$U' = f_6(x, y, z, t) \qquad (3.31)$$

The diamond bracket indicates a cross-sectional average value i.e.

$$\frac{1}{A}\int(\,)\mathrm{d}A.$$

Equation 3.28 is again simplified by the introduction of a one-dimensional dispersion coefficient, i.e.

$$\partial(AC)/\partial t + \partial(AUC)/\partial x = \partial(AE_x\partial C/\partial x)/\partial x \qquad (3.32)$$

where
$$\langle U'C'\rangle = -E_x\partial C/\partial x \qquad (3.33)$$

and the diffusion term has been neglected on the assumption that $E_x \gg \epsilon_x$ or is incorporated in E_x via equation 3.33, since $C' \ll C$. The dispersion term in equation 3.32 is, clearly, a 'catch-all' type term since it incorporates molecular and turbulent diffusive transport rates as well as advective transport due to velocity fluctuations within the flow cross-section. The use of equation 3.33 can, however, be justified theoretically[14] in steady uniform flows and by the fact that equation 3.32 has been found applicable[13] to real flow situations. A similar field-fit justification applies to equations 3.25 and 3.21, and arises because the dispersive terms are generally less important than the advective terms.

The price to be paid for simplification of equation 3.13 is thus the production of dispersion terms which will be unique to each flow situation considered. E_x, $E_{x\text{L}}$, etc. must, therefore, be determined for each flow situation encountered. Only order of magnitude estimates can be expected at best from the use of dispersion coefficient values obtained from analytical or empirical expressions derived in particular flow situations.

Equation 3.13 has been simplified so far by reducing the number of spatial co-ordinates. It is also possible to simplify the engineering problem and thus equation 3.13 by choosing the time interval over which changes within the flow system are examined. All previous equations in this chapter have used a turbulent-mean time scale. It is possible, however, to use any time interval, but some time intervals are more relevant than others. For example, the period of the tidal motion provides a convenient time interval if successive tides are identical in amplitude and period, and the boundary geometry and fresh water flow rate are also constant over the tidal period. Such conditions can only occur in a hydraulic model or the minds of men. Nevertheless, engineering situations have been studied successfully with this *single tidal-average* approach (see Chapter 7), although averaging over 2, 28 or 56 tidal cycles would be better and would ensure that tidal conditions were more nearly the same at both the start and end of the averaging period. Of course, the choice of too long a time period is undesirable due to changes in geometrical, meteorological or hydrological conditions.

A single tidal-average quantity can be produced in two ways, by integrating the relevent quantity over a single tidal period (T) either at a fixed spatial position (Eulerian average) or at a spatial position moving with the tidal flow (Lagrangian average). Both methods give identical answers in one-dimensional situations involving long simple-shaped estuaries (no tidal flats or excess meandering or branched channels) with small tidal changes in cross-sectional area compared with tidal-mean values. In two-dimensional problems, Lagrangian averaging produces larger tidal-average advective flows than Eulerian methods, but the differences are usually ignored from an engineering viewpoint since dispersion terms cannot be predicted to the same degree of accuracy. A similar argument also applies to the use of the one-dimensional Eulerian equations in estuaries with complex geometry. Consequently, Eulerian equations are generally used in solving tidal-average engineering problems although Lagrangian and even Eulerian–Lagrangian approaches are sometimes used for time-varying situations[13].

An example of the tidal-average form of equation 3.28 for a conservative substance is the equation[15]

$$\partial(A_T C_T)/\partial t + \partial(U_e A_T C_T)/\partial x = \partial(A_T E_{xT} \partial C_T/\partial x)/\partial x \qquad (3.34)$$

in which

$$U_e = U_T + [[U'_T A'_T / A_T]] \qquad (3.35)$$

$$E_{xT} \partial C_T/\partial x = -[[\langle U'C'\rangle - \langle \epsilon_x \partial \bar{c}/\partial x\rangle + U'_T C'_T]] \qquad (3.36)$$

and C_T, A_T, U_T, C'_T, A'_T, U'_T are tidal-average and deviation quantities respectively. The [[ ]] brackets represent a tidal-average value, i.e.

$$\frac{1}{T}\int_0^T (\,)\mathrm{d}t$$

and it is assumed that tidal fluctuations of A, C and ρ are small compared with their mean values, i.e. $C'_T \ll C_T$, etc.

It should be noted that the tidal-average velocity (U_T) at an estuary cross-section depends upon geometrical and tidal characteristics and the magnitude of both the freshwater discharge (Q_f) through the section and the change in the tidal-average volume of water stored upstream of that section (V_0), i.e.

$$U_T = Q_f/A_T - [[U'_T A'_T A_T]] - 1/A_T . \partial V_0/\partial t \qquad (3.37)$$

which results by longitudinal (x) integration of equation 2.17 and the use of equation 3.35. Equation 3.37 shows that U_T will only equal the fresh water velocity Q_f/A_T if successive tides are identical ($\partial V_0/\partial t = 0$) and the variations of tidal velocity and area about their mean values are out of phase by a quarter of a tidal period, as happens in pure standing wave motion (see Chapter 6).

E_{xT} is a one-dimensional tidal-averaged dispersion coefficient and must account for molecular and diffusive transport processes as well as the three-dimensional spreading effect of velocity gradients throughout the flood and ebb tide. Application of equation 3.34 to engineering situations shows E_{xT} to be large, which is not surprising since velocity gradients over the flow depth and width will cause material to be spread over a distance equal to a tidal excursion (the distance travelled by a water particle moving with the mean cross-sectional velocity for the period of the flood or ebb tide). The dispersive term of equation 3.34 also provides a major contribution to the equation and is equal to the advective term once a steady-state condition ($\partial C_T/\partial t = 0$) is reached. Use of equation 3.34 for engineering problems thus requires an accurate specification of E_{xT} and usually means that expensive tracer experiments must be undertaken in the field.

Equation 3.34 has been used to examine water quality problems in tidal rivers where the conditions assumed in its derivation ($A'_T \ll A_T$, etc.) are more likely to apply. An alternative equation which has also been used in tidal rivers[16], has a similar form to equation 3.34 but has a less satisfactory theoretical basis. The equation is:

$$\partial C_s/\partial t + Q_f/A_s . \partial C_s/\partial x = 1/A_s . \partial(A_s E_{sT} \partial C_s/\partial x)/\partial x \qquad (3.38)$$

where subscript (s) means that the equation indicates concentration values at *slack-tide* conditions, i.e. at about the time when tidal velocities are zero. E_{sT} is a dispersion coefficient which must account for the spreading action of the tide.

Equation 3.38 is clearly an approximation to actual conditions since at slack-tide U_e is zero. Equation 3.38 follows from equation 3.32 provided conditions *near* slack-water are considered and E_x is replaced by E_{sT}. Notice that E_x must be increased in magnitude in order to be comparable in size with E_{xT} since equation 3.32 does not include the longitudinal spreading action of velocity gradients throughout the tidal cycle. E_{sT} will not be equal to E_{xT} since both U_e and C_T are different from Q_f/A_s and C_s respectively.

Some engineering problems involve the continuous discharge of substances into long tidal rivers. If fresh water flow rates and boundary geometry are constant for many tidal cycles, a steady-state situation can arise when upstream dispersive transport is exactly balanced by downstream advective transport.

Equation 3.34 may then be written in the form

$$U_{es}C_{TS} = E_{xTS}\,\partial C_{TS}/\partial x \qquad (3.39)$$

where subscript *TS* indicates a steady-state approach. U_{es} represents the long term tidal-average velocity for the flow cross-section and would only tend to equal the long term river discharge if the averaging period involved a complete spring-neap tidal period.

Equations 3.13, 3.21, 3.25, 3.32, 3.34 and 3.39 can be used to assess the effects of various engineering schemes on the river and estuary environment. Application of the equations to problems involving non-conservative substances will require the addition of particular mass decay or generation terms (see Chapter 7) and problems involving heavy or buoyant discharges may require a simultaneous solution of momentum and mass continuity equations. Problems involving stratified estuary flows will require the use of appropriate diffusion–advection equations for both upper and lower layers.

Various other forms of equation 3.13 also exist. For example, a tidal-averaged form of the equation can be used to study the ocean dispersal of pollutants from barges while the dispersal of effluent from continuously discharging ocean outfalls can be studied by integrating the equation over an infinitely long time scale. Whatever the form of equation used, it is important to note that the use of equation 3.13 in time- or space-averaged situations will mean that some, if not all, of the diffusion terms must be replaced by appropriate dispersion quantities.

3.5 VARIABILITY OF DIFFUSION/DISPERSION PARAMETERS

It is impossible to give exact functional relationships or values to the diffusion and dispersion parameters appearing in the previous equation since they are unique to each physical system considered. The range and order of magnitude of the various quantities introduced in sections 3.3 and 3.4 is illustrated below (in m^2/s):

$[[\epsilon_z]]$;*	$\epsilon_x, \epsilon_y, \epsilon_z$,* E_{yd};†	E_{xd};	E_{xL}, E_x;	$E_{sT}, E_{xT}; E_{xTS}$
10^{-3}–10^{-2}	10^{-2}–10^{-1}	1–10	10–10^3	10–10^4

The wide range of diffusion/dispersion coefficients results from comparing one flow system with another and from variations in the coefficients of any one system over space and time. Variability of coefficient from one system to another arises because each flow system is the result of a unique combination of location, tidal characteristics, fluid properties, freshwater and sediment in-flows, geometrical shape, sediment type, wind and wave action and man's activities. Variability of coefficient within any one flow system is due to spatial and temporal changes in these flow system characteristics.

Changes in the magnitude of the diffusion terms ϵ_x, ϵ_y, ϵ_z are the result of the various factors outlined in section 3.2 since diffusion is directly related to the turbulent structure of the flow. Experiments with steady-flow in straight laboratory channels show that[11]

$$\epsilon_y \approx 3\epsilon_z; \quad \epsilon_x \approx \epsilon_z \tag{3.40}$$

Values of these coefficients can be increased by at least an order of magnitude in natural curved channels due to modifications in mean-flow gradients.

The value of ϵ_z is critically dependent upon the degree of vertical stratification present in a flow and the quantity undergoing diffusion. The vertical diffusion of momentum in the type of laboratory flow indicated above is usually described[11] by the equation

$$\epsilon_{mzo} = KU_* z(1-z/H) \tag{3.41}$$

which follows from equation 3.9 by assuming a linear distribution of shear stress over the flow depth, by taking $(\partial\bar{w}/\partial x) = 0$ for a parallel shear flow and by replacing $\partial\bar{u}/\partial z$ by U_*/Kz, the value appropriate to a logarithmic change of velocity with height above the channel bed. K is Von Karman's constant which is usually taken to have a value of 0·40 in clear water flows. The subscript (mzo) refers to momentum (m) transfer in the (z) direction in homogeneous conditions (o).

Equation 3.41 indicates that ϵ_{mzo} has a maximum value at mid-depth and is zero at the water surface and estuary bed, as might be expected since the surface shear stress has been taken as zero and the velocity gradient at $z = 0$ is infinite. The functional form of equation 3.41 is similar to that found in the well-mixed parts of estuaries such as the Mersey Narrows, but the magnitude of ϵ_z may well be different due to vertical stratification. For example, field observations in the Mersey Narrows[10] show that

$$[[\epsilon_{mz}]] = [[\epsilon_{mzo}]]\,(1+aRi)^b \tag{3.42}$$

where a, b are numerical constants with values of 10 and $-\frac{1}{2}$ respectively, and Ri is a quantity called the Richardson number and is defined at a point in the flow as

$$Ri = (g\,.\,\partial\bar{\rho}/\partial z)/(\bar{\rho}(\partial\bar{u}/\partial z)^2) \tag{3.43}$$

* Can be reduced to zero by large vertical density gradients.

† Assumed perpendicular to flow lines.

g is the acceleration due to gravity and $\bar{\rho}$, $\bar{u}$ are the turbulent-mean density and horizontal flow velocity respectively at a point in the flow. The [[ ]] brackets indicate a tidal-average value. For the Mersey observations $\partial\bar{\rho}/\partial z$ was replaced by $[[\partial\bar{\rho}/\partial z]]$ and $\bar{u}$ by the time-mean value of the tidal currents over half a tidal cycle.

Ri has an infinite value for complete vertical stratification and is zero for homogeneous conditions. It is also proportional to the ratio (*Rf*) of the rate of change of potential energy per unit volume to the rate at which turbulent energy is generated from the turbulent-mean flow, that is;

$$Ri = \epsilon_{mz}/\epsilon_{sz}\,.\,Rf \tag{3.44a}$$

$$Rf = g\,.\,\overline{\rho' w'}/(\rho\,.\,\overline{u'w'}\,.\,\partial\bar{u}/\partial z) \tag{3.44b}$$

and the primes and overbar refer to turbulent-fluctuation and turbulent-average values respectively. ϵ_{sz} is the vertical diffusion coefficient for salt transfer. Equation 3.44*a* follows from 3.44*b* by the use of equation 3·9 and a similar equation to 3.10 for density fluctuations (assumed due to salinity).

Equation 3.42 indicates the dramatic reduction (60 % for $Ri = 0{\cdot}5$) in tidal-average momentum transfer due to vertical salinity gradients even in well-mixed cases such as the Mersey Narrows where surface and bed salinities differ by only 2–3 parts of salt per thousand parts of water (‰). In partially-mixed estuaries such as the Tees and James (USA) where tidal-average vertical salinity differences are the order of 10–15‰ the reduction is even more pronounced (86 % for $Ri = 5$).

Equation 3.44 shows that *Ri* and *Rf* are only equal in size when $\epsilon_{mz} = \epsilon_{sz}$. Laboratory[9] and field[10] experiments show that this is not so, and that the vertical transfer of salt is much less than that for momentum. The reason for the higher transfer of momentum is that pressure fluctuations are considered capable of transferring momentum but not salt or other substances. Consequently the vertical transfer of salt reduces faster than that of momentum with increasing degree of stratification. The field results[10] indicate values of a, b in equation (3.42) (with m replaced by s) of the order of 3.33 and $-\frac{3}{2}$ respectively, i.e. a reduction in salt diffusion by a factor of 74 for an *Ri* value of 5. Such a factor seems almost unbelievable yet, if used in conjunction with a depth-average value from equation 3.41, $(KU_* H/6)$ provides a value for the vertical salt diffusion coefficient in the Tees estuary of the order of 0·6 m^2/h ($U_* = U/20$; $U = \frac{1}{2}$ m/s). Actual values are believed[13] to vary over the range 0·1–1 m^2/h.

Vertical mixing of salt by diffusive processes is negligibly small ($\epsilon_{sz} \approx 0$) for local *Ri* values between 50 and 100 since *Rf* becomes sensibly constant in this range. Vertical transfer of salt may still persist even at these large local *Ri* values due to entrainment. This is a one-way process whereby salt water is transported vertically from a low-turbulent salt layer into an overlying higher turbulent fresh water layer, by the breaking of internal waves along the salt/fresh water interface. The salt water taken from the lower layer by entrainment is replaced by horizontal advection of water within the lower layer. Diffusive mixing, on the other hand, involves the exchange of equal volumes of salt and fresh water.

Entrainment is a function of the relative degree of turbulence between upper and lower layers and consequently it is also reduced by vertical density gradients. It is zero below a critical velocity given by[19]

$$U_c = A_0(\nu g \Delta\rho/\rho)^{\frac{1}{3}} \tag{3.45}$$

where A_0 is a numerical constant with a value of 7.3 for a salt-wedge type flow and 5.6 for a stationary salt layer. $\Delta\rho$ is the difference in density between the two layers, ρ is the density of the upper fresh water layer and ν the kinematic viscosity of the lower salty layer. Equation 3.45 indicates that velocities of the order of 45 mm/s are sufficient to start the entrainment process between layers of fresh water (0‰ salinity) and sea water (34‰). Such low entrainment velocities are readily exceeded in nature, particularly in the presence of even slight surface wave action (see Chapter 4).

The variability of diffusion coefficients in the ocean and in coastal zones is much larger than in confined channels due to the random nature of wind and wave actions. Coefficients in such large bodies of water are usually considered to increase in magnitude as a function of space and time (t) since larger and larger eddies are involved in mixing processes as substances diffuse through space. A simple form of coefficient used to describe the two-dimensional radial diffusion of a vertical line of pollutant in the ocean is

$$\epsilon_r = pr^n t^m \tag{3.46}$$

where ϵ_r is a radial diffusion coefficient, r is a radial distance measured outwards from the centre of the diffusing patch and n, m, p are numerical constants. If $n = 1$ and $m = 0$, p is known as the diffusion velocity and has values between 1.5 and 15 mm/sec, with the smaller values occurring in the near-shore zone. Many other values exist for n, m (e.g. $n = \frac{4}{3}$, $m = 0$)[17] and are usually justified by assumptions concerning the transfer of energy from the large to small scale eddying motion. It is doubtful if any one functional form is better than another in view of the variable amount of energy injected by wind and wave action which cannot be accounted for in analytical descriptions of the diffusion process.

Future evaluation of diffusion coefficients may well rely on the *in situ* measurement of the amount of turbulent kinetic energy present in a flow[18], since diffusion is directly related to the size of velocity fluctuations, but considerable progress must first be made both in field instrumentation and in the development of efficient computer techniques before the method can be used generally in large three-dimensional flows involving random energy sources. Engineers will thus continue to reply upon the use of diffusion and dispersion parameters for many years to come.

The variability of diffusion coefficients is the result of alterations in turbulent fluctuations by various physical quantities, but the variability of dispersion coefficients is primarily due to changes in the structure of the turbulent-mean motion, produced by such factors as horizontal, vertical and lateral salinity or suspended sediment density gradients; changes in fresh-water flow rates; Coriolis induced circulations; wind induced currents; the mass-transporting action of waves; and natural or artificial variations in flow channel geometry. The importance of individual factors depends largely

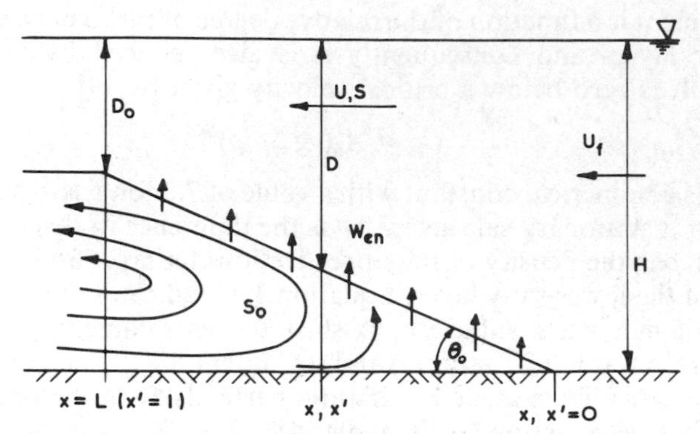

Figure 3.3. Stationary salt wedge

upon the physical characteristics of the flow system. For example, density gradients, fresh water flow rates, dredging or reclamation schemes are of prime importance in narrow tortuous estuaries. Coriolis currents, wind and wave action and large scale reclamation of tidal embayments are important in wide shallow systems. Density currents are unimportant in tidal rivers, unless there are large horizontal differences in suspended sediment concentrations.

Dispersion coefficients show the greatest degree of spatial and temporal variability in 'one-dimensional' estuaries since cross-sectional variations in horizontal advective velocity show large changes with longitudinal position and there is a large degree of mixing produced by the tidal velocities. This can be demonstrated by considering the simple example of flow over a stationary triangular-shaped salt layer of concentration (S_0) as shown in figure 3.3. The depth of the upper fresh water layer at the estuary mouth has a constant value D_0 and the bed slope is zero so that the upstream depth is constant (H).

Laboratory observations by Keulegan[19] suggest that the breaking of internal waves along a salt interface produces a continuous vertical flow of salt water. Entrainment of salt water into the fresh water zone is therefore assumed to occur at a constant vertical velocity W_{en} where

$$W_{en} = k'(U - U_c) \tag{3.47}$$

and k' is a numerical constant which varies with turbulence level and hence $\Delta\rho/\rho$ (equation 3.45) and Ri. Laboratory experiments indicate a wide range of values for k' and a dependence on Reynold's number and Ri, but many tests are unreliable due to the inclusion of diffusive mixing processes. Keulegan's[19] experiments, based on fresh water flow over a dense stationary lower layer, indicate the probable order of magnitude of k' and give values of $2{\cdot}2 \times 10^{-4}$ and $3{\cdot}5 \times 10^{-4}$ for $\Delta\rho/\rho$ values of 0·16 and zero respectively.

The variation of velocity in the fresh water layer due to entrainment can now be described by the equation

$$\partial(UD)/\partial x = k'(U - U_c) \tag{3.48}$$

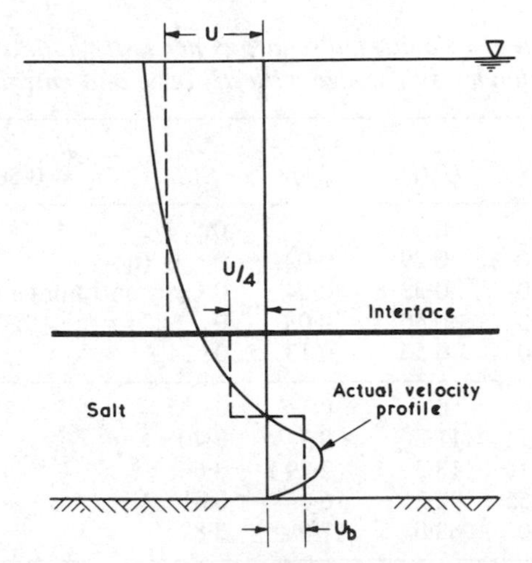

Figure 3.4. Assumed salt wedge velocity profile

where U is the depth-average velocity in the upper fresh water zone; D is the thickness of the fresh water layer at distance x from the toe of the salt wedge, that is,

$$D = H(1-\alpha_1 x') \tag{3.49a}$$

where

$$\alpha_1 = 1 - D_1/H \tag{3.49b}$$

$$x' = x/L \tag{3.49c}$$

and L is the length of the wedge.

Equation 3.48 can be solved, assuming that $U \gg U_c$ and k' is constant w.r.t. x, to give the equation

$$U = U_f(H/D)^{\beta_1} \tag{3.50}$$

$$\beta_1 = 1+k'L/(H\alpha_1) = 1+k' \cot\theta \tag{3.51}$$

The longitudinal variation of U is a function of β_1, α_1 and x', while the horizontal velocity in the salt wedge is controlled by interfacial and boundary shear forces. If the velocity distribution with depth of figure 3.4 is assumed to apply at all longitudinal positions, it is possible to illustrate the effect of entrainment on advective velocities in an estuary. The result is shown in table 3.2 for five positions along the wedge (x'), for various degrees of entrainment ($\beta_1 = 1, 5, 10$) and for a fixed α_1 and $\cot\theta_0$ value of 0·50 and 3285 respectively (similar to Mississippi river values).

Variations in longitudinal velocity (U, U_b) are seen from table 3.2 to be relatively small at low entrainment rates, but increase dramatically as entrainment is increased. Consequently, one-dimensional dispersion coefficients can be expected to show a somewhat similar longitudinal variation.

Large entrainment rates will also bring large quantities of salt water into the upper freshwater zone, as shown in table 3.2. The salinity (S) in the upper layer is assumed to be constant over the flow depth and is expressed in terms of the ocean salinity (S_0). The figures for S/S_0 in table 3.2 were obtained from

Table 3.2. *Variation of velocities, discharges per unit width and concentrations in a stationary salt wedge with distance and entrainment*

x'	U/U_f	q/q_f	U_b/U_f	q_b/q_f	S/S_0	β_1 ($\alpha_1 = 0{\cdot}50$)	$k' \times 10^3$
0	1·0	1·0	0	0	0	1	0
0·25	1·14	1·0	0·29	0·02	0	(no	Cot θ_0 = 3285
0·5	1·33	1·0	0·33	0·04	0	entrainment)	
0·75	1·6	1·0	0·40	0·08	0		
1·0	2·0	1·0	0·50	0·13	0		
0	1·0	1·0	0	0	0	5	1·22
0·25	1·95	1·71	11·8	0·74	0·40		
0·5	4·21	3·16	18·3	2·29	0·66		
0·75	10·5	6·55	32·2	6·05	0·81		
1·0	32·0	16·0	68·0	17·0	0·89		
0	1·0	1·0	1·0	0	0	10	2·74
0·25	3·8	3·33	38·2	2·39	0·63		
0·5	17·8	13·3	103·0	12·9	0·86		
0·75	110·0	68·7	389·0	72·9	0·92		
1·0	1024·0	516·0	2300·0	575·0	0·92		

equation 3.51 and the equation

$$\partial(USH(1-\alpha_1 x'))/\partial x + USH\alpha_1/L = k'US_0 \qquad (3.52)$$

which follows by integration of equation 3.21 over the flow depth, assuming laterally-homogeneous conditions with no velocity variation with depth in the upper fresh water layer ($E_{xL} = 0$), no vertical diffusive transport ($\epsilon_z = 0$), a steady state ($\partial C_L/\partial t = 0$) and a boundary condition of vertical salt transfer at the wedge interface of $k'US_0$.

Similar water circulation patterns to those indicated in table 3.2 are also found to occur in partially-mixed estuaries, but can only be seen by taking a tidal-average view of the estuary velocities, bearing in mind the difficulties mentioned earlier of adopting an ideal averaging period for real estuaries. For example, figure 3.5 shows single-tidal-average flows, salinities and depths for the Tees estuary during neap tides ($\approx$ 2.5 m range) and low river flow rates ($\approx$ 2·8 cumecs). S_0 and L have been taken as 34‰ and 26 km respectively, based on the field observations.

The middle and upper reaches of the Tees show a similar type of circulation pattern to the simple salt wedge example for a β_1 value of about 5 (table 3.2). Indeed, equation 3.51 gives a value of $\beta_1 = 4{\cdot}64$ for typical middle-estuary depths of 5 m and $\alpha_1 = 0{\cdot}5$ and a k' value of $3{\cdot}5 \times 10^{-4}$. There are, however, major differences between the simple entrainment and Tees examples. Horizontal salt transfer takes place in the real estuary due to the combined effects of turbulent mixing, lateral velocity and concentration gradients, and the spreading action of the tidal excursion. Turbulent motions also promote vertical mixing and reduce vertical density gradients. Vertical mixing, which

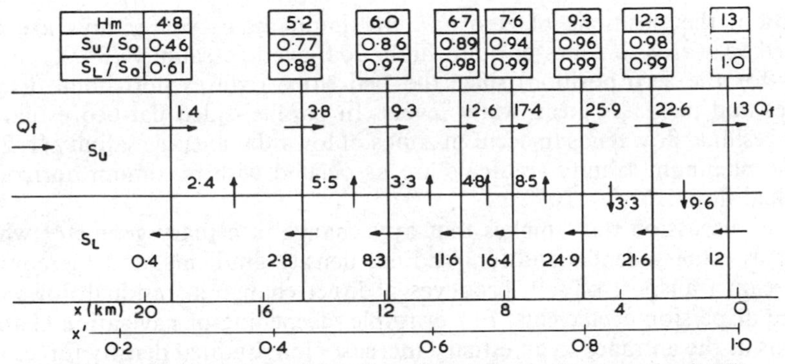

Figure 3.5. Tidal-average flows and salinities in the Tees estuary (UK)—neap tides and low river flow rates (after Farraday[13])

is more intense in the upper and seaward reaches of an estuary, will thus produce greater horizontal density gradients in the middle reaches of the estuary compared with upper and seaward reaches. Consequently, more flow deceleration occurs in the lower layer of the upper and middle reaches than in the salt wedge and produces increased vertical velocities between the layers and increased seaward flow in the upper layer. The increased advective flow of the partially-mixed estuary is simulated therefore in the entrainment example by increasing the size of the entrainment coefficient (k'). The removal of large vertical density gradients also produces a more uniform horizontal pressure gradient over the flow depth so that residual flows tend to reverse direction at about mid-depth over a large part of the saline intrusion zone. This later effect was simulated to some extent in the salt wedge example by allowing U_b to increase independently of the velocity in the top of the salt layer.

There are two other important differences between the entrainment and Tees examples. They are concerned with the pattern of vertical advective motion near the estuary entrance and the effect of a sloping estuary bed. Vertical motion occurs from upper to lower layers as shown in figure 3.5, in order to satisfy fluid continuity as the estuary surface flow decelerates near the seaward end of the estuary. Flow deceleration is produced by two effects: the acceleration of lower layer fluid by the greater horizontal density gradients in the middle reaches of the estuary compared to those near the sea, and the presence of strong tidal currents along the English coastline which have a restraining influence on the outflow of water from the estuary.

The effect of a sloping estuary bed is to reduce depths in the middle and upper reaches of the estuary and thereby to reduce salinity intrusion compared with the entrainment case. Reduced flow depths in the upper reaches also imply reduced longitudinal density forces and reduced residual flows. The large decrease in depth in the middle reaches will, however, produce large vertical flows in the middle and seaward reaches due to increased deceleration of upper and lower layer fluid (see figure 3.5). The bed slope also distorts the shape of the longitudinal salinity curve so that maximum gradients occur more in the upper parts of the estuary compared to a flat-bed situation. In addition, the positions of zero and maximum tidal-average flow are displaced seawards to zones of quite high tidal-average salinity (8–9‰ in the Tees for the zero position) since the bed slope reduces horizontal density forces and increases mean water levels. In parallel-sided flat-bed estuaries, zero-residual flow tends to occur in zones of low tidal-average salinity (1–2‰) while maximum salinity gradients are associated with maximum horizontal residual flows.

This discussion thus implies that any change in estuary geometry which changes estuary depths and longitudinal density gradients, and thereby the upstream transport of salt, produces a direct change in residual flows and hence dispersion coefficients. For example, deepening of navigation channel depths at the entrance to an estuary increases longitudinal density forces and thus salt intrusion, residual flows and dispersion coefficients (see also equation 3.56). Unfortunately, the siltation rate of fine suspended material at positions of zero residual flow is also increased as was found to occur in the Savannah estuary (USA) following navigation channel deepening[21].

The residual tidal flows have been shown to vary in size with entrainment rate, and to be affected by water depth, bed slope and longitudinal density gradients but they also vary with freshwater flow rates. The simple entrainment example (Table 3.2) shows a direct increase in upper and lower layer velocities with increased freshwater flow rates. However, consideration of the speed of propagation of turbulence-generated disturbances along the saltwater interface[20] assuming no flow in the salt wedge, shows that a critical flow is established at the mouth of a salt wedge estuary ($\beta_1 \approx 1$; $S/S_0 \ll 1$) such that

$$U^2/(\Delta\rho/\rho g D_0) = 1 \qquad (3.53)$$

Clearly an increase of velocity (U) at the estuary mouth will produce an increase in depth (D_0) in order to maintain critical conditions; that is, the salt wedge will be pushed out of the estuary. The implication of an increase in freshwater flow on tidal-average velocities is, therefore, that residual velocities at a fixed point in the saline zone will increase with small increases in flow but will be reduced at higher flow rates as the wedge is displaced seawards. This effect has also been observed in partially mixed estuaries[21] and is illustrated by figure 3.6.

The ebb predominance curves of figure 3.6 were obtained by integrating tidal velocities at particular non-dimensional depths (e.g. mid-depth = 50 %) over one or two tidal cycles, starting on one tidal state and finishing at the same value at the end of the averaging period. The ebb predominance value was then obtained by expressing the integrated area corresponding to the ebb flow as a ratio of the sum of all flood and ebb areas. For example, the ebb predominance value for a position with equal flood and ebb areas is 50 %. If flood velocities were zero, the ebb predominance figure would be 100 %. At low flow rates ebb predominance is seen from figure 3.6 to increase in upper and lower layers implying increased residual velocities. At high flow rates, residual velocities become zero in the lower layer while surface velocities show a slight increase. Eventually, residual downstream flow is established throughout the flow depth as the salt water is displaced seawards. It should be remembered that figure 3.6 is in terms of ebb-predominance and that a down-

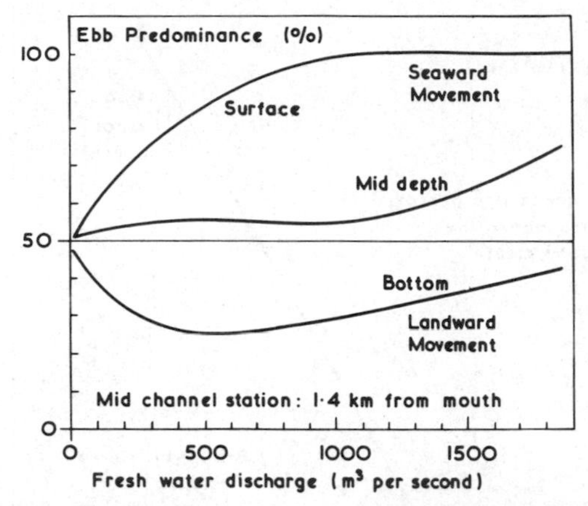

Figure 3.6. Variation of ebb-predominance curves with river discharge—Savannah Estuary (USA) (based on figure 11 from reference [21], courtesy the American Society of Civil Engineers)

stream residual (> 50 %) only implies downstream flow on a tidal-average basis and there may be considerable flood flows during the course of the tidal cycle itself.

Tidal-average dispersion coefficients, which compensate for the longitudinal dispersal produced by reversed-flow advective velocity profiles (for example, figure 3.4) can, thus, be expected (from table 3.2) to show large longitudinal variations in partially-mixed estuaries. One-dimensional tidal coefficients (E_x) will show a similar variation, particularly at times of low water when horizontal density forces dominate the flow process, and produce reversed-flow type velocity profiles.

The effect of degree of stratification and river discharge on one-dimensional steady-state dispersion coefficients is clearly illustrated by values for the Tay estuary (UK) [22] as shown in figure 3.7. Dispersion values and by implication tidal-residual flows are seen to increase continuously near the mouth of the estuary with increasing river flow. Dispersion values at the landward end of the estuary increase initially but reduce as the salt zone is displaced by the homogeneous (fresh water) tidal zone. The spatial variation of dispersion coefficients in the homogeneous tidal zone is almost negligible in comparison with the non-homogenerous estuarine zone, since instantaneous or tidal-average horizontal velocities show a relatively small longitudinal change.

Dispersion coefficient values in salt wedge conditions are less than in partially-mixed cases since residual flows are small ($\beta_1 \approx 1$) and in addition, horizontal and vertical advective transfer is less while interfacial vertical turbulent mixing is negligible. Consequently, higher concentrations of pollutants will exist over a smaller expanse of estuary compared with an equivalent sized partially-mixed estuary containing the same amount of pollutant.

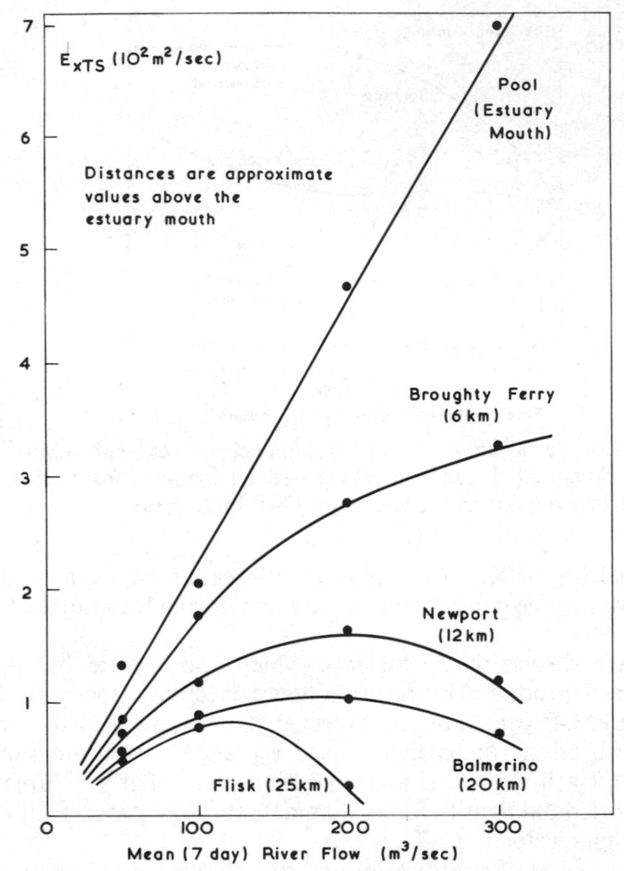

Figure 3.7. Variation of dispersion coefficient with distance and river discharge—Tay Estuary (UK) (based on figure 5 from reference[22], courtesy the Controller of Her Majesty's Stationery Office)

3.6 ESTIMATION OF DIFFUSION/DISPERSION VALUES

Previous discussion suggests that the only sensible way to determine the value of diffusion or dispersion parameters is to take field observations of the dispersal of tracer-substances contained in the flow. In the majority of pollution problems, pollutant discharges have only very local minor effects on stratification and consequently dispersion values are determined by measuring the spatial distribution of salinity. For example, equation 3.39 can be used to calculate E_{xTS} from tidal-averaged salinity results, although the equation is of limited use near estuary entrances since $\partial C_{TS}/\partial x$ cannot be determined with sufficient accuracy. Equation 3.39 gives values between 160 and 360 m²/s in the Mersey Narrows[23] for low and high river flows respectively. Values in the same range have also been found in many British estuaries including the Severn and Tay estuaries, and Southampton Water. Values near the mouth of the Thames Estuary at low river flows are about half the Mersey low flow values.

Lateral or one-dimensional tidal dispersion values (E_{xL}, E_x) can be determined in narrow estuaries by the use of equations 3.20 and 3.33 respectively. For example, equation 3.33 has been used to compute dispersion values in the Tees estuary[13] at a cross-section located some 14½ km from the sea where the mid-tide width and depth was 100 m and 6 m respectively. Values found at mid-flood and mid-ebb tide were 86 and 108 m^2/s respectively.

Salinity cannot be used as a tracer in homogeneous tidal zones and it is usual to use a fluorescent or radioactive tracer which has the advantage of being detected at very low concentration levels. For example, Rhodamine WT dye or radioactive bromine 82 is often used for such purposes. Fluorescent tracers are more commonly used in estuarial waters since tracing can be undertaken with the minimum of skilled labour and with readily available commercial equipment. Radioactive tracers require the use of highly skilled personnel because of the health hazard and are very useful in coastal and ocean situations since only small quantities of tracer need be used.

Fluorescent tracers and equation 3.34 have been used to determine E_{xT} in the homogeneous tidal zone of the Patuxent River (USA) at a distance of between 50 and 65 km from the estuary mouth. A value of 50 m^2/s was found[24]. Similar fluorescent tests have been used in the Irish Sea[25] (UK) to determine tidal-average dispersion coefficients in x, y and z directions. Equation 3.13 was used with $\bar{u}$, $\bar{v}$ and $\bar{w}$ replaced by constant tidal-average values and ϵ_x, ϵ_y, ϵ_z replaced by appropriate dispersion values. Average dispersion values over a five day tracing period were 33, 4 and 2×10^{-3} m^2/s in the x, y, z directions respectively. These values varied by $\pm$60 to 70% due to changing meteorological, tidal and salinity conditions over the test period. Fluorescent tracers have also been used[26] in the Solent (Isle of Wight, UK) to determine E_{xd} and E_{yd} from equation (3.25). Values of 5 m^2/s and 0·5 m^2/s were found for the x and y directions.

Tests have been performed[27] some half a kilometre off the English south coast, in water 15 m deep. Diffusion coefficients in the top 2 m of the unstratified flow were found to follow equation 3.46 with p(m/h) given by the equation

$$p = 7{\cdot}8 + 0{\cdot}00065W_s \tag{3.54}$$

where W_s is the wind speed in m/h at standard height (19 m).

Tracer experiments are extremely time-consuming and expensive, and only provide information valid for the period of the experiment. They cannot be used to predict the effect of changes in salinity or pollutant concentrations in an estuary following navigation channel deepening, or bank reclamation, or the construction of tidal barrages or large man-made changes in river flow rate. Such problems are best tackled by using a hydraulic model combined with a numerical or analytical solution of an appropriate diffusion–advection equation. For example, a hydraulic model may be used to study salinity and tidal changes resulting from the constructions of a tidal barrage while an equation such as 3.21, with appropriate loss and generation terms, is used to study pollutant dispersal.

In view of the expense of field or hydraulic model tests, it is hardly surprising to find that many attempts have been made to produce analytical relationships for both dispersion and salinity values. Such methods are not universally applicable because of simplifying analytical or experimental assumptions. For example, many techniques use results from parallel-sided laboratory flume tests or from rivers and estuaries with small tidal ranges and narrow simple-shaped geometry. The techniques may prove useful if field data is used to determine fitting parameters or if dispersive transport rates are small compared with advective ones, which implies low flow rates or small concentration gradients. As a general rule, results from analytical techniques should not be accepted unless the method is shown capable of reproducing existing conditions. Confidence in any method will also decrease as problem boundary conditions deviate more and more from existing conditions.

One-dimensional dispersion coefficients in the homogeneous tidal zone of rivers and estuaries are often calculated from the equation[31]

$$E_x = 63nUR^{5/6}(\mathrm{m^2/s}) \tag{3.55}$$

where n is Manning's roughness coefficient and R is the flow hydraulic radius. Equation 3.55 is based on steady uniform flow in a straight pipe involving a logarithmic distribution of flow velocities. If tidal flows are small (< 1 m/s) and the estuary has a simple cross-sectional shape along its length, it is possible to use a constant value of E_x for the whole tidal cycle based on equation 3.55 with appropriate tidal-mean parameter values.

The value of the numerical constant in equation 3.55 is also found to vary with the cross-sectional shape of field velocity profiles. For example, equation 3.55 has been increased by a factor of three when used on tidal rivers on the east coast of the USA[28]. The constant may well be increased by a further factor of ten in narrow but tortuous conditions as occur, for example, in the upper reaches of the Wear Estuary (UK).

Wide tidal rivers ($\bar{B}/R > 10$) are more likely to have larger lateral flow variations relative to cross-sectional mean values than in the vertical direction, and the numerical constant in equation 3.55 may well increase by a further factor of ten[30]. The increased effect of width may, however, be offset by a reduction in dispersion due to the period (T) of the tidal motion, since it can be argued that an infinitely small tidal period produces no dispersion while an infinitely long period is equivalent to steady uni-directional flow for which equation 3.55 might be applied. The degree of any tidal-reduction is difficult to establish with any certainty due to the many interacting factors present in the dispersion process. Holley *et al.*[29] have, however, indicated a possible functional relationship which may enable the degree of reduction to be established for one-dimensional dispersion coefficients in the homogeneous fresh water tidal zone of simple-shaped estuaries.

The dispersion coefficient (E_x) in the non-homogeneous tidal zone may be estimated according to Thatcher and Harleman from the equations

$$E_x = K_0(\partial S'/\partial X) + E_{x\infty} \tag{3.56a}$$

$$K_0 = 0{\cdot}002U_0L_2(Es)^{-0{\cdot}25} \tag{3.56b}$$

$$S' = S/S_0;\ X = x/L_2 \tag{3.56c}$$

Es is the estuary number defined in Chapter 1, with g replaced by $(\Delta\rho g)/\rho_m$; $\Delta\rho$ is the maximum change in cross-sectional mean density of the estuary water from the fresh water to the seaward end of the estuary; ρ_m is the mean longitudinal density of the estuary water when $\Delta\rho$ is determined; S_0 is the ocean salinity (e.g. 34‰) and S is the cross-sectional mean salinity at a point along the estuary, which has a length (L); $E_{x\infty}$ is the value of equation 3.55, increased as appropriate (e.g. by 3) to allow for lateral effects; and U_0 is the maximum cross-sectional mean tidal velocity at the estuary entrance.

The correlation of E_x with estuary number is not too surprising since it can be related to the stratification parameter (G/J) mentioned in Chapter 1. The G/J number is itself inversely proportional to Richardson number, and thus to the degree of vertical stratification and entrainment.

Equations 3.56 and 3.32 have been successfully used by Thatcher and Harleman[28] to predict tidal salinity distributions along the Delaware, Potomac and Hudson estuaries on the east coast of America, although results were marginally better in the case of the Hudson when the numerical constant in equation 3.56 was reduced to 0·0015.

Equation 3.56 is likely to be less useful in estuaries with very wide or irregularly shaped seaward zones, since lateral salinity variations can be high and will require the introduction of large dispersive transport rates near the seaward boundary of the estuary. Equation 3.56(a) shows relatively small values near the ocean boundary.

3.7 SUMMARY

Each estuary or tidal river has a unique internal velocity structure which, for engineering purposes, is assumed to be composed of a periodic turbulent-mean flow component and a random turbulent one directly related to the magnitude of mean-flow gradients. The mixing process which is responsible for the distribution of quantities such as heat, dissolved gas and solids, and suspended sediment is shown to consist of a uni-directional movement by turbulent mean flows, called advection, and a three-dimensional spreading action produced by the turbulent flow components, called diffusion by analogy with molecular motions.

Mixing processes are shown to be accelerated by mean flow velocity gradients and by horizontal density differences but to be reduced by vertical density gradients due to salinity or suspended sediment. If little tidal energy is available, vertical diffusion of salt is suppressed almost completely, producing salt-wedge conditions. However, vertical advective transport of salt is still possible due to entrainment; that is, the breaking of internal waves along the salt/fresh water interface. Engineering works can also modify mixing processes by producing changes in mean flow velocities as well as in their spatial and temporal gradients.

The effect of engineering works on the distribution of dissolved and suspended matter can be investigated, in theory, at any point in a flow system by the use of an appropriate continuity of mass (diffusion–advection) equation. Unfortunately, the volume of data required to solve such equations for large three-dimensional unsteady flows is prohibitive, even after simplifying them by relating turbulent transport terms to local turbulent-mean gradients via diffusion coefficients. Practical problems are solved, therefore, by averaging the equations over time and space and this leads to the introduction in the simplified equations of a set of parameters called dispersion coefficients which contain both advective and diffusive components and which are unique to each flow system. Attempts have been made to relate dispersion coefficients to simple parameters characteristic of the tidal motion but the results can only be expected to provide 'order-of-magnitude' estimates since each flow system has a unique velocity structure. Dispersion coefficient values should therefore be obtained from field observations, which may involve fluorescent or radioactive tracers.

3.8 NOTATION

a, b Numerical constants or lateral co-ordinates of the sides of a flow system
c Concentration of any substance defined as the mass of substance per unit mass of solution or as a subscript, indicates a critical entrainment value
d Subscript, indicates a depth-average quantity
$f(\,)$ Indicates a general function
g Acceleration due to gravity
i Subscript indicating lateral or vertical directions
k' Entrainment coefficient
mz Subscript indicating momentum transfer in the z-direction
mzo Subscript indicating momentum (m) transfer in z direction for homogeneous (o) conditions
n Manning's roughness coefficient or C_L'/C_L
n, m, p Numerical constants
$\dot{q}$ Rate of addition of heat in degrees/unit time
q Depth-average water discharge per unit width in upper layer of a salt wedge
q_b Depth-average water discharge per unit width in lower layer of a salt wedge
r Radial direction as subscript
s Salinity of water defined as the mass of salt per unit mass of seawater
s Subscript indicates values apply at slack tide.
sz Subscript indicates salinity transfer in the vertical direction
tr Turbulent average mass transport rate per unit area in the x-direction
t_1 Averaging time for turbulent fluctuations
u, l Subscripts relating to an upper layer or lower layer value
$\bar{u}, v, \bar{w}$ As Chapter 2
u', v', w' As Chapter 2
w_f Free fall velocity of sediment particles
xd, yd Subscript indicates a depth-average has been taken in relation to x-, y-directions respectively

xL Subscript indicating reference to the x-direction and that a lateral average has been taken

xx, xy, zz Subscripts indicating mass transport in the first-letter direction due to turbulent-mean concentration gradients in the second-letter direction

x, y, z Subscripts indicating that the symbol applies to directions x, y, z respectively

x' Non-dimensional length along salt wedge ($= x/L$)

xT Subscript indicates relevance to the x-direction and that a tidal average has been taken

sT Subscript indicates relevance to slack-time conditions and that a tidal average has been taken

xTS Subscript indicates relevance to x-direction and that a steady state condition has been reached

A Cross-sectional area of flow

A_0 Numerical constant

B Average lateral estuary width

C Temporal or spatial average concentration value

C' Concentration variation from temporal or spatial average concentration value

C_p Specific heat at constant pressure

D, D_0 Molecular dispersion coefficient or flow depth; freshwater flow depth at the ocean entrance for salt wedge

E Dispersion coefficient

Es Estuary number

$F_{x,y,z}$ Rate of transfer of mass per unit area due to molecular motion in -, y,x- z-directions

H Flow depth

K, K_0 Von Karman's constant, or a dispersion parameter

K_1 Decay rate constant

L, L_1, L_2 Salinity intrusion length, mean free path of molecules, and estuary length from its mouth, respectively

L Subscript indicates a lateral average

Q_f Freshwater flow rate

R, R_0 Hydraulic radius, tidal range

Ri Richardson number

Rf Flux Richardson number

S Cross-sectional average salinity value

S' Ratio of cross-sectional salinity to ocean salinity (S/S_0)

S_0 Indicates ocean salinity value

T Tidal period or as subscript a single tidal average has been taken

TS Indicates a steady value as a subscript

U_e Effective single-tidal cross-sectional average velocity

U_{es} Long term tidal-average cross-sectional flow velocity

U_* Shear velocity ($\sqrt{(\tau_0/\rho)}$)

U_0 Maximum tidal velocity at the estuary entrance

U, V, W Spatial or temporal average velocity values

U', V', W' Velocity variations from spatial or temporally-averaged velocity values

V_1 Tidal-average velocity of motion of molecules
V_0 Volume of water stored upstream of a section
W_s Wind speed in m/h
W_{en} Entrainment velocity for a salt wedge
X Non-dimensional length along an estuary, (x/L_2)
$\sigma_u, \sigma_v, \sigma_w$ Standard deviation of turbulent velocity fluctuations u', v', w'
ρ Fluid density
θ Temperature of fluid
α Molecular thermal diffusion coefficient $= \kappa/\rho C_p$
κ Thermal conductivity of a fluid
ϵ Coefficient of eddy viscosity
β' Proportionality coefficient, e.g. between momentum and sediment transfer coefficients
[ ] Represents a lateral-average
[[ ]] Represents a tidal-average
{ } Represents a depth-average
∞ Indicates a steady flow value as a subscript
$\Delta\rho$ Indicates a density difference
< > Indicates a cross-sectional average value
μ Coefficient of molecular viscosity
τ_{zx} Shear stress in the x-direction on a plane perpendicular to the z-direction
‰ Parts per thousand parts of fluid by weight
ν Kinematic viscosity of salt water
α_1, β_1 Parameters given by equations 3.49*b* and 3.51

REFERENCES

[1] Elata, C. and Ippen, A. T., The dynamics of open channel flow with suspensions of neutrally bouyant particles, *Tech. Rep. No. 45. Hyd. Lab.*, Department of Engineering, M.I.T. (1961)

[2] Pechenkin, M. W., Optical method for experimental studies on turbulent flows, *Proc. 12th Congress IAHR*, **2**, Fort Collins, Colorado, U.S.A. (1967)

[3] Richardson, E. V. and McQuivey, R. S., Measurement of turbulence in water, *J. Hyd. Div., ASCE*, **94**, HY2 (1968)

[4] Greated, C., Measurement of Reynolds stresses using an improved laser flow meter, *J. Physics* (*E: Sci. Inst.*), **3** (1970)

[5] Bowden, K. F., and Howe, M. R., Observations of turbulence in a tidal current, *J. Fluid Mech.* **17**, (1963)

[6] Boussinesq, J., Essai sur la théorie des eaux counsantes. *Mémoires présentés par divers savants à l'Académie des Sciences*, **23** (1877)

[7] Reynolds, O., On the dynamical theory of incompressible fluids and the determination of the criterion, *Phil. Trans. R. Soc.*, **186** (1894)

[8] Jobson, H. E. and Sayre, N. W., Vertical transfer in an open channel flow *J. Hyd. Div., ASCE*, **96**, HY3 (1970)

[9] Ellison, T. E. and Turner, J. S., Turbulent entrainment in stratified flows, *J. Fluid Mech.*, **6** (1959)

[10] Bowden, K. F. and Gilligan, R. M., Characteristic features of estuarine circulation as represented in the Mersey estuary, *Lim and Ocean*, **16**, 3 (1971)
[11] Hinze, J. O., *Turbulence*, McGraw-Hill Book Co. (1959)
[12] Pritchard, D. W., The equations of mass continuity and salt continuity in estuaries, *J. Marine Res.*, **13** (1958)
[13] Farraday, R. V., *Finite element models for partially-mixed estuaries*, Ph.D. thesis, Manchester University (1973)
[14] Taylor, G. I. The dispersion of matter in turbulent flow through a pipe, *Proc. Roy. Soc. A.*, **223** (1954)
[15] Okubo, A., Equations describing the diffusion of an introduced pollutant in a one-dimensional estuary, *Studies on Oceanography, Tokyo* (1964)
[16] O'Connor, D. J., Estuarine distribution of non-conservative substances *J. San. Eng. Div. ASCE*, **91**, SA1, Feb. (1965)
[17] Okubo, A., Horizontal diffusion from an instantaneous point source due to oceanic turbulence, *Tech. Rep. 32, Chesapeake Bay Inst., Johns Hopkins University* (1962)
[18] Launder, B. E. and Spalding, D. B. *Mathematical models of turbulence*, Academic Press, London (1972)
[19] Keulegan, G. H., Interfacial instability and mixing in stratified flows. *J. Res. Nat. Bur. Stds.*, **43** (1949)
[20] Shijf, J. B. and Schönfeld, J. C. Theoretical considerations on the motion of salt and freshwater, *Proc. Minnesota Int. Hyd. Convent.*, Minneapolis (1953)
[21] Simmons, H. B. Some effects of upland discharge on estuarine hydraulics, *Proc. Am. Soc. Civil Eng.*, **81**, Paper 792 (1955)
[22] Williams, D. J. A. and West, J. R., A one-dimensional representation of mixing in the Tay Estuary, Paper 11, Mathematical and hydraulic modelling of estuarine pollution, *Water Pollution Research*, Tech. Paper No. 13, HMSO (1973)
[23] Hughes, P., Tidal mixing in the Narrows of the Mersey Estuary, *Geo. J. Roy. Ast. Soc.*, **1** (1958)
[24] Carter, H. H. and Okubo, A., Longitudinal dispersion in non-uniform flow, *Tech. Rep. 68, Chesapeake Bay Inst., Johns Hopkins University*, November (1970)
[25] Talbot, J. W., Transport and dispersion of soluble material, Appendix 11, *Out of Sight out of Mind*, **2**, Dept. of Environment, HMSO (1972)
[26] Adey, R. A. and Brebbia, C. A. Finite element solution for effluent dispersion, Paper 18, *Numerical methods in Fluid Mechanics*, Pentech Press, London (1974)
[27] Barrett, M. J., Munro, D. and Agg, A. R., Radiotracer dispersion studies in the vicinity of a sea outfall, *Int. Ass. Water Poll. Res* (1969)
[28] Thatcher, M. L. and Harleman, D. R. F., A mathematical model for the prediction of unsteady salinity intrusion in estuaries, *Rep. 144, Dept. of Civil Engineering, M.I.T.*, Cambridge, Mass., USA. Feb. (1972)
[29] Holley, E. R., Harleman, D. R. F. and Fischer, H. B., Dispersion in homogeneous estuary flow. *J. Hyd. Div., ASCE*, **96**, HY8 (1970)

[30] Sooky, A. A., Longitudinal dispersion in open channels, *Proc. Hyd. Div., ASCE*, **95**, HY4, April (1969)
[31] Harleman, D. R. F., Ch. 12, *Estuary and coastline hydrodynamics*, ed. A. T. Ippen, McGraw-Hill, New York (1966)
[32] Lighthill, M. J., Turbulence, *Osborne Reynolds Centenary Symposium*, University of Manchester, September (1968)

4

Sediment movements

Many of the world's major cities have grown to their present size because they are centred around a large navigable tidal river or estuary. Failure to preserve adequate navigation channels has often caused a city to decline in importance or even be abandoned altogether. For example, the city of Chester is situated at the head of the Dee Estuary in England and was a major port in Roman times, but is today hardly accessible to small pleasure craft owing to the progressive siltation of the estuary channels. Engineering schemes concerned with tidal rivers and estuaries must, therefore, be carefully investigated in order to avoid disastrous changes in the size and location of port approach channels.

Unfortunately, complete theoretical prediction of navigation channel changes in an unsteady tidal environment is unreliable and will remain so until the mechanics of the transport process are more fully understood and can be expressed by analytical equations. Channel changes can, however, be estimated for engineering works, but must be based on field observations supplemented perhaps by hydraulic model tests (see Chapter 8). Because of this reliance on field data, it is essential that the engineer should understand the various sedimentation and erosion processes at work in a tidal environment and their relationship to the total environment.

This chapter is an attempt to make the reader aware of the possible sources and types of sediment which contribute to tidal shoaling problems, as well as the various sediment transport processes at work in a tidal environment and the factors on which they depend. Attention is generally confined to the more difficult estuarine zone.

4.1 SEDIMENT SOURCES

Engineering problems associated with the tidal environment often require an estimate of the total quantity of sediment available for deposition. Failure to identify actual sources of sediment can lead to the design of inadequate engineering works. The most common examples of such failures in the past have been associated with the return of dredged spoil from deposit grounds (see Chapters 9 and 10). Problems have arisen in many estuaries; for example,

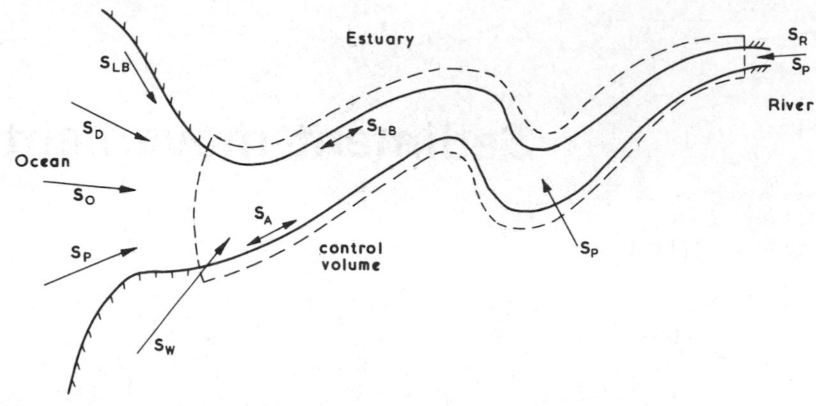

Figure 4.1. Sources of sediment for an estuary

the Mersey, Thames, Ribble and Humber estuaries in Britain, and the Savannah and Delaware estuaries in the United States.

Possible sediment sources for an estuary or for particular shoals within an estuary are illustrated in figure 4.1.

These can be summarized as follows:

(i) Land erosion by rivers and streams (S_R).
(ii) Disposal of domestic and industrial effluents and solid wastes (S_P).
(iii) Littoral drift and/or bank erosion (S_{LB})
(iv) Wind erosion of coastal dunes and drying inter-tidal shoals (S_W).
(v) Erosion of the near-shore Continental Shelf (S_O)
(vi) Return of dredged spoil (S_D)
(vii) Decomposition and excretions of marine and river plants and animals (S_A).

The above sources will contribute a wide variety of material to estuarine shoals. The cutting back of rivers and streams, together with land surface drainage produces large quantities of sand, silt and clay size material as does tide and wave action on the sea-bed and on beaches and cliffs. Rivers also supply organic matter and dissolved mineral salts. Wind action on dunes and inter-tidal banks produces fine sand while land erosion contributes silt and clay material. Gravel may be transported along beaches to the entrance to an estuary where wave action is present. It is unlikely to be supplied to an estuary in significant quantities by river and ocean currents.

Pollutant effluents contribute electrochemically charged fine colloidal organic matter, which may also be mixed with solid waste products such as wood-fibres and grease. Refuse material, as well as sludge from sewage works, may also be contributed by direct barge disposal in the ocean. Organic material is also supplied from dead marine and river flora and fauna and will combine ionically with the inorganic clay sediments, as indeed, will most of the organic matter discharged to an estuary.

Living creatures, such as marine bivalves and filter-feeding zooplankton, can provide material indirectly. They have the ability to filter clay particles from water and to eject them later as larger sized agglomerate particles, which are capable of settling from suspension. Unicellular plants, such as diatoms, may then cover these particles with a slimy layer and so inhibit erosion. Other plants, such as cordgrass, can reduce flow and wave action on the top of estuarine deposits and thereby encourage sedimentation of fine material.

The most important sediment sources are undoubtedly the river (S_R), ocean (S_O) and littoral (S_{LB}) sources, although dredged material (S_D) may be as important as river sources, if disposal areas are located close to estuary entrances or persistent shoaling zones. Pollutant sources may be important in the long term history of an estuary, since suspended material is unlikely to be removed from an estuary unless it remains permanently in the near surface waters, or the tidal-average water circulation pattern (see figure 3.6) is flushed completely out of the estuary by large fresh water flows. The relative importance of plants and animals (S_A) is largely unknown. It is probable, however, that they contribute to long term estuarine changes as in the case of the Dee Estuary, where there has been a continual growth of salt marsh over many centuries. It has also been suggested that filter-feeding zooplankton play a major part in the sedimentation of fine material in coastal and estuarine areas as, for example, in Upper Chesapeake Bay (USA)[1].

The quantity of material contributed by the more important sources can be large. For example, the Savannah River supplies some 6–7 M tonnes of fine sediment to its estuary each year[2], while littoral movement of sand along the beach at Madras in India amounts to some $\frac{3}{4}$M tonnes per year[3]. Erosion of the bed of the Irish Sea by wave and current action is thought to supply some 4 M tonnes of sand and silt to the Mersey Estuary each year.[15] Fine material, returning from spoil disposal areas, may also amount to several million tonnes per year, as was the case formerly in the Thames and Savannah estuaries.

Sediment deposits in estuaries, therefore, consist of various proportions of gravel, sand, silt, clay and organic matter. Gravel and sand are often found at the seaward ends where wave action and residual currents remove the finer fractions, while fine sand, silt, clay and organic matter (often collectively referred to as mud), is found in the upper reaches of an estuary near the limit of tidal-average salinity-intrusion. Mud may also be found in coastal deposits, particularly if salt water is flushed completely out of the estuary during high fresh water flows or severe wave action stirs up offshore deposits so that tidal-average coastal currents can move the material shorewards[3].

Typical physical characteristics of muddy deposits from the middle and upper reaches of the Mersey and Severn estuaries in England[4] are shown in table 4.1. Muds from other parts of the world have similar compositions; for example, San Francisco Bay (USA), the Gironde Estuary (France) the Hooghly Estuary (India), the Chao Phya Estuary (Thailand), and the Demerara Estuary (South America).

The precise composition of deposits in any particular estuary depends upon the particular geological and social history of the estuary and its hinterland, as well as the sediment transport characteristics of the tidal currents. Consequently, each estuary must be considered unique even though it has many tidal and sedimentary features in common with other estuaries.

Table 4.1. *Physical characteristics and critical erosion shear stresses of British estuarine muds*[4]
(Courtesy of the Institution of Civil Engineers)

	Origin of material			
	Bromborough (Mersey Estuary)	Gloucester (Severn Estuary)	Portishead (Severn Estuary)	Hayle (Cornwall)
Specific gravity	2·39±0·02	2·28±0·01	2·46±0·04	2·54±0·01
Organic content (%)	2·08	0·87	0·88	4·06
Mean surface diameter of particles (μ)	9·3 ±0·5	8·9 ±0·1	3·6 ±0·3	11·9±0·8
% clay	10	9	31	6
% silt	40	65	61	64
% sand	50	26	8	30
$\tau_c(N/m^2)$	1·6	3·2	5·1	16*

* Estimated value

4.2 SEDIMENT TRANSPORT PROCESSES

The erosion, transport and deposition of sediments in tidal flows depends on the physical and chemical properties of the bed sediments and the surrounding fluid, as well as the turbulent nature of flow. Bed sediment properties have a major influence on the movement of sediment. Gravel, sand and silt particles move as solid individual (discrete) grains, but clay particles adhere together in sponge-like clusters (flocs) which change in size and shape during sediment motion. Consequently, the following sections treat the transport of discrete particles separately from that of flocculated material, but interaction between the two groups is discussed at appropriate points since most natural sediment deposits are a mixture of discrete and flocculated particles.

4.3 DISCRETE PARTICLE SEDIMENTS

Non-cohesive sediment deposits are composed of individual grains of various minerals. The precise composition depends on the geology of the estuary and its surroundings. Many deposits contain large proportions of quartz, silica and felspar, which have similar densities (2550–2750 kg/m^3). However, deposits also contain lightweight calcium carbonate shell fragments and small amounts (usually < 2 % of the total) of heavy minerals such as zircon (density ρ_b = 4600 kg/m^3) and magnetite (5200 kg/m^3). The presence of particular heavy minerals can be very useful and may help to establish the source of particular sediments (see Chapter 10).

The mineral composition of individual sediment grains is also largely responsible for the wide variety of particle shapes found in natural deposits. Igneous material, such as felspar, produces spheroidal-shaped grains, while the metamorphic slates and shales produce flat disc-like particles. All particles moving on the estuary bed are, of course, smoothed and rounded by the abrasive action of other grains during sediment motion. A particular estuarine deposit thus usually contains a large number of particles of widely differing density, shape and form, with the harder materials, such as quartz, forming the bulk of the deposit. The average properties of the deposit, such as porosity and bulk density, may, therefore, show only small variations from one point to another. Particle shape, size and density are of fundamental importance from the sediment transport view point since they influence both the erosion and deposition of material (see also sections 4·4 and 4·5).

The size and distribution of particles in a particular deposit is largely controlled by the ability of the tidal streams to move and transport individual sediment grains. Thus, silt and fine sand deposits are found along estuary margins while medium to coarse sand is found in navigation channels. Gravel may appear locally in areas of high velocity and turbulence, such as occur on the outside of channel bends and around the end of revetments and training walls. However, the presence of grains of widely differing density or size can produce a protective layer on the surface of a deposit which prevents the erosion of any underlying fine material.

Studies of sediment transport problems should therefore include details of the size, size distribution, density and shape characteristics of individual sediment grains as well as the bulk properties of the total deposit. It is also useful in coastal situations to determine the statistical properties of bed grain-size distributions. Attempts have been made[5] to relate these parameters to the hydraulic characteristics of the flow environment. It is doubtful, however, if the effort required is justified in an estuarine environment, except perhaps for the median grain size (d_{50}: 50 % of particles finer than this value), since the shape of grain-size distribution curves is largely controlled by the amount of silt, clay and organic matter present in the test sample.

4.4 INITIATION OF MOTION

Erosion starts at the flow boundary when the applied moments, about a point of contact between sediment grains (figure 4.2 – point (A)), are greater than the restoring moments. The forces acting on an individual sand grain on the point of movement are the steady attraction of gravity and the fluctuating hydrodynamic forces, which include viscous forces, pressure gradients due to flow separation, forces due to seepage into or out of the bed and impacts due to particle motions. The hydrodynamic forces fluctuate with time due to the birth, motion and decay of eddies within the flow and the period of the tidal motion. Sediment movement thus tends to be an intermittent process at low flow rates and to take place in separate places over the bed surface. It is usual, therefore, in engineering problems, to introduce the concept of a critical erosion velocity or bed shear stress below which there is insignificant sediment motion. These flow parameters then may be correlated with those of the bed sediment, so that the design of non-erodable channels can be accomplished by the use of easily determined quantities.

An example of a velocity–grain size correlation is shown in figure 4.3 in

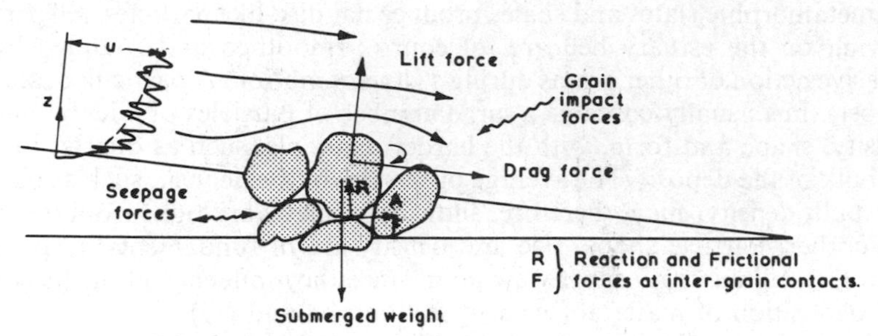

Figure 4.2. Forces acting on a sediment grain

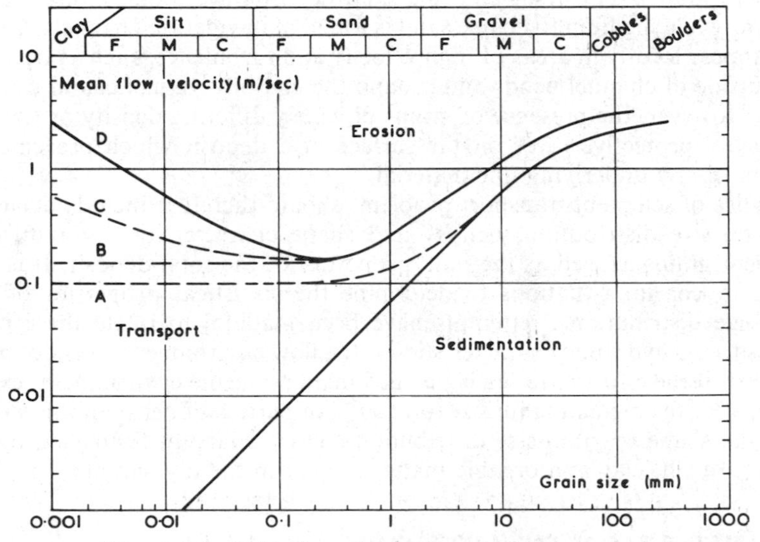

Figure 4.3. Critical erosion/sedimentation boundaries (after Hjulström[6] and Postma[7])

terms of a depth-average velocity, which is assumed to be 40 % greater than the flow velocity around the sediment grains. The solid lines are based on sediment-free uni-directional flow laboratory experiments [6] involving a level bed of loose, single-sized, rounded grains. Lines $A-D$ are general trends based on field and laboratory tests [7] and indicate (curve A) a more realistic transportation/sedimentation boundary as well as the effects of clay particles on erosion. Figure 4.3 illustrates several interesting points observed in nature. For example, the most easily eroded particles are seen to be fine to medium sand, while fine silt and clay material can require a velocity which is as large as that for gravel and cobbles. It is not surprising, therefore, to find that fine marine sands form the bulk of deposits in many estuaries or that tidal deposits accumulate slowly over the years in estuaries containing a high suspended load of silt and clay material.

Figure 4.3 also indicates that sediment can be carried in suspension at much

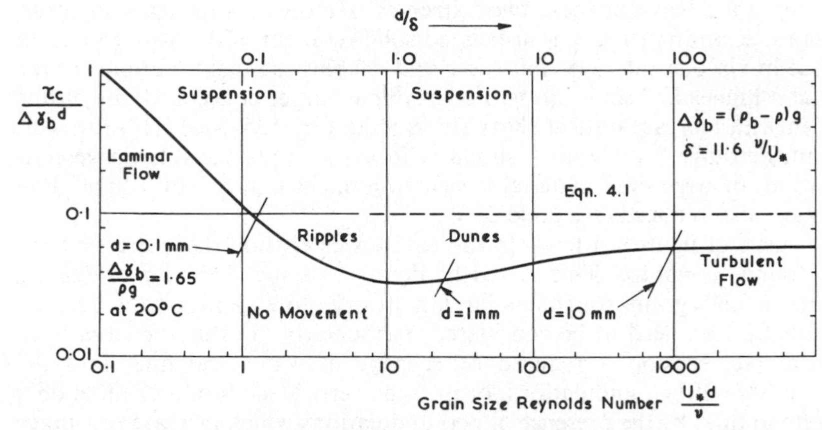

Figure 4.4. Critical shear stress/grain size correlation diagram (after Shields[9])

lower velocities than those required for erosion. Consequently, fine sand particles may be transported into an estuary on the flood tide to be deposited at slack water into cohesive deposits, but the requirement of a higher erosion velocity on the ebb tide means that sand particles progressively accumulate in cohesive deposits. This in turn produces an armouring effect on the deposit's surface which allows a longer time for the underlying material to consolidate and become more resistant to erosion.

Initial motion in steady flows is often related to the critical shear stress (τ_c) acting on the whole sediment deposit so that the difficulty of specifying a velocity at the grain level is avoided. The shear stress concept is particularly useful in laboratory situations where it can be determined with sufficient accuracy. Unfortunately the calculation of shear stress values in field situations involving unsteady non-homogeneous flows requires an excessive amount of field data; the critical near-bed velocity concept is just as useful in the present state of knowledge of sediment transport processes.

An example of a particularly simple critical shear stress relation is the equation[8]:

$$\tau_c = 166\bar{d}\,(\text{g/m}^2) \tag{4.1}$$

where $\bar{d}$ is a weighted-mean bed grain size in millimetres. Equation 4.1 has been found to be an adequate description of initial sediment motion for particles larger than 0·3 mm for sediment-free, steady, unidirectional flows over a level bed.

A further example of a critical shear stress–grain size correlation is shown in figure 4.4, which is based on Shields' steady-flow laboratory experiments over a level bed of loose, uniform-sized material[9]. Figure 4.4 indicates that the motion of small particles is controlled mainly by viscous forces acting on the whole surface of the sediment deposit. The motion of large particles, greater than 6 mm in diameter, is controlled almost completely by turbulent lift and drag forces produced by pressure differences across particles projecting above the viscous sub-layer ($d/\delta > 1$). Sand-sized particles lie in the transition zone between these two extremes. However, a decrease in water temperature, or an increase in suspended solids content of the flow, causes an increase in viscous sub-layer thickness and enables sediment motion to take place at a higher bed shear stress. For example, larger concentrations of fine sediment can increase critical shear stress values by 125–300 %[10] for sediment in the range 1–0·1 mm. A similar effect is also produced by large concentrations of large sized material in suspension, since turbulent lift and drag forces are also reduced (see table 3.1).

The curve of figure 4.4 tends to indicate lower erosion shear stresses than those found in nature. This is partly because of the difficulty of scaling turbulence and grain roughness in laboratory flows and partly because additional forces need to be considered, particularly in estuarine situations. For example, erosion is reduced by seepage of water and fines into the upstream side of bed undulations, or into the vertical sides of a channel on a rising flood tide; by the presence of bed undulations which increase resistance to flow; by the deposit of flocculated material at low flow rates and by a reduction in lift and drag forces due to flow acceleration. On the other hand, erosion is increased by the seepage of water from the downstream side of bed undulations and from the sides of a channel during a falling ebb tide; by the presence of underwater slopes; by wave action increasing the hydrodynamic forces and by an increase in lift and drag forces produced by flow deceleration.

4.5 ESTABLISHED MOTION AND RESISTANCE TO FLOW

Once the instantaneous applied forces exceed the resistance, sediment grains will roll or slide along the estuary bed, and bed load sediment transport occurs. If the instantaneous lift forces exceed the submerged grain weight, sediment particles may be plucked from the bed and carried into suspension with the assistance of the turbulent eddies within the flow. Suspended load sediment transport then results and is the dominant transport mode of fine sediments. If grain inertia is high or turbulence levels are low, the suspended particle may fall back into the estuary bed after travelling a short distance. This transport mode is referred to as saltation and is thought of as part of the bed load transport.

As sediment grains become mobile, an initially flat bed deforms into a series of undulations which in turn affect sediment transport rates and cause an increase in resistance to flow. A progressive increase in velocity results in development of bed forms similar to those found in steady flow (see figure 4.5). If bed material is sand less than 600 μm diameter, ripples form on the bed. These are three-dimensional waves of small dimensions. At higher flow speeds, ripples give way to dunes which are much larger and often coalesce into largely two-dimensional long-crested waves. At still higher flow speeds dunes may be washed away, giving rise to a plane bed at Froude numbers

$$F = U/(gH)^{\frac{1}{2}}$$

near unity. Such high Froude numbers seldom occur in estuaries except in local drainage gullies or in shallow regions near the head of estuaries in which

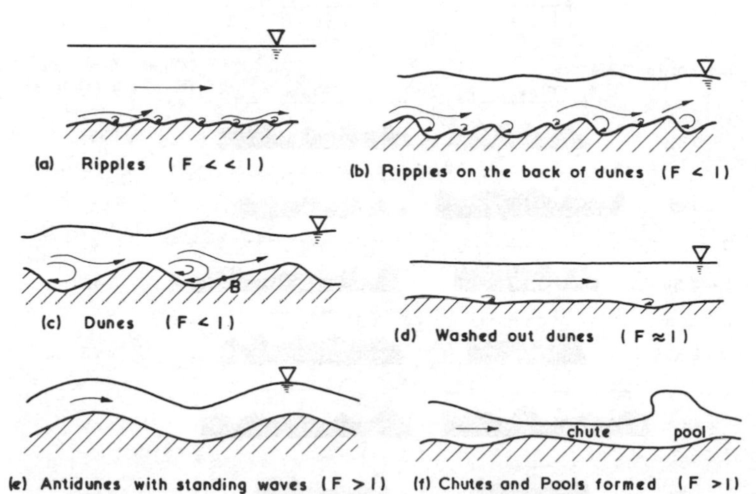

Figure 4.5. Bed form classification (based on figure 2 of reference[11], courtesy the American Society of Civil Engineers)

the tidal range is large. In both cases, critical or super-critical flow can only occur during falling tides.

The resistance to flow during the development of bed forms depends on the work done in transporting sediment and on energy losses due to the resistance of the bed itself. The bed resistance includes the effect of textural roughness due to grains on the bed (grain drag), and drag due to the shape of the ripples or dunes (form drag). It has been found that bed resistance can be greater during slowly increasing (quasi-steady) velocities than during slowly decreasing velocities[47]. This is probably caused by the extra work done on the bed during the form-building period.

In tidal flows, bed forms have a limited time to develop and are only likely to approximate to fully-developed steady forms in estuaries that have strong currents. They may then have shapes typical of bed forms found in steady flows, with relatively gentle upstream slopes but steep downstream slopes at the angle of repose of the bed material (see figure 4.6). These bed forms offer a high resistance due to flow separation and eddy formation downstream of each wave. When flow reverses, however, the same profile has a low form drag. Once sediment transport begins in the reversed direction, the bed profile gradually re-forms, beginning at the crest where velocities are highest. In strong currents, reversal of the bed form may take place quickly, but in weak currents it may not occur at all. An extreme case would occur in part of an estuary in which the flood tide was only just strong enough to develop dunes. The ebb tide may then be insufficiently strong to cause much sediment transport. Frictional coefficients during the ebb tide would then be lower than during the flood, as occurs during dry-season conditions. At the other extreme, freshets in rivers can dominate the flow in the upper reaches of estuaries and in this case, ebb-formed waves would persist throughout the tidal cycle, giving rise to high resistance coefficients during ebb tides and low

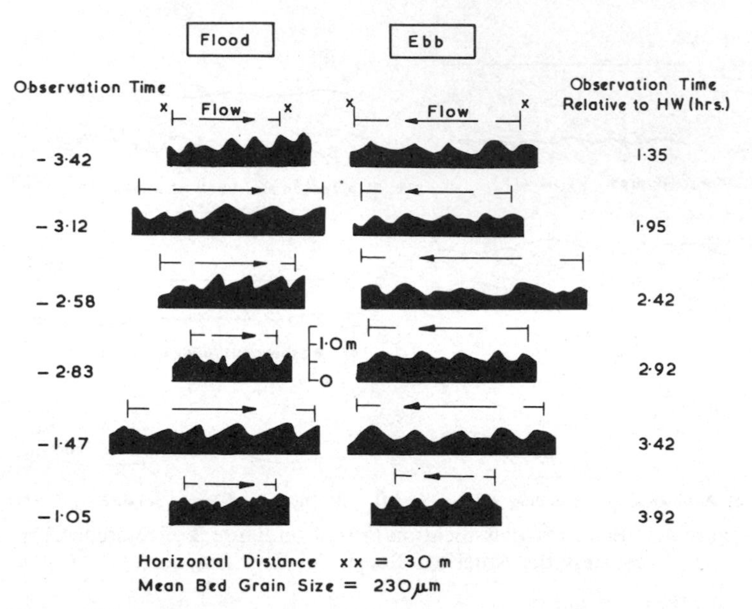

Figure 4.6. Bed forms in the Ribble Estuary (based on observations by the staff of the Hydraulic Research Station, Wallingford, UK[22])

coefficients during flood tides. These asymmetric effects usually occur towards the upper reaches of estuaries, since the influence of fresh water flow near the mouth is often small compared with tidal currents and flood and ebb tidal flows are more nearly equal in strength. However, asymmetric effects can occur at any point in an estuary in ebb- or flood-dominated channels.

Classification of bed forms and bed resistance is only possible for steady flows in broad terms. Nevertheless guidance can be sought from the results of extensive work that has been done on this subject. Fortunately it is usually possible to solve estuarial problems without having a precise knowledge of bed resistance. In operating mathematical or physical models, for example, adjustments can be made to friction during the process of calibration. This process can, however, be shortened and simplified if good estimates of friction can be applied in the first trial.

In steady, uniform flow situations, several methods of classification have been suggested[11, 12]. All depend on the properties of the flow and the grains, although no satisfactory system has been produced to date. Froude number has been used to classify uniform open channel flows as in figure 4.5, which shows that for $0 < F < 1$, ripples and dunes form as velocities increase, but as F approaches unity, dunes give way to plane-bed sediment transport, while at higher Froude numbers dunes re-form and move upstream; they are called anti-dunes. Typical estuary Froude numbers range up to 0·3 in main channels and 0·5 in branch channels; higher values may be found in drainage gullies and near the head of estuaries.

The properties of the sediment, such as effective grain size, shape and gradation must also be included in any classification system. A useful parameter is the effective fall velocity of the particles, which is a fundamental property concerned with sediment transport. However, it is not a measure of the rate at which sediment will be lifted from the bed so it is not a property embracing all aspects of sand transport. Simons and Richardson have correlated bed forms with stream power ($\tau_{z_0} U$) and a 'fall diameter', defined as the diameter of a sphere having a specific gravity of 2·65 and the same settling velocity as a particle when each is allowed to settle in quiescent water of infinite extent at a temperature of 24° C. One great advantage of such a definition is that it can be applied to a wide range of particle shapes and densities, including very light material which is often used in small-scale hydraulic models.

Ripples and dunes are similar in general shape, but differ in size and in their dependence on fall diameter and flow conditions (figure 4.5). Ripples are a major factor in the determination of bed resistance in models, whereas dunes are more significant in real rivers. Any differences in their modes of formation must therefore be taken into account when examining scale effects in models. Ripples are small waves of bed material with gently sloping upstream faces and steep downstream faces. They form in relatively fine grained materials (fall diameter less than 600 μm) but not in coarser materials. They occur when shear stresses are little higher than would be necessary to initiate movement of the bed material. At higher shear stresses dunes form, sometimes with ripples on their backs.

When ripples are present, resistance to flow is large and reduces with depth but is independent of grain size. Resistance is, in fact, dependent on the form of the ripples much more than on their surface roughness. Sediment transport on a rippled bed is thus less than on a flat bed because only a small part of the shear stress is applied directly to the sediment grains. The distribution of ripple heights is best described by statistical means[13], as for water waves, but the size of ripples to be expected in particular situations is extremely difficult to estimate even in uni-directional flow situations. The ratio of average ripple height to wavelength ($\Delta\bar{H}/\bar{\lambda}$) is usually found to be in the range 0·04–0·15, the smaller values being associated with the finer sand sizes. Average ripple wavelengths ($\bar{\lambda}$) are thought to be independent of bed grain size[14] and to have a magnitude of some 1000 d for $d \leqslant 0{\cdot}18$ mm.

Dunes are relatively large wave-forms of bed material which form at higher shear stresses or stream powers than ripples, possibly at grain size Reynolds numbers ($U_* d/\nu$) of twenty[14] (see figure 4.4). Unlike ripples, dunes can form in sands coarser than those with a fall diameter of 600 μm. Dunes can grow to large amplitudes and can increase in height as flow depth increases. Dunes of height 1 m to 2m are common in estuaries, but heights exceeding 20 m have been measured in the North Sea.

Simons and Richardson[11] found that resistance to flow increased with increase of depth for coarser sands ($d_{50} > 300\ \mu$m) but decreased with increase of depth for finer sands. Moreover, when dunes were relatively flat, with a height less than ten times their length, the resistance to flow was dependent on grain diameter.

Bed form characteristics in the dune range can be described by statistical curves[13, 16] involving parameters related to the size of the flow situation.

For example, steady flow laboratory experiments[16] with a medium sized sand show that bed forms have a similar statistical distribution to ocean wind waves. For fully developed forms,

$$\Delta\overline{H} = 2{\cdot}3\sigma_0; \quad \Delta H_m = 6\sigma_0; \quad \lambda_m = 2{\cdot}6\overline{\lambda} \tag{4.2}$$

where $\Delta\overline{H}$, ΔH_m, $\overline{\lambda}$, λ_m are the mean and maximum bed form height and wavelength respectively; σ_0 is the standard deviation of bed surface changes at a fixed point with time and is related to the depth of flow and the flow Froude number[13, 17], although no precise relationship has been established applicable to all flow conditions. A particularly simple relationship for σ_0 has the form[13]

$$\sigma_0 = 0{\cdot}12R \tag{4.3}$$

where R is the channel hydraulic radius.

Confirmation of the importance of turbulence levels in bed form growth is provided by observations in streams[18] and estuaries[19], which show a reduction in bed form height at low water temperatures and in the presence of large concentrations of suspended matter.

Dune wavelengths also show a scale effect and many laboratory and river situations[14] seem to follow the relationship

$$\overline{\lambda} = aH \tag{4.4}$$

where (a) is a coefficient ≈ 5 with some tendency to decrease with increasing particle size[20].

Analysis of irrotational flow over sinusoidal shaped bed forms[21] suggests that dune wavelengths are also a direct function of flow Froude number. However, the theoretical analysis only agrees with equation 4.4 when conditions are nearly critical ($F \approx 1$); the value of the coefficient then being equal to 2π.

Equations 4.2, 4.3, 4.4 suggest that very large bed features can develop in large flow systems. The size of tidal bed forms is, however, often smaller than the values given by them. For example, observed[22] mean bed form heights and wavelengths in the Ribble Estuary (UK) are of the order of 0·46 m and 7 m respectively (figure 4.6), while equations 4.2–4.4 indicate values of some 2·4 m and 43 m respectively. The major reason for the discrepancy is probably the limited time available during a tidal cycle for bed form growth. Clearly bed form development time is restricted to the time interval between successive slack tides during which flow velocities exceed critical values and by the time taken for bed forms to re-adjust from previously developed shapes.

Steady-flow laboratory experiments at moderate Froude numbers (0·3–0·5) with a typical estuary-size sand ($d_{50} = 250\ \mu$m) suggest that bed forms reach steady-state values in shallow water areas in the upper reaches of an estuary where energy slopes are steep enough to generate large velocities, or where the erosion period of the flood tide is so small that successive ebb tides provide a long development time ($t \rightarrow \infty$ in table 4.2). Development will be less than steady-state values in deep water or in moderately large velocity areas where (t) is small, or in low velocity areas where both U and (t) are small. For

Table 4.2. *Development of bed features with time (Jain and Kennedy*[17])

Ut/H	0	1500	3500	6500	10,000	22,000
$\sigma_1/\sigma_0(\%)$	0	20	40	60	80	95

NOTE: t is the time of development from an initially flat bed; σ_1 is the standard deviation of bed elevations at time (t); wavelengths show a similar rate of growth to σ_1/σ_0.

Table 4.3. *Friction factors due to bed forms*

Type	Grain	Ripples	Dunes	Flat bed	Anti-dunes	Chute/pools	Depth (m)
Reference [11]	—	0·05–0·13	0·05–0·16	0·02–0·03	0·02–0·07	0·07–0·09	0·1–0·3
Equation 6.42	0·02[b]	0·06–0·11[c]	0·13[a]	—	—	—	0·3
Equation 6.42	0·01[b]	0·02–0·03[c]	0·13[a]	—	—	—	10

[a] Fully developed conditions; [b] k_b = 400 μm; [c] $\Delta H_m/\Delta \bar{H}$ taken as 6/2·3 = 2·6.

example, mean bed form heights and wavelengths reach some 60 % and 90 % respectively of steady-state values (as given by equations 4.2–4.4), in the upper tidal reaches of the Weser Estuary (Germany)[19] where flood tide transport is small, but only some 20 % and 16 % respectively in the middle reaches of the Ribble Estuary. It is interesting to note that table 4.2 also suggests a 20 % development value for the Ribble since U, t and H have typical values of 1·5 m/s, 7200 s, and 6.85 m respectively.

The results shown in table 4.2 also suggest that dune bed forms will increase in height and wavelength during a flood tide until flow velocities fall below some critical value near high water. Consequently, bed form development and flow resistance will be out of phase with tidal velocities, and maximum values will be reached more towards high water.

Unfortunately, it is impossible to quantify flow resistance in the present state of knowledge of sediment transport processes in oscillatory flows. Field experience does suggest, however, that in parts of many estuaries friction coefficients approximate to steady flow values in similar channels (see also Chapter 8). Typical friction coefficient values (f) found in shallow (0·1–0·3 m) steady uniform laboratory flows[11] are shown in table 4.3 and are compared with the results from equations 4.2, 4.3 and 6.42 with the effective roughness height (k_b) taken as half the maximum bed form height.

The effect of depth on grain and ripple resistance is clearly seen from Table 4.3. The use of an effective bed roughness of half the maximum bed form height also seems to give a reasonable measure of bed friction coefficient. Dunes do not show an effect with depth since bed form height has been related directly to flow depth. In tidal conditions, partial development of the dunes could occur, which greatly reduces the bed friction value. For example, (f) is reduced by over 50 % (to 0·062) for dunes which have developed to only

20 % of steady flow values, as in the Ribble Estuary. The effect of flow depth is now implicitly contained in the calculation of (f) through the variables of table 4.2.

4.6 BED SEDIMENT MOTION

Changes in river and estuary bed levels due to changes in tidal flow characteristics can be related to sediment transport rates by an application of continuity of mass principles (see Chapter 2) to an elemental area of bed surface. The resulting equation is:

$$\frac{\partial q_x}{\partial x} + \frac{\partial q_y}{\partial y} + \gamma_b(1-m)\frac{\partial z_0}{\partial t} = P - D \tag{4·5}$$

where q_x, q_y represent bed load transport rates in weight per unit width in the longitudinal (x) and lateral (y) directions respectively; m is the porosity of the bed material; γ_b is the specific weight of the sediment ($= \rho_b g$) and z_0 is the elevation of the sediment surface above a horizontal datum. P is the quantity of sediment removed into suspension from the moving bed layer, in weight per unit area of bed surface per unit time, and D is the quantity of suspended sediment deposited onto the mobile bed, in weight per unit area per unit time. P and D tend to zero for large sized particles in the absence of suspended load, while they are equal to each other for steady-state conditions such as occur in unidirectional river flows.

Changes in elevation of the mobile bed surface could be predicted from equation 4.5 if sediment discharge rates and sediment pick-up and deposition parameters were known as functions of space and time. Unfortunately the precise form of the parameters is unknown in a tidal environment and engineering problems involving bed sediment transport must, in general, be solved by field observations and/or a hydraulic model study. Unfortunately, hydraulic models are less useful for predicting the movement of suspended material, due to problems in scaling the turbulence characteristics of a flow. Hydraulic models can, however, provide useful pointers to particular areas of siltation and can show changes in residual flow patterns. Some local sedimentation problems can be tackled by specifying analytical distributions of suspended sediment. This technique is useful, for example, in calculating sedimentation rates in port approach channels.

4.7 SUSPENDED SEDIMENT MOTION

Many estuary deposits contain large proportions of fine sand-sized material (60–200 μm) which is readily set in motion by the tidal currents. Such particles have a relatively small submerged weight and therefore are easily entrained within the flow by turbulent eddy motions, particularly from the crests of bed forms. The primary transport mode of fine sediment is, in fact, as suspended load and such sediment may amount to some 75–95 % of the total load.

The distribution of suspended material in an unsteady environment can be determined for low concentrations of small uniform-sized particles by a sediment mass continuity equation (see Chapter 3), i.e.

$$\frac{\partial C}{\partial t}+\bar{u}\frac{\partial C}{\partial x}+\bar{v}\frac{\partial C}{\partial y}+(\bar{w}-w_f)\frac{\partial C}{\partial z}=\frac{\partial}{\partial y}\left(\epsilon_y\frac{\partial C}{\partial y}\right)+\frac{\partial}{\partial z}\left(\epsilon_z\frac{\partial C}{\partial z}\right) \tag{4.6}$$

in which C is the turbulent-average concentration of sediment in mass per unit volume of sediment/water mixture: ϵ_y, ϵ_z are lateral and vertical sediment diffusion coefficients and w_f is the particle fall velocity. Equation 4.6 can also be modified to include large volume concentrations of various sized sediment mixtures[46].

Equation 4.6 must be solved in conjunction with equation 4.5, with appropriate boundary conditions at the water surface and at either end of the flow system. The difficulties of specifying flow and diffusion coefficients together with appropriate boundary conditions at all points in an unsteady environment prohibit a general solution to equation 4.6 at present. The equation can, however, be simplified and used to illustrate the reaction of suspended sediment to a changing environment.

Consider a very wide laterally-well-mixed estuary with a bed of fine to medium sand. Equation 4.6 then reduces to the form

$$\frac{\partial C}{\partial t}=\frac{\partial}{\partial z}\left(\epsilon_z\frac{\partial C}{\partial z}+w_fC\right)-\bar{u}\frac{\partial C}{\partial x} \tag{4.7}$$

in which the vertical water velocity ($\bar{w}$) has been neglected in comparison with w_f. This is a reasonable approximation for sand-sized particles where fall velocities are greater than vertical water velocities.

Equation 4.7 simply means that the change of suspended concentration with time at a point within the flow is produced by the vertical rate of change in vertical advective (w_fC) and diffusive ($\epsilon_z\,\partial C/\partial z$) transport rates, together with the longitudinal difference in horizontal advective ($\bar{u}C$) transport. The effect of longitudinal transport would be negligible if estuarine deposits and turbulence conditions were nearly uniform in the longitudinal direction, since $\partial C/\partial x$ would then be nearly zero. The concentration of sand in the flow would then change as a result of the difference between vertical diffusive and advective transport rates. This implies that an equilibrium situation can be reached at all levels in the flow when diffusive and advective transports are equal. Equation 4.7 may then be integrated over the flow depth to give the result

$$\frac{C}{C_a}=\exp\left[-\int_a^z\frac{w_f}{\epsilon_z}\mathrm{d}z\right] \tag{4.8}$$

in which C_a is the sediment concentration at level a above the sediment bed and it is assumed that equilibrium conditions ($\partial C/\partial t = 0$) exist at all levels above the bed. This condition is satisfied in uni-directional river flows where sufficient time is available for the sediment to be diffused over the flow depth and may be approximately satisfied in tidal flows for large sized particles or shallow flow depths.

Equation 4.8 can be evaluated by specifying the variation of w_f and ϵ_z over the flow depth. The particle fall velocity is usually taken to be independent of turbulent conditions in low concentrations ($< 1\,\%$ by volume) and therefore to be independent of flow depth. The form of the vertical sediment diffusion coefficient depends on the depth variation of turbulence characteristics and the presence of vertical temperature, salinity or suspended solids density gradients which are, in turn, reflected in the depth variation of horizontal water velocity and shear stress (see Chapter 3).

Analytical expressions for the vertical distribution of sediment can be produced, for example, for a linear distribution of shear stress and a parabolic or logarithmic distribution of horizontal water velocity. ϵ_z is then either constant over the flow depth or shows a parabolic form as given in equation 3.41. Integration of equation 4.8, in the absence of vertical density gradient then gives the results:

$$C/C_a = \exp\left[-w_f(z-a)/\bar{\epsilon}_z\right] \tag{4.9a}$$

$$C/C_a = [(H/a-1)/(H/a-1)]^Z \tag{4.9b}$$

$$Z = w_f/\beta K U_* \tag{4.9c}$$

where β is a correction factor to allow for sediment rather than momentum transfer ($= \epsilon_z/\epsilon_{mzo}$) and the overbar indicates a constant value over the flow depth.

Equations 4.9 indicate that sediment concentrations are greatest at the river bed and decrease towards the water surface. Very fine sediment ($w_f \rightarrow 0$) is seen to be almost uniformly distributed throughout the flow depth ($C \rightarrow C_a$) while large sized material is concentrated near the river bed and has a small suspended load. Highly turbulent flows (ϵ_z, $U_* \rightarrow \infty$) also cause a more uniform distribution of sediment over the flow depth ($C \rightarrow C_a$).

The suspension exponent (Z) is often given the value $Z = 5$ at the boundary between bed load and suspended load, while the transition to significant suspended load motion occurs[23] at Z values between 0·2 and 2·0; that is w_f/U_* values of 0·08–0·8 for $\beta = 1$. The size of sediment found predominantly as suspended load is therefore likely to be less than about 80–300 μm for typical estuary values of $\beta = 1$, $U = 1$ m/s and $U/U_* = 20$. The transition to significant suspended load transport also appears to correspond to the point in the tidal cycle at which the standard deviation of vertical velocity fluctuations (σ_w) is comparable in magnitude to the fall velocity of the sediment grains. For example the range of $\sigma_w (\approx \sigma_u/2)$ over the flow depth is about 0·5–1·5U_* (see Chapter 3), and may be compared with values of w_f for $Z = 0{\cdot}2$ to 2 of 0·08 to 0·8U_*. The Z criterion for significant suspended load also suggests that the majority of fine material ($< 100\ \mu$m) is carried directly into suspension from the bed once critical erosion conditions are exceeded. For example, figure 4.3 shows that erosion begins when $U \geqslant 0{\cdot}2$ m/s ($U_* \geqslant 10$ mm/s) and implies that $w_f/U_* \leqslant 0{\cdot}8$ (based on $Z \leqslant 2$) since $w_f \leqslant 8$ mm/s for $d \leqslant 100\ \mu$m.

The chief difference between equations 4·9(a) and (b) is the rate of change of sediment concentration near the flow boundaries, as is shown in figure 4.7, in which ϵ_z has been taken as the depth-average value of equation 3.41 and $Z = 0{\cdot}34$ has been used.

Clearly, equation 4.9(b) represents quite a good fit to field values and is in

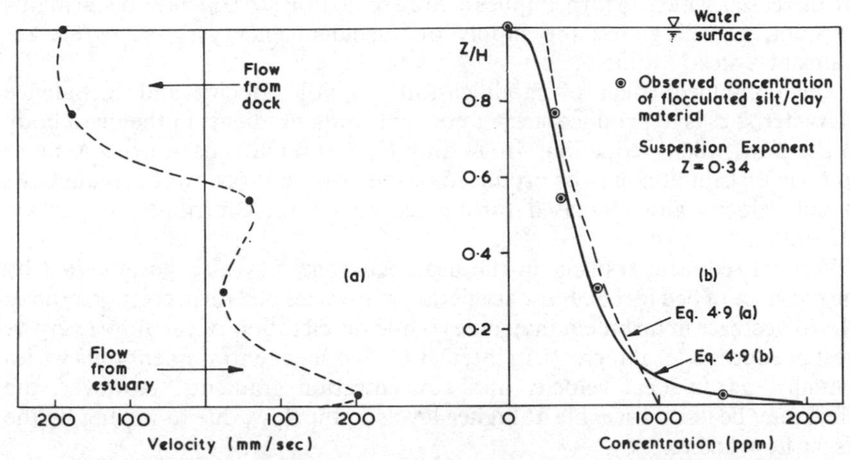

Figure 4.7. Observed velocity and suspended sediment profiles in a density-exchange tidal flow in Gladstone Lock (Mersey Estuary, UK) and the comparison with equation 4.9(*a*) and (*b*)

line with results found in parts of other English estuaries, where the vertical distribution of fine sand particles ($< 150\ \mu$m) has also been found to follow equation 4.9(*b*)[24]. Much of the discrepancy between observed values and predicted values can be attributed to the magnitude and form of diffusion coefficient used to determine equation 4.9(*b*). The assumption (in equation 3.41) of a linear shear stress distribution and a logarithmic horizontal velocity distribution over the flow depth is clearly incorrect as can be seen from figure 4.7(*a*). However, quite good agreement between equation 4.9(*b*) and the field observations would be obtained by making a small change in the value of Z. This is probably one of the main reasons why equation 4.9(*b*) has been found to apply to river situations, in which the suspension exponent (Z) is invariably different from its value in equation 4.9(*c*). Agreement with field situations is also improved by applying the equation to each grain size fraction present on the river bed and then combining the results in direct proportion to the amount present.

Other reasons for the difference between equation 4.9(*c*) and field values of (Z) is the modification of particle fall velocity and flow turbulence by vertical density gradients and in particular by the presence of suspended sediment in the flow. Large concentrations of suspended solids lead to a reduction in particle fall velocity, due to an increased frequency of particle collision and an increase in viscosity of the sediment-water mixture. For example, a 1 % (by volume) concentration of sediment or lowering the water temperature from 27° C to 4° C reduces particle fall velocities by 20–60 % for grain sizes between 1 mm and 0·1 mm.

Large sediment concentrations cause turbulent fluctuations to be reduced, as described in Chapter 3. This reduction in turbulence velocity fluctuations also implies a reduction in shear velocity and in vertical momentum and sediment transfer. Consequently, flow discharges and local velocity gradients are increased, which, in turn, implies a large reduction ($\propto U_*^3$) in Von Karman's constant, assuming that the supply of turbulent energy ($\approx \tau_{zx} \partial\bar{u}/\partial y$) remains at a steady value.

The combined effect of modifications in fall velocity and turbulence characteristics is to produce steeper concentration gradients in the main body of the flow, where reductions in K and U_* outweigh those in w_f. A more uniform distribution may be produced in the near bed zone, where reductions in fall velocity and increased turbulence may both contribute to particle suspension.

Vertical sediment transfer in the near-bed zone may also be modified by the presence of bed forms. Flow acceleration towards bed form crests produces a local decrease in turbulent intensities while deceleration of the flow over the crest produces a local increase in intensities. Sediment entrainment thus varies spatially, as do local velocity and concentration gradients. However, the effect may be less noticeable at higher levels in the flow, due to mixing by the larger flow eddies.

The vertical distribution of sediment in a tidal environment is also affected by vertical salinity gradients. These could be accounted for in shallow partially-mixed estuaries by including a Richardson number in the expression for ϵ_z and would have the effect of reducing vertical sediment transfer. Spatial and temporal accelerations also affect vertical mixing processes (see Chapter 3) but it is impossible at present to quantify these effects for practical flows.

Equation 4.8 is unlikely to apply to deep tidal systems containing small sized sediment since the tidal cycle may not be long enough to allow equilibrium conditions to be reached, at all levels within the flow. This point is illustrated by solving equation 4.7 by numerical techniques with $\partial C/\partial x$ taken equal to zero[25] and with appropriate bed and surface boundary conditions. Parameter values similar to those existing on a mean spring flood tide in the Mersey Narrows (figure 1.9 Station H) have been used. This implies a tidal range of 9 m, mean water depths of 22 m and a maximum flow of some 2 m/s. ϵ_z was determined by a similar equation to 3.41 and U_* was allowed to vary with time in a similar manner to observed near-bed velocities. The rate of sediment pick-up (P) from the estuary bed was allowed to vary throughout the flood tide and was calculated from Einstein's transport theory[23]. No corrections were used for the effects of bed forms on bed load transport and sediment pick-up or on vertical salinity gradients. Erosion was assumed to take place at a critical shear stress given by figure 4.4.

The result of the calculation, which was performed for two grain sizes (76 μm and 195 μm), is shown in figure 4.8 for two levels above the estuary bed. The results have also been expressed as a percentage of the maximum tidal equilibrium concentration (C_{eq}(max)) at each level above the bed. The solid line in figure 4.8 represents the equilibrium values which would result if steady-state conditions were reached simultaneously at all levels in the flow, and are given by similar equations to 4.9(*b*) with appropriate Z values for each instant during the tide. The dashed line represents the numerical solution and thus non-equilibrium concentrations (C_t).

The equilibrium concentrations (C_{eq}) in figure 4.8 increase from slack low water to reach maximum values at times of maximum tidal velocity, and

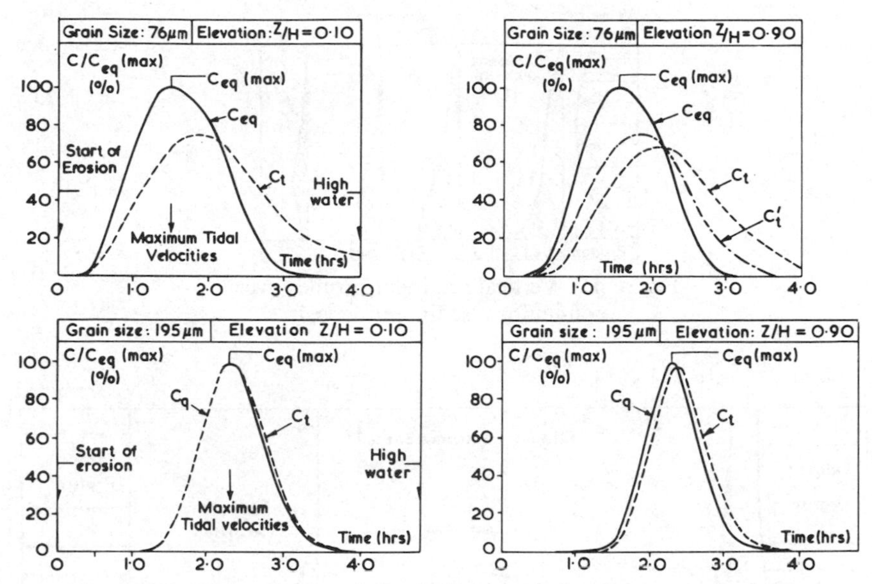

Figure 4.8. Theoretical variation of suspended sediment concentrations during a flood tide appropriate to conditions at the entrance to the Mersey Narrows

reduce towards slack high water. This is an inevitable consequence of relating U_* to the flow velocities. The non-equilibrium concentrations depend on the value of Z and thus sediment grain size. The larger sized material has sufficient time to achieve equilibrium values, which is to be expected since the majority of the equilibrium profile is situated near the estuary bed. The finer material takes longer to reach an equilibrium profile and reaches a maximum concentration towards high water slack. The small fall velocity of the particles prevents complete settling of the sediment at slack water.

Figure 4.8 suggests that equilibrium profiles can be established in deep vertically homogeneous tidal flows with a uniform longitudinal sediment cover and spatially constant turbulence conditions, provided that turbulence levels are small or particle fall velocities are high enough to produce large Z values. If turbulence levels are high or fall velocities are small, non-equilibrium profiles will result. A useful guide to the formation of equilibrium profiles is that the parameter $(w_f t/H)$ has a value of about 2; (t) being the time available for erosion. For example, a 195 μm particle requires an erosion time of some 0·6 hours for $w_f = 24$ mm/s and $H = 25$ m, while a 76 μm particle ($w_f = 4$ mm/s) requires about 3·6 hours for the same depth.

The example shown in figure 4.8 also gives an idea of the order of error produced in the case of the Mersey by predicting near-surface concentrations (C'_t) from observed near-bed values (C_a taken equal to C_t at $z/H = 0{\cdot}10$) and equilibrium concentration curves. The method produces good results at each point in the tidal cycle for coarse, but not for the finer, material. The C'_t method is better, however, than using equation 4.9(b) with C_a determined

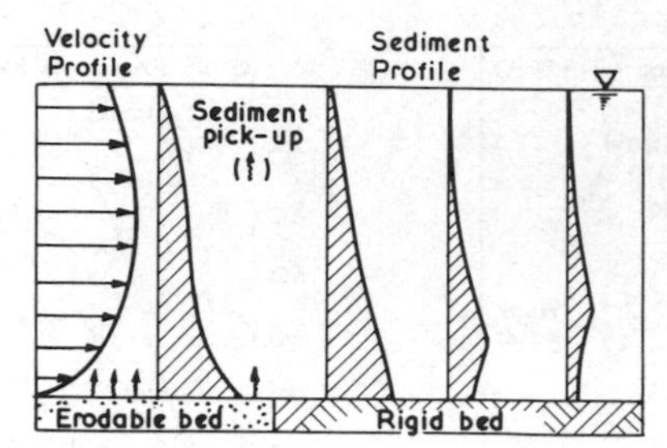

Figure 4.9. Vertical sediment profiles over a non-uniform sediment deposit

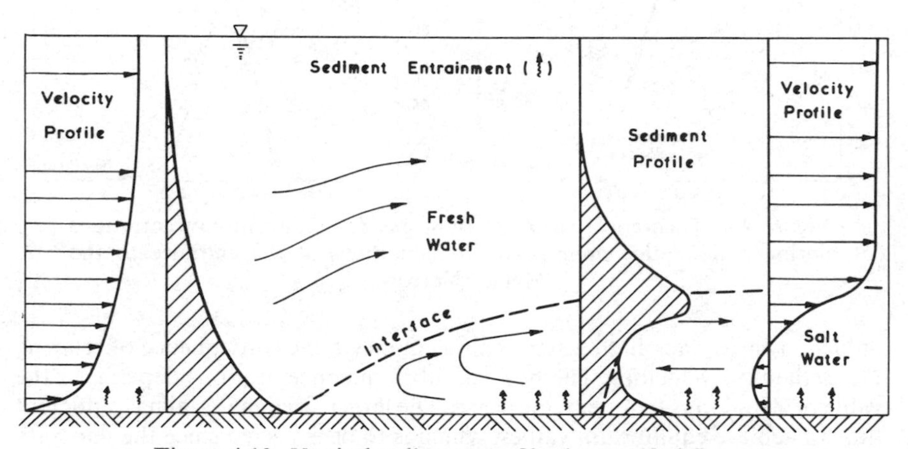

Figure 4.10. Vertical sediment profiles in stratified flows

from some analytical bed load formula[23] and will give better results in shallower flow systems. The C_l' method has, for example, been used with success in predicting sedimentation rates in dredged channels at the entrance to the Thames estuary[24].

Many estuaries contain spatially non-uniform sediment deposits and consequently non-equilibrium vertical profiles are readily produced. Figure 4.9 illustrates one type of vertical profile produced above a particular non-uniform sediment deposit, which consists of an erodable area with a rigid bed downstream of it. Even more complicated profiles are produced in stratified environments as is shown in figure 4.10, in which sediment entrainment from the estuary bed is shown taking place at all points except near the tip of the salt wedge. The dashed sediment profile results if entrainment does not occur in the salt layer.

The non-equilibrium profiles of figures 4.9 and 4.10 could be predicted by the solution of equation 4.6 or, in simplified two-dimensional situations, by the use of equation 4.7, provided that sediment concentrations and particle sizes are small and the boundary conditions of the problem are known. However, a major uncertainty exists in defining the spatial and temporal

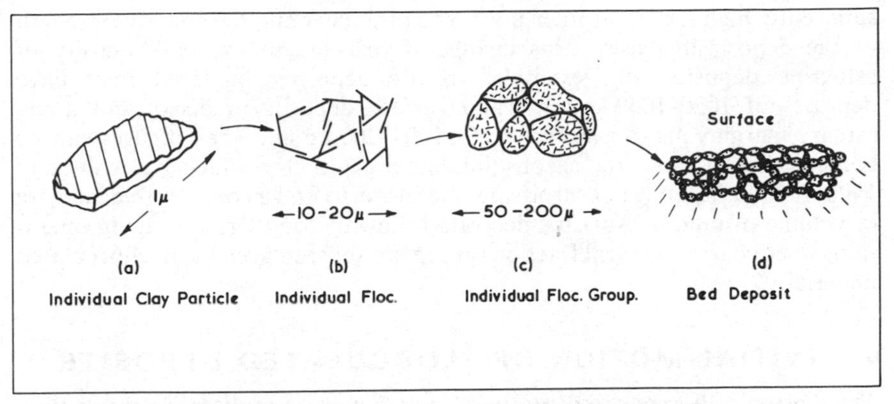

Figure 4.11. Typical arrangements and sizes of flocs and floc groups

entrainment of sediment at the estuary boundaries. Consequently, field observations of horizontal velocity and suspended solids are usually required to determine total suspended loads in a tidal environment. These may be calculated from the expressions:

$$q_{ssx} = \int_0^H cu\,\mathrm{d}z; \quad q_{ssy} = \int_0^H cv\,\mathrm{d}z \tag{4.10}$$

where q_{ssx}, q_{ssy} are longitudinal and lateral suspended sediment transport rates in weight per unit width respectively, c is instantaneous sediment concentration (in weight per unit volume), and u, v are velocity values, all at height z above the estuary bed. Often turbulent-mean values are used in equation 4.10 on the assumption that longitudinal diffusive transport is small.

4.8 FLOCCULATED SEDIMENTS

These consist mainly of clay minerals such as illite, kaolinite, montmorillonite and chlorite which are derived principally from the chemical weathering of rocks and by the reduction of one clay mineral to another. The clay minerals themselves are composed of hydrated silicates of iron, aluminium and magnesium and may also be mixed with electrochemically charged organic matter from chemical plants and sewage works. An estuarine deposit may also, of course, contain non-cohesive material.

Individual clay minerals have a similar density to silica and quartz but have a highly variable shape. Plate-like forms are very common (figure 4.11 *a*) but tube and needle-like shapes also exist. Small groups of these particles (flocs) with a more rounded appearance (figure 4.11 *b*) are also present in a deposit, due to the interaction of the ionically charged clay minerals with dissolved salts in the estuarine water. Individual flocs also unite into larger floc groups (figure 4.11 *c*) which may be rounded in form, although long chain-like or branch-like units are also found in suspension at times of slack water.

Deposits formed from floc groups have considerable porosity which slowly changes as the deposit consolidates. If the sediment deposit is subjected to sufficiently high loads, the inter-floc voids disappear and the cohesive strength of the deposit increases. This change is reflected in the bulk density of estuarine deposits. For example, freshly deposited material may have densities of 1020–1050 kg/m^3, while surface deposits in docks and along estuary margins may have values of 1100–1200 kg/m^3. These figures can be compared with those for a consolidated organic clay which has values of 1400–1600 kg/m^3. It is not surprising, therefore, to find a considerable increase in volume of muddy estuarine deposits following long term dredging operations since coarse sediment fractions are invariably replaced by fine flocculated material.

4.9 INITIAL MOTION OF FLOCCULATED DEPOSITS

The erosion of flocculated deposits is a much more complicated process than that of discrete particles. The forces indicated in figure 4.2 must be considered together with electrostatic forces between individual particles (see Chapter 1) and electrochemical forces between groups of flocs. The erosion resistance of a deposit depends largely on the cementing action of chemicals on surface particles, the ionic bond strength of surface flocs and the degree of consolidation of the deposit. This latter quality depends in turn on the mineral composition of the sediment, the shape and packing of individual flocs and floc groups, the presence of easily compressible organic material and sand drainage paths, the physical and chemical properties of the fluid environment and time. Figure 4.3 indicates the rapid increase in mean velocity required to initiate erosion with increasing degree of consolidation: Line B represents unconsolidated material and may correspond to a moisture content of 90 % while lines C and D are approximate representations for partially consolidated and fully consolidated material respectively [7].

The velocities of figure 4.3 for particle sizes of 1–2 μm are very similar in magnitude to those calculated for various concentrations of San Francisco Bay mud (see table 4.4) on the basis of laboratory viscometer tests. The viscometer results also indicate that a silt/clay suspension behaves as a Bingham fluid. This exhibits Newtonian behaviour (stress proportional to rate of strain, $\tau_{zx} \propto \partial \bar{u}/\partial z$) only at shear stresses in excess of a particular value. Once this value is exceeded, the sediment bed will flow as a fluid with an increased viscosity corresponding to its sediment content.

Clay material deposited in saline water may also exhibit a total loss in shear strength if exposed to a freshwater environment for sufficient time, so that the saline pore water is replaced by fresh water. A slight disturbance may then be sufficient to set the bed in motion, as occurs with certain 'quick' Norwegian clays. Clay deposits sometimes exhibit thixotropic behaviour and show an increase in shear strength (and thus in resistance to erosion) during slack water periods. In this way sediment deposits may accumulate on the bed during neap tide periods only to be eroded again during the next spring tide cycle.

Many attempts have been made to correlate the 'erosion velocity' of consolidated material with such bulk sediment parameters as plasticity index, liquid limit, voids ratio, mean particle size, and the percentage of clay present in the deposit. The most significant correlations usually involve the first three

Table 4.4. *Limiting shear stresses resisted by sediment in a 30 ft channel*[26]
(Courtesy of the American Geophysical Union)

Average velocity (m/s)	Shear stress (N/m²)	Sediment concentration (kg/m³)	Bulk density (kg/m³)
0·305 (1)	0·098	17	1011
0·61 (2)	0·343	59	1051
0·915 (3)	0·737*	127	1092
1·22 (4)	1·26*	217	1148

* Extrapolated values. Bracketed figures in ft/sec.

parameters, but other correlations involving the overall shear strength of the deposit, as measured by a shear vane, have also been suggested. Laboratory evidence as to the most important variables is contradictory, and correlation is unlikely to be improved until the flow and sediment parameters are more precisely defined. In particular, the effects of near-bed turbulence, *in situ* soil structure and the surface cementing action of sand particles and chemicals, such as iron oxide, are not well understood. Determination of critical erosion velocities must be attempted, therefore, by use of empirical formulae, based on laboratory and field results. Laboratory experiments involving steady uniform flow over a bed of reconstituted *in situ* material provide a useful guide for recently deposited or partially-consolidated estuarine deposits, but cannot represent estuarine conditions precisely because the *in situ* turbulent structure and soil fabric is not reproduced. Such laboratory tests on recently deposited (unconsolidated) material have given critical shear stresses in the range 0·3–0·6 N/m² for bed sediment concentrations of some 200–300 kg/m³. Table 4.4 shows similar but somewhat lower τ_c values than the laboratory results. Partial consolidation of a deposit can, however, increase τ_c values by at least an order of magnitude (see Table 4.1, which is based on steady unidirectional flow experiments over a sediment bed produced by allowing mud to settle from suspension for a period of between 1–3 days and then scraping off the surface bed layer).

4.10 ESTABLISHED MOTION OF FLOCCULATED SEDIMENTS

The general mode of transport of flocculated sediments depends on the degree of consolidation of the estuarine deposit. Freshly deposited sediment from a previous tide has little chance to consolidate over slack water and indeed may never do so, if continuous sand-drainage paths are absent. Consequently, such deposits may be set in motion on relatively shallow slopes once critical shear stresses have been exceeded. Moreover, sediments may continue to flow on much flatter slopes once motion is initiated, perhaps by shock loading or wave action. Bed flows of freshly deposited sediment are known to occur locally in the Thames, Severn and Gironde (France) estuaries. Indeed, observations in the latter estuary indicate flow velocities of the order of 0·5 m/s with sediment concentrations of some 50 kg/m^3[27].

If the sediment does not lose shear strength on initial motion, mud flows only take place locally where bed slopes are steep enough, as for example along the sides of dredged cuts and navigation channels. The normal mode of transport of fine sediment is therefore by suspension. Transport is initiated when the applied forces break cohesive bonds between floc groups. These are easily taken into suspension by the turbulent eddies because of their low density.

If the material is unconsolidated, erosion proceeds layer by layer through the deposit at a constant rate and is independent of the strength of the bed as measured by shear vane tests. The latter are representative of the resistance of a large number of inter-floc bonds distributed around the vane circumference and over its depth and thus of the macroscopic strength of the bed rather than of the resistance of the surface layers. Bed resistance may increase with depth in partially consolidated material and a point may be reached at which erosion stops.

Armoured mud deposits require higher erosion stresses than normal. When critical values are exceeded the surface skin tends to rupture randomly producing large (20–30 mm) pieces of cemented mud which are rolled into balls by the flow. These are called mud pellets and can usually be neglected when considering bed-load transport. They may also be produced by the erosion of alternate thin layers of mud and sand. The less resistant flocculated material beneath the surface skin is eroded very quickly since shear stresses are higher than those required for normal failure.

Resistance to erosion is also affected by the physical properties of the flowing fluid. Salinity is considered to have only a small effect on erosion rates at salt concentrations above 1 kg/m^3; below this value, bond strength is rapidly reduced and can lead to the erosion of quantities of partially consolidated flocculated material during times of freshets. Water temperature probably has a slight effect on ionic bond strength with a tendency for greater erosion resistance in warmer water. An increase in viscosity of the water has a two-fold effect. Sub-layer thickness is increased, leading to erosion at higher velocities, while particle fall velocities are reduced leading to high concentrations of suspended sediment. Many British estuaries are found to have much higher suspended load concentrations in cold weather as suggested by a reduction in viscosity, but additional factors such as increased freshwater flows (and thus sediment supply) and de-flocculation effects are probably more important.

Quantification of the erosion rate of flocculated materials is difficult in view of the variables outlined above. Even laboratory evidence is conflicting. Some small scale laboratory experiments[28, 29] suggest that partially consolidated material with concentrations of some 200–300 kg/m^3 is eroded in direct proportion to applied shear stresses and that the process is independent of suspended load concentration. The functional relationship is

$$\frac{dm}{dt} = M\left(\frac{\tau}{\tau_c} - 1\right) \tag{4.11}$$

where m is the mass of sediment removed per unit bed area per unit time, and M is a 'constant' with units of mass per unit bed area per unit time. Equation (4.11) does not provide for simultaneous erosion and deposition and implies that cohesive material reacts much more slowly to changes in environmental conditions than sandy material; many river bed-load transport formulae suggest that sand transport rates are proportional to $\tau_{z_0}^{2\cdot5}$ or $\tau_{z_0}^{3}$. Other small scale laboratory experiments[26] suggest that an equilibrium concentration is reached in the flow, as for discrete particles, at which scour and deposition rates are equal. Some of the conflict between the laboratory results may well arise from the type of material used in the various tests. Equation 4.11 is likely to be true for flocculated clay material while a balance between erosion and deposition is likely to be true for silt sized material. Experiments conducted with a mixture of silt and clay material will tend to give mixed results.

The constant M is a function of the degree of consolidation of the bed and erosion depth within the deposit as well as the other parameters detailed earlier. It can be determined by flume experiments; for example, Thames and Gironde mud have values of 1·7 and 2·0 $g/m^2/s$ respectively[35].

Equation 4.11 suggests that the concentration of fine material near the bed of a well mixed estuary increases and decreases in phase with the tidal currents, provided that a sufficient depth of uniform bed material is available for erosion; that the deposit extends over a length comparable to the tidal excursion (the distance travelled by a fluid particle over the course of a flood or ebb tide); that the variation of longitudinal velocity is small; and that M remains constant. These conditions seem to apply to Upper Chesapeake Bay, as indicated by figure 4.12, even though some degree of stratification is present. The presence of fine suspended material which does not settle at slack water is also indicated in figure 4.12; surface concentrations show little tidal variation.

Once individual flocs and floc groups are carried into suspension, they become reduced in size by the shearing action of local velocity gradients. This reduction process continues until an equilibrium situation is reached, at which point the rate of floc destruction is balanced by the formation of new flocs from particle collisions. The average size of floc present in the flow thus varies throughout the flow depth and the tidal cycle. Nevertheless, the vertical distribution of flocculated estuarine sediment over a large proportion of the flow depth is sometimes found to follow equation 4.9*b* (see figure 4.7), but with Z values typical of coarse silts and fine sand particles. This implies that a nearly constant effective floc size has been reached over the flow depth and that the effective particle fall velocity is comparable to silt and sand-size material. This comparison breaks down close to the flow boundary since a proportion of the flocculated material is re-entrained into the flow: equation 4.9*b* assumes no re-entrainment.

The determination of effective particle size and, more importantly, sediment fall velocity for flocculated material is extremely difficult since the size is a function of clay mineral type, the physical and chemical properties of the surrounding fluid, the number of particles present in the flow and its mean and turbulent structure. Laboratory experiments are of little use since field

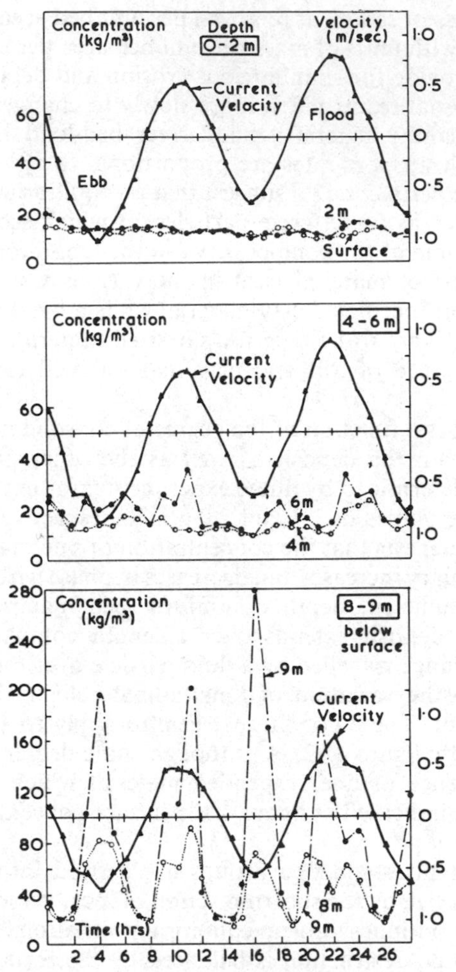

Figure 4.12. Distribution of suspended silt/clay material in Upper Chesapeake Bay, USA (after Shubel[1])

turbulence and shear stresses cannot be adequately reproduced. Indeed, field observations[39] in the Thames Estuary suggested that particle sedimentation rates can be at least an order of magnitude greater than implied by laboratory results.

Once velocities reduce towards slack water, floc destruction is reduced while particle collisions are increased as the result of the differential settling rates of the flocs and other sediment particles present in the flow. Particle size thus tends to increase but the situation very close to the sediment boundary is not too clear. Some laboratory tests[26, 30] suggest that sediment cannot settle onto the bed if the applied bed shear stress is above a certain limiting value (τ_d). The argument for the existence of τ_d is usually based on the idea that the zone just above the sediment boundary is the area of greatest turbulent intensity and that large flocs entering this zone are broken apart into smaller units and immediately re-entrained into the main body of the flow. Other tests[31] suggest that a minimum shear stress (τ_{min}) exists below which all material deposits and that the minimum shear stress required to keep all material in suspension is considerably larger than τ_d. The argument in favour of a larger τ_d is based on the idea that the flow contains a wide range of floc sizes, and that the larger flocs are more resistant to applied turbulent forces than the small flocs, and thus require a larger applied shear stress than τ_{min}.

Laboratory experiments involving steady recirculating flows in straight[26] and in circular flumes in which flow was produced by rotating paddles[7], indicate that the limiting shear stress (τ_d) is of the order of 0·06–0·08 N/m^2 and is less than τ_e (see table 4.4) so that neither erosion nor accretion occurs if $\tau_d < \tau_{z_0} < \tau_e$. The laboratory results also imply depth average velocities of 0·16–0·18 m/s (for $U/U_* = 20$) and are the basis for line A in figure 4.3. Experiments in circular flumes in which flow was produced by rotation of the flume and a surface shear ring[31] suggest that the minimum shear stress (τ_{min}) has typical values (between 0·06–0·12 N/m^2) for kaolinite suspensions and San Francisco Bay mud and is clearly similar in magnitude to τ_d. The shear stress required to keep various proportions of material in suspension was also given by the laboratory tests[31]. The results suggest that the majority of material (98 %) remains in suspension at bed shear stresses of 1.2 N/m^2 or at a depth mean velocity of 0·70 m/s (assuming $U/U_* = 20$) and that $\tau_{min} = 0·1$ N/m^2. This calculated velocity value is much greater than the straight flume results but appears to agree with field observations taken in the Chao Phya Estuary in Thailand[32] (see figure 4.13). It should be noted, however, that both the field and some of the moving circular flume results were obtained with sediment containing silt-sized material, which will certainly deposit at shear stress values in excess of τ_{min}, if the sediment grains are cohesionless.

Once shear stress values fall below τ_d or τ_{min} large concentrations (> 10 kg/m^3) of clay type material flocculate due to Brownian motion alone and form a low density fluid-mud layer which settles very slowly at the estuary bed. A greater rate of flocculation occurs at concentrations in the range 0·3–10 kg/m^3 and produces a rapid deposition of sediment. At concentrations less than 0·2–0·3 kg/m^3 flocculation rates are again very slow due to the small number of particle collisions. However, rapid deposition can be promoted at these low concentrations if particle collisions are marginally increased by means of additional turbulence. Deposition by this mechanism has been observed in Mare Island Strait in the United States[26] due to the turbulence generated by piled structures.

A quantitative statement of deposition rates is difficult to make, particularly in view of the conflicting laboratory evidence. The precise deposition mechanism of flocculated material clearly requires further study. Present knowledge does, however, highlight the radically different behaviour of cohesive and discrete particles. The implication of the laboratory tests is also important from the dredging activity point of view, since flocculated sediment is readily stirred up by dredging plant. Once sediment is in suspension it will travel large distances if $\tau_{z_0} > \tau_d$. Dredging at one point in a flow system

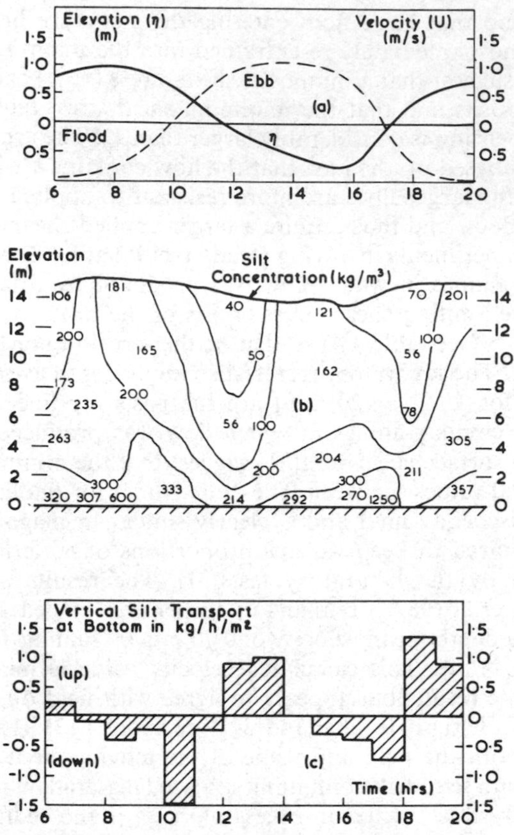

Figure 4.13. Flow and sediment transport characteristics at station (+51 km) Chao Phya estuary, Thailand (based on figure 6 from reference[32], courtesy the American Society of Civil Engineers)

may then add to shoaling rates at points located many kilometres from the dredging area. Disturbance of the larger sized cohesionless material by dredging activity affects a much smaller area, since large concentrations are rapidly reduced by particles settling from suspension, but fine sand and silt will still travel large distances before settling out, particularly if discharged into the surface of deep, high speed flows.

Previous discussions of the erosion and deposition of flocculated sediments, including equation 4.11, suggest that variation of near-bed concentrations with time above a thick bed of flocculated material will show four maximum values per tidal cycle provided concentrations are in excess of 0·3 kg/m³. Two of the maximum values occur near times of maximum tidal velocities due to bed scour and two occur near times of slack water due to flocculation and particle deposition. Figures 1.12 and 4.13*b* show good examples of an erosion peak on the flood tide and a flocculation peak on the ebb tide. Once concentrations are less than 0·3 kg/m³, the flocculation maxima near slack water are absent, as is found in Chesapeake Bay (figure 4.12).

Longitudinal tidal-averaged sediment concentration profiles in the near-bed zone may also show two distinct maximum zones provided the longitudinal sediment cover is continuous and flow is confined to a single channel. One maximum value is associated with maximum estuarine flow velocities, which may occur near the estuary mouth, while the second maximum value is associated with the salinity-associated sedimentation area located near the landward end of the estuary. Both maximum zones coincide near the sea for very large fresh water flow rates, but only the salinity-intrusion maximum exists in estuaries with a shallow bed slope and a limited amount of flocculated bed material. Spatially non-uniform bed deposits and multi-channel estuaries, however, produce variations on this simple longitudinal pattern.

The prediction of estuarine sediment transport rates for flocculated silt and clay material is a difficult, if not impossible, task in the present state of knowledge. Solutions to engineering problems must continue to be based on in situ field measurements because, in physical models, turbulence cannot be simulated completely and in mathematical models it is difficult to specify w_f, ϵ_z and, in particular, the appropriate bed boundary conditions (P, D). Consequently, field measurements should be made as extensive as time and money allow in view of the large number of seasonal variables involved in the problem.

4.11 WAVE ACTION EFFECTS

Wave action can have a major effect upon sediment motion in estuarine and coastal waters. Waves with large orbital bed velocities provide a very effective stirring mechanism while orbital motion throughout the water depth, together with breaking waves, encourages mixing of suspended material. Indeed, attempts have been made to include the orbital wave motion in vertical eddy diffusion coefficients.

Field and laboratory experiments suggest that sand is set in motion on a level bed at approximately the same near-bed velocity as in unidirectional flow ($\approx 0{\cdot}2$ m/s for 300 μm material). Bed slope would cause motion to begin at a lower bed velocity and would promote the sorting of bottom sediment; on beaches the fine fraction moves onshore near the bed but offshore at mid-depth due to mass transport currents[33]. The height of wave initiating sediment motion can be estimated approximately from the orbital bed velocity (u_b) and the critical erosion velocity of uni-directional flow. For example, the bed velocity for a swell wave is given by the equation[33].

$$u_b = (\pi H_1/T_1)/\sinh\,(2\pi H_1/L_1') \qquad (4.12)$$

where H_1, L_1', T_1 are wave height, length and period respectively. Other wave-motion/sediment movement co-relations have been achieved by using Shields' results[34] (figure 4.4) or by laboratory experiments in oscillating water tunnels[35].

Equation 4.12 and figure 4.3 or 4.4 indicate that nearly all waves are capable of moving sand-sized sediment in shallow water (< 3 m deep) but waves in excess of 1 m height and 6 sec period are required in deeper water,

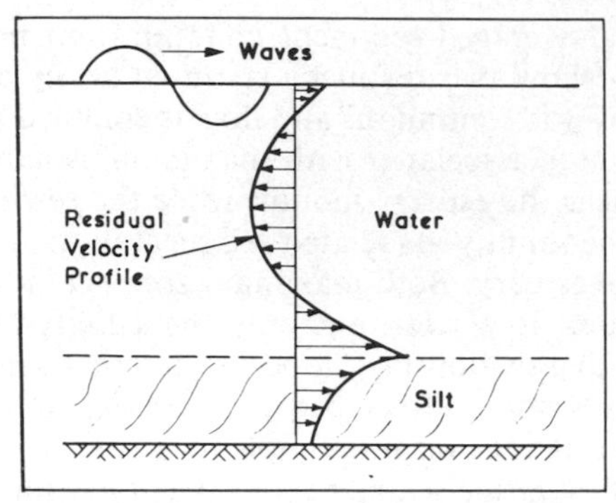

Figure 4.14. Mass transporting action of waves on water and silt (after Lhermite[44])

which is typical of navigation channel depths (10 m). Clearly, it is the long period waves which are important for strong bed-stirring action.

Once cohesionless sediment is set in motion, small ripples are formed which are a few centimeters in height and which have wavelengths of the order of the near-bed water particle motion[36]. These increase bed friction, but disappear at high near-bed velocities (> 1 m/s).

Waves combined with weak tidal currents are particularly effective in causing erosion since fine material ($< 150\ \mu$m) can be transported at quite low velocities (0·15 m/s in figure 4.3). Quantification of this process is not possible at present, although modifications of steady-flow sediment-transport formulae have been suggested[37]. A simple view of the problem, which ignores vertical velocities and flow accelerations, is obtained by vector addition of the two horizontal near-bed velocity fields followed by an averaging process over the wave period. This method eventually results in the equation[38]

$$U_{*T} = [U_*^2 + p_1^2 K^2 u_b^2]^{0 \cdot 5} \tag{4.13}$$

where p_1 is a constant ($= 0 \cdot 45$) and U_{*T} is the combined wave and current shear stress.

Equation (4.13) with $U_* = 0$ gives similar values to equation 4.12 (provided equation 4.1 is used for τ_c) and also indicates that a wave of 4 m height and of 7 s period produces the same shear stress in 20 m of water as a 2 m/s tidal current, assuming that $U/U_* = 20$.

The effect of wave action on flocculated sediments depends largely on the degree of consolidation of the deposit. Well-consolidated material, as found in drying inter-tidal flats, is highly resistant to erosion; but unconsolidated material is easily eroded, since near-bed flow velocities may readily exceed critical erosion values (see figure 4.3). Wave action may also penetrate into the deposit once critical values are exceeded if the deposit completely looses its shear resistance. Sediment transport may then occur even in the absence of tidal currents by wave mass transporting currents[44] (see figure 4.14). Waves, therefore, must be considered in any sediment transport problem in large open estuaries and coastal areas, but their effect tends to be of minor importance in deep, narrow estuaries and rivers.

4.12 STABILITY OF ESTUARINE CHANNELS

Estuarine channels, whether natural or artificial, are continually changing. For example, they may either meander or divide into two or more smaller channels, so becoming braided, or change their alignment, width and depth. Changes in cross-sectional area are usually most marked in those sections of an estuary which have small bed slopes, large suspended loads and moderately large tidal ranges. Lateral movement is also rapid if the estuary is wide, so that cross-sectional circulation currents can develop either through the non-isotropic nature of the turbulent motion, the earth's rotation or local centrifugal forces arising from channel curvature.

The stability of any particular channel, or indeed of the bed of the whole estuary, depends on the sediment transporting ability of the tidal streams as well as on the supply of sediment from external sources. Clearly, stability is a desirable feature of navigation channels but is unusual without external assistance. Equations 4.5 and 4.6 indicate why this is so. Integration of the equations over the channel cross-section and with time indicate that the condition for stability of a channel is

$$\int_{t_1}^{t_2} \frac{\partial}{\partial x}(Q_{Tx})\mathrm{d}t + \Delta Q_{Ty} + \Delta C_t = 0 \qquad (4.14)$$

where x, y are co-ordinates along and perpendicular to the channel axis respectively; Q_{Tx} is the total (bed and suspended) transport rate over the channel cross-section: ΔQ_{Ty} is the net lateral total sediment transport per unit channel length in the time interval $\Delta t(= t_2 - t_1)$; ΔC_t is the change in cross-sectional average suspended concentration over time Δt.

The time interval used in equation 4.14 is of particular importance. It must be long enough to cover all major variations in those parameters which effect sediment motion as well as the supply of sediment. In view of the variability of fresh water inflows, tidal conditions and weather, it is best to consider as long a period as possible. For example, the time interval corresponding to a neaps–springs–neaps cycle might be used if sediment inflows and weather conditions are approximately constant. However, if marked seasonal or yearly changes occur in fresh water inflows and weather conditions, longer time intervals must be selected, perhaps of the order of 5–20 years.

Consider the hypothetical situation in which tidal conditions and fresh water inflows are constant for several tidal cycles and lateral sediment inflow is zero.

Application of equation 4.14 over a time scale of one tidal cycle then produces the simple criterion for stability:

$$\frac{\partial}{\partial x}[[Q_{Tx}]] = 0 \qquad (4.15)$$

in which the double square bracket indicates a tidal average.

Equation 4.15 indicates that a channel section is stable, in the case of a homogeneous tidal flow, if Q_{Tx} is constant along the channel length. A stable channel is thus obtained if the maximum effective bed shear stress is below the

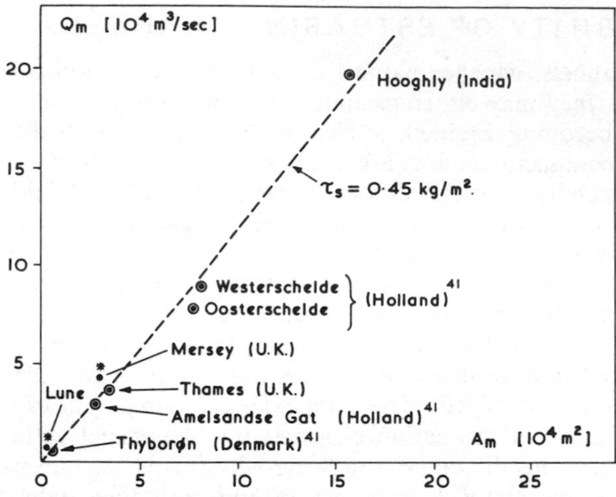

Figure 4.15. Variation of estuarine and tidal-inlet cross-sectional area at mean tide level with maximum tidal discharge

initial value for sediment motion and sediment inflows are absent. If effective tidal shear stresses exceed critical values then equation 4.15 is only satisfied in the special case of an 'ideal' estuary, assuming that Q_{Tx} is proportional to the cross-sectional mean flow velocity. The 'ideal' estuary[40] has a level bed and an exponential shape such that the tidal amplitude (A_0) and maximum tidal velocity are constant along the estuary length. Tidal velocities also show a constant phase lag (ϕ) behind water surface elevations, which implies that frictional effects are constant along the estuary and that the tide has a small amplitude in comparison with the mean depth. This is summarised in equation form as:

$$\eta = A_0 \cos(\sigma t - kx) \tag{4.16a}$$

$$U = -(A_0 g/c_0 \sin\phi) \sin(\sigma t - kx - \phi) \tag{4.16b}$$

$$B = B_0 \exp(-kx \cot\phi) \tag{4.16c}$$

where η is the water surface height relative to mean water level; A_0 is half the tidal range and is assumed equally distributed about the mean water level; $\sigma = 2\pi/T$; $k = 2\pi/L_0$; c_0 is the tidal wave speed, $(gH)^{\frac{1}{2}}$ for a progressive wave (see Chapter 6); B is the surface width of the estuary and B_0 is the value at $x = 0$; x is measured positive in the upstream direction along the estuary axis and is zero at the entrance.

The concept of an 'ideal'estuary implies that a unique relationship exists between maximum tidal discharges (Q_m) and channel cross-sectional area (A_m) at all points along the estuary. Many real estuaries with sandy bottom sediments (0·1–0·5 mm diameter) show a similar correlation near the ocean, as indicated in figure 4.15 in which the broken line corresponds to the equation:

$$Q_m = A_m C \sqrt{(\tau_s/\rho)} \tag{4.17}$$

Q_m is the maximum tidal discharge on a mean spring tide; τ_s is an average cross-sectional 'stability shear stress' which is thought to have an average value of 0·45 kg/m^2 for many tidal inlets, but which may vary from 0·35 to 0·50 kg/m^2 with increasing sediment load and littoral transport[41]; C is Chezy's friction coefficient, which in many equilibrium shaped tidal inlets can be approximated by the empirical equation[41]

$$C = 30+5 \log (A_m) \text{ m}^{\frac{1}{2}}/\text{s} \tag{4.18}$$

for A_m in m^2.

Equation (4.17) provides a simple design criterion for unrestrained estuaries and is likely to be most useful for the improvement of straight tidal inlets in medium to coarse sized sediments subject to little suspended load movement. It would be less useful for estuaries with large fluctuating fresh water inflows, large suspended loads or for estuaries restrained laterally by inerodable boundaries. Such estuaries will not conform to equation 4.17 unless a larger value of τ_s is used. The effect of a hard lateral boundary is seen in figure 4.15 for the cases of the Lune and Mersey Estuaries (UK).

Figure 4.15 also indicates that the maximum cross-sectional velocity in a stable channel in the absence of wave action, etc. is about 1 m/s, which is also the rough figure often used to indicate the start of considerable bed sediment movement.

An alternative to equation 4.17 is based on the concept of a net tidal stability velocity for the estuary entrance. This is obtained by dividing the total ebb or flood tidal volume by the total tidal period (44,700 s for a semi-diurnal tide) and the estuary cross-sectional area at mean tide level. The result, in Dutch estuaries[42] which have sandy deposits (0·15–0·2 mm grain diameters) is a velocity of 0·55 m/s. This is expected to rise to 0·7 m/s due to the deposition of mud following the construction of the Deltaworks[42].

Real estuaries show significant differences from the 'ideal' estuary outlined in equation 4.16. For example, an 'ideal' estuary in 'daily equilibrium' shows a direct correlation between upstream water volume and tidal elevation or cross-sectional area at every longitudinal cross-section. This would not be so in an estuary with significant fresh water sediment inflows, because sediment volumes tend to increase with time due to the trapping effect of salinity/density circulations. The 'ideal' estuary equations also indicate that the durations of flood and ebb tide are equal at all points. This is not so in real estuaries, in which spatially varying frictional forces and estuary shape, together with secondary forces due to meteorological conditions, Coriolis force and density gradients, modify the periods of flood and ebb tidal motion, particularly near the estuary bed (see Chapter 1). A tidal-average of the near-bed velocity at a point in a real estuary, even in the absence of density effects, usually indicates a net landward or seaward movement of a fluid or suspended solid particle.

A theoretical investigation of the direction of tidal-average motion due to vertical variations in horizontal momentum exchange ($\propto u\,\partial u/\partial x$, $w\,\partial u/\partial z$) has been made by Abbott[43] for a simplified estuary in which the bed was assumed to be horizontal, the flow two-dimensional (no lateral variation), the bed boundary-layer width independent of distance and the eddy viscosity constant. It indicated that the net movement of a particle just above the bed boundary-layer is zero, if

$$\frac{\mathrm{d}}{\mathrm{d}x}(U_0 e^{\phi(x)}) = 0 \tag{4.19}$$

where $\phi(x)$ is the phase lag between surface velocities (U_s) and tidal elevations (it is equal to $\sigma x/c_0$ for a progressive tidal wave – see Chapter 6) and the variation of surface velocity is assumed to be related to maximum tidal velocities by the equation:

$$U_s = U_0 \cos(\sigma t - \phi(x)) \tag{4.20}$$

Landward or seaward movement of a particle was found to occur if the differential term in equation 4.19 was positive or negative respectively; the positive direction of x being taken in the upstream direction.

Equation 4.19 represents a stability criterion for sediment movement close to the flow boundary and indicates that stability is only possible in estuaries in which the surface velocities show an exponential decrease in the landward direction. Application of equation 4.19 to the Thames Estuary by Abbott[43] showed that it was satisfied at only three points, one of which corresponded to the traditional shoal area known as the Mud Reaches: transport to this zone was indicated from both upstream and downstream directions. However, it must be remembered that the residual flow pattern in the Thames estuary is also influenced by longitudinal salinity/density currents which also produce a point of zero net-tidal movement.

The dependence of stability on estuary-shape and density current circulations is also seen from equation 4.15 by assuming that Q_{Tx} is a function of τ_{z_0}. Equation 2.20 shows that the Eulerian tidal-average of τ_{z_0} ($\propto i_f$) depends on three quantities; the tidal-average longitudinal density force, which is always directed upstream; the tidal-average horizontal change in flow momentum, which depends on boundary roughness and estuary shape; and the tidal-average pressure force arising from a sloping water surface. In an 'ideal' shaped estuary, all quantities are assumed to be zero, so satisfying equation 4.15, but in a real estuary all terms are important. Near the estuary head, the slope term exceeds the density term (assuming that the momentum term is negligible) while near the sea the opposite is true. Equality of forces ($[[\tau]] = 0$) is only possible, therefore, at some intermediate position which then represents a shoal area. The momentum term is especially important in flow systems with abrupt changes of flow cross-section or flow direction. Consequently, equality of forces in homogeneous systems is likely to be confined to a few particular points. For example, the Grimsby middle shoal near the mouth of the Humber Estuary (UK) occurs in an area with large changes of geometry. The tidal-average form of equation (2.20) is, however too simple to predict the location of estuary shoal areas. They can usually be found from *in situ* current observations by determining locations of 50 % ebb-predominance (see Chapter 3), that is, the points where the tidal-average velocity is zero (see also Chapter 10). Experience on many different estuaries suggests that an Eulerian average at a point close to the estuary bed is sufficient to locate shoal zones even although Lagrangian averaging is a closer representation of particle motion.

Short term stability does not appear to be possible, therefore, in estuaries with small bed slopes and large suspended load movements, consequently training works or dredging may be necessary to keep navigation channel depths within acceptable limits. However, long term stability is possible because extreme fresh water flows and weather conditions provide mechanisms for the removal of sediment from an estuary. It is interesting to note that bed levels at the entrance to the Rotterdam waterway in Holland can be lowered or raised by a metre or so following periods of prolonged strong offshore or onshore winds in the North Sea[45].

It is important to note that engineering works can modify the long term stability of an estuary by interference with natural flushing processes. For example, construction of reservoirs on rivers feeding into an estuary has a two-fold effect. Sediment supply is reduced by direct trapping in the reservoir, which causes a reduction in estuary sedimentation rates and the development of a new deeper long-term equilibrium situation. The second effect is a reduction in extreme peak river flows, which can have various effects on sedimentation processes in the estuary. Peak flows may be reduced but may still be large enough to flush the tidal-average density circulation system out of the estuary (see figure 3.6). A new deeper equilibrium situation will then occur as described above. Moderate reductions in peak flow will allow increased saline penetration and an increase in estuary siltation rates, provided sediment supplies are not drastically reduced. Erosion of downstream areas can occur, followed by progressive accretion from the landward end of the estuary. The estuary bed will continue to accrete until the saline zone is pushed seawards and the new peak flow rates are able to provide flushing action once more. A large reduction in peak flow allows increased saline penetration and progressive siltation from the landward end of the estuary but at a reduced rate.

Once dredging is adopted as a maintenance measure, it becomes necessary to dispose of the dredged material. Often this material is deposited at sea in shallow water some miles from the estuary entrance, while in other cases, it is deposited within the estuary adjacent to the main navigation channel. In both cases, sediment can be eroded by tide and wave action and returned to the navigation channel shoaling zones. Successful estuary management depends upon recognising the limitations of such disposal methods. Equation 4.14 provides a means of examining the likely return of dredged spoil in estuarine accretion zones.

Consider the estuary shown in figures 4.1. Application of the general form of equation 4.24 to the control volume leads to the equation:

$$\sum_{t_1}^{t_2}\left(S_{\mathrm{O}}+S_{\mathrm{D}}+\sum_{1}^{n} S_{\mathrm{P}}+\sum_{1}^{m} S_{\mathrm{LB}}+\sum_{1}^{i} S_{\mathrm{A}}+S_{\mathrm{R}}-S_{\mathrm{E}}\right) = p_2 \sum_{t_1}^{t_2} D+\Delta V \quad (4.21)$$

where S is a volumetric rate of sediment supply; (O – E) are subscripts referring to sediment source shown in figure 4.1: E is introduced to allow for sediment loss from the control volume by tidal flushing action. D is the volume of material dredged from the control volume in time interval

($\Delta t = t_2 - t_1$). p_2 is a correction factor to account for the difference between the volume of the dredged spoil in the dredgers hold and the volume of sediment forming the boundaries of the control volume. ΔV is the change in volume within the control volume in time Δt. Σ indicates the summation of the various sources, either over the time interval Δt, which must be chosen in the same way as that for equation 4.14, or the number of repetitive sources n, m, or i.

Dredged quantities in many estuaries consist of a silt/clay fraction from estuary docks and a sand fraction from estuary navigation channels. In this case equation 4.21 can be applied to each size fraction separately; the division being made on the basis of *in situ* samples. The quantity of dredged material returning to the control volume can be estimated from equation 4.21, provided that the various terms can be quantified[15]. The right-hand side of equation 4.21 can readily be determined from the use of bed samples, survey charts and dredging records as well as an echo sounding or radioactive density gauge traverse of the estuary bed. However, the left-hand side requires a field sampling programme which must be extensive enough to cover variations in fresh water flow, tidal range and weather conditions.

4.13 SUMMARY

Sediment deposits in a tidal environment are derived from a variety of sources, chief of which are rivers, the near-shore sea bed and the littoral zone. Pollutants and animal/plant life may make important long term contributions. Transport modes of sediment deposits depend on the presence of cohesive sediment. Cohesionless grains move as individual solid particles whose shape and density is determined by abrasion during transport and by parent source. They form moving bed undulations (bed load) which produce a large resistance to flow so that water and sediment transport rates are less than flat bed values for the same total bed shear stress. Fine material is easily eroded and travels primarily in suspension (suspended load), exchanging particles continuously with the bed so that a particular steady uniform flow conveys a fixed amount of sediment. The erosion of cohesive material depends initially upon the degree of consolidation of bed deposits. Unconsolidated material is easily eroded and may subsequently move as a fluid. Cohesive sediments travel generally as suspended load in soft sponge-like clusters (flocs). Effective particle size changes continuously in response to environmental turbulence, and deposition only occurs below some limiting bed shear stress at a rate dependent upon flow concentration. Consequently very large quantities of clay material can be carried in suspension. While tidal flow structure and the spatial distribution of sediment deposits controls the quantity of sediment in motion, it is the tidal-average flow structure which dictates the location of accretion zones. The long term stability of estuaries, however, depends on the flushing ability of large fresh water flow rates and can be seriously affected by engineering works. Dredging plant may be required to maintain stability.

While much is known about the mechanics of sediment movement, it is impossible to predict sediment transport rates with sufficient accuracy on a purely theoretical basis. Analytical equations linked with field data can be useful for simple situations but, in general, use must be made of field and/or hydraulic model tests.

NOTATION

c_0	Tidal wave speed $= (gH)^{\frac{1}{2}}$
d	Sediment grain size
$\bar{d}$	Weight mean sediment grain size
d_{50}	Median diameter of a sample of sediment (50% of sample finer than this size)
g	Acceleration due to gravity
i	Numerical constant
i_f	Friction slope as defined in Chapter 2
k	Wave number $= 2\pi/L_0$
m	Mass of sediment eroded from a cohesive deposit or porosity of bed deposit or numerical constant
mzo	Subscript indicating momentum transfer in homogeneous conditions
n, i	Numerical constants
p_1	Numerical coefficient in Bijker's formulae
p_2	Correction factor to account for difference between dredge volume and *in situ* volume
q_x, q_y	Sediment transport rate in weight per unit width in x-, y-directions
q_{ssx}, q_{ssy}	Suspended sediment transport rate in weight per unit width in x-, y-directions
t	time coordinate, or time
$\bar{u}, \bar{v}, \bar{w}$	Turbulent mean velocity components in x-, y-, z-directions
w_f	Particle fall velocity
u_b	Maximum orbital wave velocity at the bed level
z_b	Elevation of mobile sediment surface above a horizontal datum
A_0	Half tidal amplitude for ideal estuary tide
A_m	Cross-sectional area at maximum tidal discharge on a mean spring tide
B	Surface width of ideal estuary at longitudinal position x
B_0	Surface width of ideal estuary at $x = 0$ (mouth)
C	Chezy coefficient or suspended sediment concentration
C_a	Steady-state suspended sediment concentration value at level (a) above a mobile sediment deposit
C_t	Non-equilibrium suspended sediment concentration
C'_t	Near-surface non-equilibrium suspended sediment concentration obtained from actual near-bed concentration and equilibrium curves
D	Quantity of suspended sediment removed from the bed in weight per unit area of bed surface per unit time
H	Water depth at point or for a cross-section
H_1	Wave height
K	Von Karman's constant
L'_1, L_0	Wavelength of wind wave, tidal wave
M	Erosion constant for partially consolidated cohesive sediment
F	Cross-sectional Froude number $(= U/(gH)^{\frac{1}{2}})$

P Quantity of suspended sediment depositing on to a sediment bed in weight per unit area of bed surface per unit time
Q_m Maximum tidal cross-sectional discharge
Q_{Tx} Total cross-sectional sediment transport rate
R Hydraulic radius
S General sediment source
T Tidal period
T_1 Wave period
U_* Shear velocity
U_s Lateral-mean tidal velocity at the water surface
U_0 Maximum lateral-mean tidal velocity at the water surface
U Cross-sectional mean velocity
U_{*T} Shear velocity due to waves and currents
Z Suspension exponent $(=w_f/\beta K U_*)$
β Ratio of sediment to momentum transfer coefficient $(=\epsilon_z/\epsilon_{mzo})$
γ_b Specific weight of sediment $(=\rho_b g)$
δ Laminar sub-layer thickness
ϵ_y, ϵ_z Sediment diffusion coefficients in the y-, z-directions
ρ_b Density of sediment grains
λ, λ_m Mean and maximum bed form length respectively
μm Micrometre (10^{-6} m)
ν Kinematic viscosity of fluid flow
σ_1 Standard deviation of bed form elevations at time t
σ_0 Steady-state standard deviation of bed form elevations
σ_u, σ_v Standard deviation of turbulent velocity fluctuations in the u-, w-directions
σ Tidal wave number $= 2\pi/T$
η Tidal amplitude
ϕ Phase angle of ideal estuary tide
$\phi(x)$ Phase angle of tidal motion which varies with distance (x)
τ_{zx} Fluid shear stress in x-direction on a plane perpendicular to z-direction
τ_{z_0} Bed shear stress
τ_c Critical bed shear stress for sediment
τ_d Bed shear stress required to keep all fine grained material in suspension
τ_{min} Bed shear stress below which all fine material settles out
τ_s Stability shear stress
$\Delta\overline{H}$, ΔH_m Mean and maximum bed form height
ΔC_t Cross-sectional average suspended sediment concentration change in time Δt
ΔQ_{Ty} Lateral sediment inflow to a channel per unit length in time Δt
ΔV Change in volume of an estuary control volume
Δt Time interval
A; D; O; E; W; LB; P, R Subscripts referring to various sediment sources

REFERENCES

[1] Shubel, J. R., Suspended sediment of Northern Chesapeake Bay, *Tech. Rep. 35, Chesapeake Bay Inst. Johns Hopkins University*, March (1968)
[2] Wicker, C. F., Maintenance of Delaware Estuary Ship Channel, *XXI International Navigation Congress*, Stockholm (1965). Section II, Subject 3, PIANC, Brussels
[3] Johnson, J. W. Dynamics of nearshore sediment movement, *Bull. Am. Soc. Pet. Geol.*, **40**, 9 Sept. (1956)
[4] Peirce, T. J., Jarman, R. T. and de Turville, C. M., An experimental study of silt scouring, *Proc. Inst. Civ. Eng.*, **49**, March (1970)
[5] Folk, R. L. and Ward, W. C., Brazos River Bar. A study in the significance of grain size parameters, *J. Sed. Pt.*, **27**, no. 1 (1957)
[6] Hjulström, F., The morphological activity of rivers as illustrated by rivers Fyris, *Bull. Geol. Inst., Uppsala*, **25** (1935)
[7] Postma, H., Sediment transport and sedimentation in the estuarine environment, *Estuaries*, Ed. Lauff, *Am. Ass. Adv. Sci.*, Pub. 83 (1967)
[8] Leliavsky, S., *An introduction to fluvial hyraulics*, Constable, London (1955)
[9] Shields, A., Anwendung der Ähnlichkeitsmechanik und turbulenzforschung auf die geischiebebewegung mitteil Preuss, *Versuchsanst*, Wasser, Erd. Schiffsbau, Berlin, no. 26 (1936)
[10] Lane, E. W., Progress report on studies on the design of stable channels of the Bureau of Reclamation, *Proc. ASCE*, **79** (1953)
[11] Simons, D. B. and Richardson, E. V., Forms of bed roughness in alluvial channels, *Proc. ASCE*, **87**, HY1 (1961)
[12] Liu, H. K., Mechanics of sediment-ripple formation, *Proc. ASCE*, **83**, Feb. (1957)
[13] Nordin, C. F., A stationary gaussian model of sand waves, *Stochastic Hydraulics*, ed. Chao-Lin Chiu, Sch. of Eng., Pub. No. 4, University of Pittsburgh, U.S.A
[14] Yalin, M. S., Geometrical properties of sand waves, *J. Hyd. Div. ASCE*, **90**, May (1964)
[15] Halliwell, A. R. and O'Connor, B. A., Quantifying spoil disposal practices, *Proc. 14th Coastal Engineering Conference*, ch. 153, vol. III, June (1974)
[16] Ashida, K. and Tanaka, Y. A statistical study of sand waves, *Proceedings 12th Cong. IAHR*, **2**, Colorado State University, Fort Collins, Colorado, USA, September (1967)
[17] Jain, S. C. and Kennedy, J. F., The growth of sand waves, *Stochastic Hydraulics*, ed. Chao-Lin Chiu, Sch. of Eng., Pub. no. 4, University of Pittsburgh, USA
[18] Colby, B. R. and Scott, C. H., Effects of water temperature on the discharge of bed material, *U.S. Geol. Survey*, Prof. Paper 462-G (1965)
[19] Nasner, H., On the behaviour of tidal dunes in estuaries, *Proc. 15th Congress IAHR*, **1** (1973)
[20] Simons, D. B., Richardson, F. V. and Nordin, C. F., Jr, Sedimentary structures generated by flow in alluvial channels, *Report CER 64*, DBS/EVR/CFN 15, Colorado State University (1964)

[21] Kennedy, J. F., The mechanics of dunes and antidunes in erodable-bed channels, *J. Fluid Mech.*, **16**, Pt. 4, August (1963)

[22] An investigation of sand movements in the Ribble Estuary using radioactive tracers, *Rep. No. Ex.* 280, Hydraulic Research Station, Wallingford, England, July (1965).

[23] Einstein, H. A., The bed load function for sediment transportation in open channel flows, *US Dept. Agric. Soil Conservation Ser.*, Report No. 1026 (1950)

[24] Thorn, M. F. C., Deep tidal flow over a fine sand bed, Paper A 27, *Proc. 16th Cong. IAHR*, São Paulo, Brazil, July (1975)

[25] O'Connor, B. A., Mathematical model for suspended sediment distribution, **4**, *Proc. 14th Congress IAHR*, Paris (1971)

[26] Einstein, H. A. and Krone, R. B., Experiments to determine modes of cohesive sediment transport in salt water, *J. Geo. Res.*, **67**, no. 4, April (1962)

[27] Bonnefille, R., Etude de l'aménagement de l'estuaire de la Gironde, report no. 10, *Division Hydraulique Maritime Department Laboratoire National d'Hydraulique*, Chatou (1970)

[28] Cormault, P., Détermination expérimentale du débit solide d'érosion de sédiments fins cohésifs, **4**, *Proc. 14th Congress IAHR*, Paris (1971)

[29] Partheniades, E., *A study of erosion and deposition of cohesive soils in salt water*, Ph.D. Thesis, University of California (1962)

[30] Krone, R. B., Silt transport studies utilising radio-isotopes, 1st and 2nd Annual Reports, *Hyd. Eng. Lab.*, *University of California*, (1956–9)

[31] Mehta, A. J. and Partheniades, E., Effect of physico-chemical properties of fine suspended sediment on the degree of deposition, *Proc. Int. Symp. of River Mech. IAHR*, **1**, Bangkok, Thailand (1973)

[32] Allersma, E., Hoekstra, A. J. and Bijker, E. W., Transport patterns in the Chao Phya Estuary, *Proc. 12th Coastal Eng. Conf.*, ch. 37, (1966)

[33] Coastal Engineering Research Centre, Shore Protection Planning and Design, *Tech. Memo. No. 4*, US Gov. Print. Office (1966)

[34] Bonnefille, R. and Perkecker, L., Le début d'entrainement des sédiments sous l'action de la houle, *Bulletin du Centre de Recherches et d'Essais de Chatou*, no. 15 (1966)

[35] Rance, P. J. and Warren, N. F., The threshold of movement of coarse material in oscillatory flow, *Proc. 11th Coastal Eng. Conf.*, **1**, ch. 12 (1968)

[36] Homma, M., Horikawa, K. and Kajima, B., A study of suspended sediment due to wave action, *Coastal Engineering in Japan*, **8** (1965)

[37] Einstein, H. A. A basic description of sediment transport on beaches, *Proc. Advanced Seminar. Math. Res. Centre*, University of Wisconsin (1971). Ed. R. E. Meyer, Academic Press, New York/London (1972)

[38] Bijker, E. W., Some considerations about scales for coastal models with movable bed, *Delft Hydraulic Lab.*, *Pub. no.* 50, Nov. (1967)

[39] Owen, M. W., The effect of turbulence on the settling velocity of silt flocs, **4**, *Proc. 14th Congress IAHR*, Paris (1971)

[40] Pillsbury, G., *Tidal Hydraulics*, Corps. of Engineers, Vicksburg, USA (1956)

[41] Bruun, P. and Gerritsen, F. *Stability of Coastal Inlets*, North-Holland Pub. Co., Amsterdam

[42] Dronkers, J. J., Research for the coastal area of the delta region of the Netherlands, *Proc. 12th Coastal Eng. Conf.*, ch. 108, Washington (1970)

[43] Abbott, M. R., Boundary layer effects in estuaries, *J. Mar. Res.*, **18**, 2 (1960).

[44] Lhermite, P., *Bulletin d'Information, Comité d'Océanographique et d'Etudes des Côtes*, **10**, no. 5 (1958)

[45] Terwindt, J. H. J., de Long, J. D. and Van der Wilk, E., Sediment movement and sediment properties in the tidal area of the lower Rhine (Rotterdam Waterway), *Verhandelingen van het Koninklijk Nederlands Geologisch Mijnbouwkundig Genvotschap* (*Geol. Serie*), deel 21–2 (1963)

[46] Hunt, J. N., On the turbulent transport of a heterogeneous sediment, *Quart. J. Mech. App. Maths.*, **XXII**. Pt 2, May (1969).

[47] Raudkivi, A. J., Analysis of resistance in fluvial channels, *J. Hyd. Div. ASCE*, **93**, HY5, Sept. (1967)

5

The study of tidal systems: field measurements

Introduction to tidal studies

Estuaries have been shown to be regions in which many factors interact. Disturbance of any of them can affect others. One of the first aims of any study must be to build up a picture of the behaviour of the estuary and all its component parts. There are many ways of doing this and some of the major ones form the subject of the next four chapters.

Various aspects of surveying, measurement of surface gradient and measurement of water quality at points in the flow are discussed later in this chapter, but first, a brief comparison of the various methods of study is given, including those that can be used for prediction of behaviour when conditions are changed.

Comparison of methods of study

Whenever major works are contemplated, it is essential to know the immediate consequences of their construction. It is also desirable to forecast as much as possible about the long-term consequences, although these are more difficult to establish in any detail. There are several methods of study, none of which is capable of providing all the information that is needed. Usually, at least two of them must be used to complement each other if reasonable forecasts are to be made. Four methods have been used widely and are described in succeeding chapters. They are as follows.

Mathematical analysis

The equations of momentum and continuity can be solved analytically if they are simplified to linearised forms. Practical applications include analysis of the motion of tides in estuaries and the diffusion of pollutants. In each case, it is only possible to obtain solutions of grossly simplified situations. These solutions can help towards understanding the physical behaviour of systems in general, but they give little help towards solution of real problems. They are discussed briefly in Chapter 6.

Numerical computations

Digital, analogue and hybrid computers are widely used for solution of the non-linear equations of tidal motion. The amount of detail that can be included is increasing rapidly as more powerful computers come into use, but it is still only possible to solve rather simple representations of real estuaries, e.g. by averaging velocities over at least one dimension. However, with numerical computations there is considerable freedom in the choice of friction parameters, diffusion coefficients and effects such as Coriolis force that are difficult to represent in physical models. The scope of numerical computations is described in Chapters 6 and 7.

Simulation

Some of the essential parts of the physical process can be simulated by electric circuitry or by use of small-scale hydraulic models. Electric simulators are a form of analogue computer, but whereas the latter operate by representation of the equations of motion, the former operate by direct simulation of the hydraulic system by electronic components (Ishiguro [1], Harder [2]). They have not been widely used and have been superseded by digital computers.

Physical models are used to represent the hydraulic system continuously in three dimensions. They are the only means of simulation that can do so. All other methods require at least some of the parameters to be lumped together at intervals of space and/or time. Electric analogues, simulators and physical models can all be operated continuously with time. Physical models can thus give a useful three-dimensional picture of the behaviour of an estuary as a whole. They suffer, however, from scale effects, particularly with regard to sediment transport, turbulence and mixing processes. Moreover, it is rarely practical to include effects such as the Coriolis force. Compared with other methods of study, they are slow and expensive to construct, calibrate and operate. Physical modelling is considered in more detail in Chapter 8.

Field measurements

Field measurements are an essential part of all tidal studies. They can often provide all the information needed for solution of a local problem, such as the speed and direction of flow at a site for a mooring buoy. In general, however, it is necessary to observe the behaviour of a system over a wide area and in several of its aspects. In practice, the size of estuaries is such that it is impossible to obtain more than a very sparse sampling of conditions of tide and flow. The variability of conditions of tide, meteorology and fresh water flow leads to a requirement for protracted observations at any one site. A further limitation to the value of field measurements is that they can only be used to study existing conditions; they cannot be used to predict the consequences of change, yet prediction is the main objective of most studies. For these reasons, field measurements are usually used in conjunction with other methods.

5.1 FIELD MEASUREMENTS

To begin with, charts used for navigation can provide a great deal of background information. Mariners need to know such things as tidal rise and fall, directions and strengths of currents, the nature of the sea bed and the general bed topography. Data about tides, currents and the nature of the sea bed presented on charts are sparse and simple, but are often sufficient to enable the salient features to be understood. They can be of great help in preparing plans for a campaign of field investigations.

Charts can only provide a broad background, however. Much essential detail is not shown on them and the accuracy, even of soundings, is insufficient for scientific purposes. Navigators are concerned with least depths; these tend to be given more weight than average depths in charts. Regions used by vessels may be surveyed quite frequently. Soundings outside the main channels, often shown in quite full detail, may be revised only occasionally—in some cases at intervals approaching 100 years. The publishers of navigational charts can usually give details about the sources of information and may also be able to provide copies of the original surveys upon which the different parts of a chart were based.

The rate of accretion and scour is likely to be of special interest to engineers. Quite apart from any immediate objective, knowledge of rates of sediment movement is as essential to an hydraulic engineer as knowledge of rainfall patterns to an hydrologist. Complete and accurate charts obtained under controlled conditions are therefore needed. Surveys should be made at regular intervals so that knowledge of any changes can be built up. Care must be taken to ensure that the surveys give useful and accurate information, bearing in mind that channels in estuaries can undergo rapid change.

Information about local current velocities is also needed in most cases. The data shown on charts are intended for navigators and show the variation of speed and direction of the near-surface water at frequent time intervals during typical spring and neap tides. This is helpful for initial planning but is inadequate for serious studies which require information about currents close to the bed related to specific conditions of tide and river flow. A problem with current velocity measurements, as with any other measurements of water quality, is to decide on the most suitable location of instruments and on the frequency of measurement for each purpose. Data given on charts can be helpful for this purpose.

The most significant factors that affect field measurements are the size of estuaries and the variability of the forces acting in them. River flows, wind and waves are subject to random variation so that flow conditions never recur in exactly the same patterns and can never be predicted precisely. Systematic study of an estuary by measurements taken at different times is thereby made very difficult. The ideal of simultaneous measurements taken at many sites cannot be attained because of the large resources of skilled manpower and scientific equipment that would be needed. The alternative is to try to build up a relationship between cause and effect at several sites throughout an estuarine system as done, for example, by Inglis and Allen[3] in the Thames Estuary. The problem is then to co-relate conditions at all of these sites. Good co-relation is sometimes possible, but random variations of fresh water flow at the landward end and wind and waves at the seaward end may prevent it. Simultaneous measurements of water surface levels at several points during field measurements is one essential requirement because it enables the strength of tide during different periods of observation to be compared. It is unlikely to be sufficient if the main interest is in rates of movement of sediment or in the net transport of water or pollutants. It is better, in such cases, to monitor the important variables at one reference point in the estuary whenever measurements are being taken elsewhere, so providing a basis for comparison between all measurements. The siting of the monitoring station will depend on the local situation. It should clearly be in the main stream of flow and away from the effect of local eddies so that it gives an indication of typical response to the variables affecting the estuary at any time. Moreover, if any particular factor is being examined, such as the variation of net transport of water and solids over the depth of the stream, the monitoring site should be chosen so that the main factors will have a discernible effect; neither so far landward that saline water seldom reaches it, nor so far seaward that the influence of fresh water is negligible.

It is sometimes necessary to obtain simultaneous measurements at different cross-sections; for example, to estimate dispersion coefficients. There is then a further complication arising from the irregular cross-sections and curved plan shapes of estuary channels which cause appreciable variations in flow over a cross-section. Measurements of discharge through a cross-section can only be done with reasonable accuracy by measuring velocity at several points in plan and at several depths below the surface at each point. This usually requires the use of up to five boats simultaneously and presents severe problems if the work is done in a busy port area. Before an operation on this scale is begun it is worth considering whether the information could be gleaned in any other way – for example by numerical computation based on the equation of continuity or, in the case of sediment transport, by the use of tracers attached to typical particles.

The various schemes of field measurements depend so much on the characteristics of individual estuaries and on the particular problems that require solution that their description is beyond the scope of this book. Instead, some of the main things that affect the accuracy of investigations are described. The subjects discussed are measurement of water surface and bed level and, by implication, estimation of sediment volume changes; measurement of water state and quality at points in the flow; and measurements of rates of sediment transport.

5.2 MEASUREMENTS OF SURFACE AND BED LEVEL

5.2.1 Measurements of water surface level

A datum for measurements of level is a basic requirement. In ports, a fixed underwater datum level may be provided by a horizontal surface such as a dock sill. This datum is only useful locally. It must be transferred to other locations just as land levels must be related to known data. Local datum levels have to be transferred to any locality via the water surface. This is usually done by tide gauges which record surface level at points along the margins of estuaries. When the water surface level has been established at a tide gauge it still has to be transferred to the required point of measurement, which may be some distance away. It is then necessary to measure depths relative to the water surface from a floating vessel which, in its turn, has to be used as a local datum of level.

Tide gauges consist of pressure transducers fixed at a known level, or of devices such as floats which record the water surface level relative to the land. A transducer records only the local pressure intensity at a point below the water surface. The pressure can be affected by flow curvature at the point of measurement and by changes in the density of liquid above the transducer. Floats must be placed in vertical chambers which protect them from currents and wave action. These chambers are also possible sources of inaccuracies.

Tide gauges record levels at the shore. These levels may not be typical of water level over a cross-section of the tidal channel passing through them. The accuracy of tide gauges therefore depends on their location within the estuary and on their method of installation. If they are sited on bends they will record the super-elevated water surface due to motion of the water. In extreme cases, the difference in level between opposite banks of a bend may be as much as 0·6 m. For example, if it is assumed that the speed of flow U is uniform over the stream depth, the local radial surface slope at radius r is $\partial z/\partial r = U^2/gr$, where z is water surface level at radius r. Then

$$z_2 - z_1 = \int_{r_1}^{r_2} \frac{U^2}{gr}\,\mathrm{d}r$$

If it is assumed that the speed of flow is equal at all radii

$$z_2 - z_1 = \frac{U^2}{g}\ln\frac{r_2}{r_1}. \tag{5.1}$$

In an actual case, $r_2/r_1 = 2$ and $z_1 - z_2 = 0{\cdot}07U^2$. If U reaches 3 m/s during spring tides, $z_2 - z_1$ can reach 0·6 m when flow speed is greatest. Differences of level of this order have been observed between opposite banks of the River Hooghly at Hooghly Point (figure 10.13).

Equation 5.1 gives a first approximation of the order of magnitude of the difference in water level across a bend, but numerical integration using actual velocities would be required for an accurate estimate to be made. The use of mean velocity in place of actual local velocity causes the super-elevation to be underestimated because the effect depends on U^2.

Tide gauge errors may also occur due to velocity head effect and to density effects[4].

Velocity head effect can occur when a tide gauge records level in a flowing stream. A poorly designed gauge may have its ports facing upstream or downstream or its location may be in a region affected by velocity head, such as upstream or downstream of a solid jetty. If the stream speed is U, the velocity head can reach $U^2/2g$; 0·2 m if the speed is about 2 m/s. Properly designed and located gauge wells should obviate these errors.

Density effects can occur in a tide gauge consisting of a tube open below low water level, such as a float chamber. During the passage of a tide, salinity may vary from zero to a value approaching that of the open sea. At low tide, the water outside the tube may be fresh. During rising tide the density of the water outside will gradually increase. The tube will fill progressively with water of increasing density. At high tide, the tube will be filled with water having mean density nearly equal to the mean density of the water that has passed it, but the water outside will have the maximum salinity. If the density of water at high tide is 1025 kg/m^3 and the mean density in the tube is 1012 kg/m^3 the height of water column inside the gauge well will be (1025/1012) or 1·013 times that outside; an error of 0·13 m on a 10 m tide. An even greater error can occur if heavy rainfall is able to fall into the gauge well; it could then read 1·025 times too high. In addition, errors arise from badly maintained tide gauge recording mechanisms and from the sedimentation of suspended solids in the tide gauge well. In particular, siltation can lead to a complete blockage of well openings. Regular maintenance is therefore essential.

5.2.2 Surveys of bed level

These must be done from the water surface, either by physical measurement using sounding lines (lead lines) or by use of echo sounders. Sounding lines provide a single point measurement of depth whereas echo sounders provide a continuous line of soundings. In both cases, measurements are made from a floating vessel which is usually moving both horizontally and vertically. Errors in measurement of apparent depth can arise from the technique of measurement itself; from errors in transferring a datum of level from shore to ship; from errors in estimating instrument level in the ship relative to the undisturbed water surface; and from errors in estimating the true location of the point of measurement on the bed.

When estimating the instrument level in relation to the water surface, allowance must be made for variations of the mean draught and trim of the vessel according to the load in the vessel and its distribution. Instruments sited near the centre of the vessel will suffer least in this respect. Particular care should be taken when instruments are mounted over the side of small craft.

High-speed craft are sometimes used to complete a survey in the shortest possible time. This is desirable if rapid bed changes are taking place. The change of trim caused by relative motion between vessel and water must then be taken into account. This effect is likely to be small on 'displacement' craft in deep water, but in shallow water, vessels squat deeper when moving fairly fast[5]. The effect on craft that rely on some lift from their hulls or on hovercraft is too big to ignore. In these cases, the only way to estimate the depth of instruments below the water plane is by pressure transducers sited on the instruments. Even then, care must be taken that the pressure recorded is relative to a large area of water surface rather than a part of the wave field associated with the vessel. Velocity head effects at the transducer must also be avoided. High accuracy can only be achieved by sounding from a slowly moving vessel.

Sounding by line has been superseded by echo sounding, but many old charts must be used for comparative estimates of siltation or erosion. Most charts prepared before 1940 were based on soundings by line. A *sounding line* consists of a calibrated line with a weight or 'lead' attached. Sounding by line presents no problems if used from a stationary vessel on a firm bottom in still water, but in all other cases indeterminate errors arise. When used from a moving survey vessel, the 'leadsman' works from a platform suspended as near as possible to the water surface and swings the lead ahead, with the intention that it would reach the bottom in time for the line to be just vertical as the platform passes over it. Accuracy is limited by the calibration of the line and by the accuracy with which a leadsman can read the intersection of line with water surface from a distance of over 2 m. This is usually assumed to be between $\pm 0{\cdot}075$ m and $\pm 0{\cdot}15$ m, depending on the skill of the leadsman. A moving vessel or flowing water will cause the line to depart from the vertical. If the bottom is at all soft, the weight on the end of the line will penetrate into the bed and the leadsman will lose the sense of feel upon which he depends in estimating the moment of contact with the bed. Weights having a large plan area can reduce the penetration into soft mud. A standard technique is necessary if successive surveys are to be compared. These errors are important when navigational charts are being prepared because they all tend to over-estimate the true depth, and they must be taken into account when quantitative estimations of sediment movement are attempted.

Note that soundings by line can only give point values. They are useless for detection of irregularities of the bed.

Echo sounders work by emitting a short series of pressure pulses towards the bed from a transmitter which is attached to the hull of a surveying vessel. The pressure pulses are reflected back towards the transmitter from any solid object; in fact, any well-defined gradient of density can cause partial reflection to occur. A receiver placed close to the transmitter detects the returning train of pulses and a recorder provides a visible trace, the position of which depends on the time taken for pressure pulses to return back to the receiver. This time interval multiplied by the speed of pressure waves in water gives the distance travelled by the returning wave and hence the distance to the nearest point of reflection. The accuracy depends on knowledge of the speed of propagation of pressure pulses in water which, in turn, depends on the density ρ and bulk modulus K. For a small-amplitude wave, the wave-speed $c = (K/\rho)^{\frac{1}{2}}$. These values can either be assumed from knowledge of temperature, silt content and salinity or measured. In practice, echo sounders mounted in vessels are calibrated empirically by measuring the apparent distance to an object such as a dock sill or a horizontal bar or disc suspended at a known depth below the transmitter. This must be done at the start of each period of surveying and whenever significant changes of water density occur, e.g. during rising and falling tide.

Echo sounders operate at frequencies between 15 and 210 kHz corresponding to wavelengths in water between 85 and 6 mm respectively. The shorter wavelengths are reflected almost completely from any sharp density change. They show the surface of soft mud or even an interface between layers of water of different density. The longer wavelengths are only partially reflected by

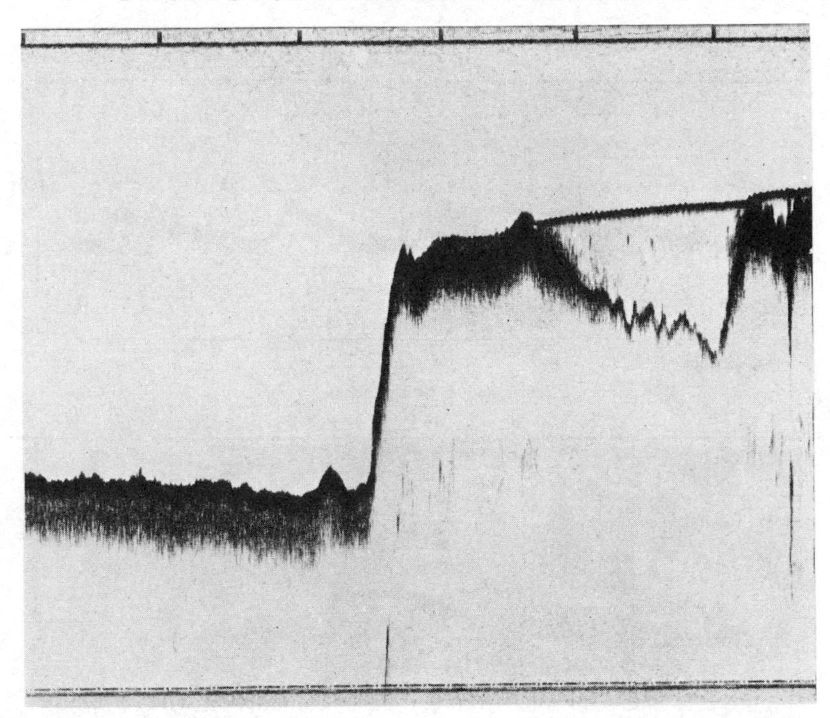

Figure 5.1. Echo trace showing mud deposited in a dredged channel at Tilbury, Thames Estuary (courtesy Kelvin Hughes, a division of Smiths Industries Limited)

moderate density changes but are totally reflected by solid rock. They are capable of returning multiple echoes from each of several density interfaces, beginning with a surface of soft mud, which may be poorly defined, followed by firmer layers that occur below the surface (figure 5.1). The interpretation of such echoes is not easy. A mud layer that can return a clear echo may be so soft that it offers no significant hindrance to a ship. Such mud layers can form in newly dredged channels or in natural channels, for example during neap tides. If quantitative estimates of solid volumes are to be determined from echo soundings, it is essential that the bed should be sampled to determine the density of solids present. This, in itself, is not a simple matter where soft mud is concerned.

Some modern echo sounders have been designed to operate with two frequencies simultaneously. It is claimed that the higher frequency (210 kHz) will penetrate aerated water under the hull of a moving ship and will give a clear echo from the top of a mud layer. Solid rock or pebbles reflect both high and low frequencies, resulting in a darker trace than that given by mud alone. In addition to the record on a paper chart, a digital output can be provided, sensitive to ± 50 mm. This can then be recorded in a form suitable for computer analysis.

The accuracy with which echo soundings can be read from recordings taken from a firm bottom is better than $\pm 0{\cdot}07$ m. If the record includes vertical

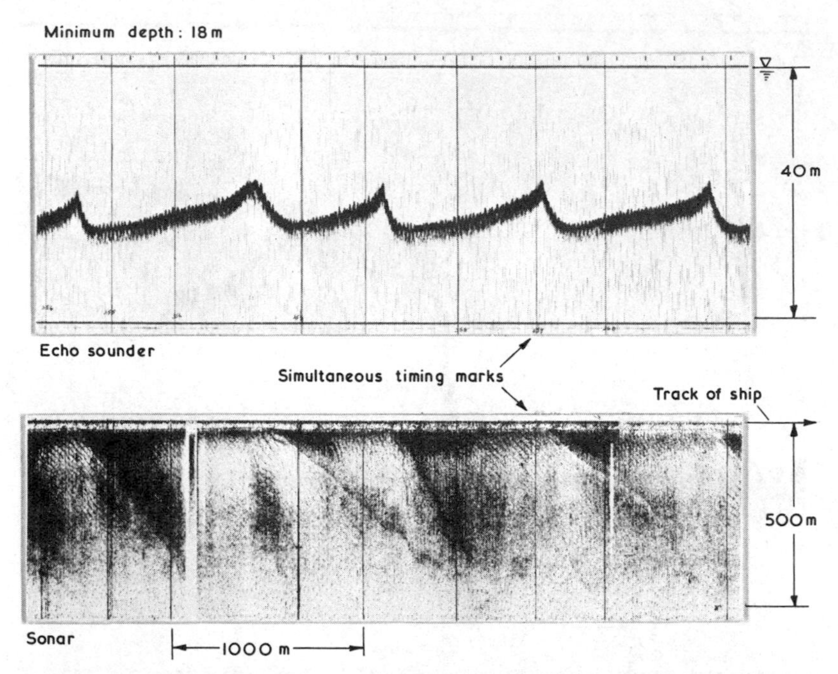

Figure 5.2. Side-scan sonar record used in conjunction with an echo sounder (courtesy Kelvin Hughes, a division of Smiths Industries Limited)

motion of the transmitter relative to a land datum, this must be taken into account. It may be difficult to interpret records obtained over a sandy bottom where dunes occur if the surveying vessel is affected by waves.

Side-scan sonar is an instrument based on the echo-sounder principle, in which a train of pulses is emitted from a transmitter in a variable direction. At the start of the train, the narrow beam of pulses is directed vertically towards the bed, but the beam is swept in an arc towards the horizontal as the train is emitted. The frequency of pulses is higher than normal for echo sounding; this allows the pulses to be concentrated in a pencil-like beam. It also causes a sharp echo to be returned from bed features aligned nearly normal to it. A side-scan sonar chart gives a pictorial view of the bed to one side of the track of the vessel in which it is mounted. Changes of reflectivity of the bed are recorded. A steep feature such as a wreck or trench may cause a shadow. Some knowledge of bed features is useful when interpreting a chart.

The side-scan sonar is an indispensable aid to surveying over a sandy or rocky bed. It provides continuity of information between lines of echo soundings, and can show up areas of dunes and obstacles that might otherwise be missed. In normal use, it does not give accurate soundings or distances to bed features and is usually operated simultaneously with an echo sounder [6].

Figure 5.2 is a typical record of an echo-sounder chart of sand waves with a side-scan sonar chart taken simultaneously. The latter shows the orientation of the sand waves.

5.3 MEASUREMENT OF WATER STATE AND QUALITY

This includes all measurements required at a point within the flowing water. They include the following.

(1) Flow speed and direction.
(2) Concentration of suspended solids.
(3) Density.
(4) Conductivity.
(5) Temperature.
(6) Substance in solution, including mineral salts and dissolved oxygen (DO).
(7) Biochemical Oxygen Demand (BOD).
(8) pH value.

The density can be determined from concentration of suspended solids, temperature and salinity, the latter being inferred from measurement of conductivity.

Measurements of all these properties are affected by problems of location in space of the point of observation or sampling. They are also affected by the rate of variation of the properties with time during observations. In most cases, it is necessary to integrate a series of point measurements over depth or over a cross-section to obtain an estimate of mean concentration or total discharge. These topics are common to all point measurements and are discussed in this chapter, but the techniques of individual measurements are not; they depend on a wide variety of instrumentation supplied by many manufacturers and are beyond the scope of this book.

Location in space has two elements. One is the problem of positioning an instrument where required; the other is the accurate measurement of its location once in position.

5.3.1 Positioning of instruments

Instruments suspended beneath a boat or buoy are affected by hydrodynamic drag on the instrument assembly and on the suspension cable. This drag is bound to cause displacement from the vertical. If the displacement is small, the length of cable below the water surface is only a little greater than the depth to which the instrument is lowered. Drag on instrument assemblies can be much greater than the weight of the assembly, however, with the result that special measures must be taken to prevent excessive streaming of the instruments downstream of the vessel.

Addition of weight to the bottom of the instrument assembly is an obvious palliative, but this increases the load that must be raised or lowered to bring the instrument to each required level. If the weight were to become very large, it would impose a prohibitive load on the instrument suspension.

A more efficient alternative is to suspend a very heavy weight just above the bed by means of a wire. The instrument assembly and its cables can then be attached to this wire by sliding links as in figure 5.3. The depth of the

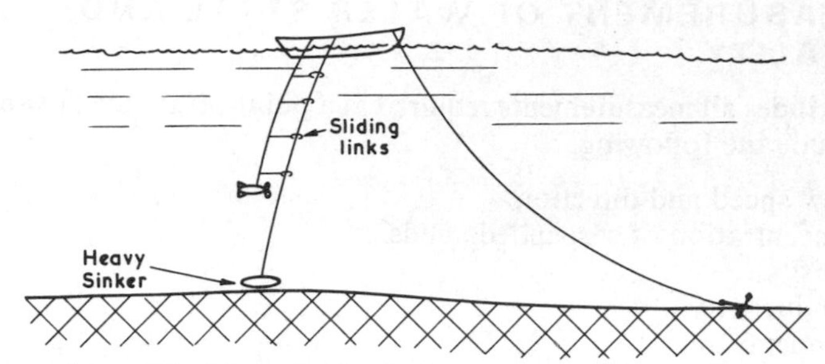

Figure 5.3. Method of reducing streaming of instruments

heavy weight below the vessel will need frequent adjustment to compensate for tidal rise and fall.

The effect of current drag on various types of mooring and suspension systems can be analysed using methods evolved by Wilson[7]. Some idea of the magnitude of the problem may be obtained by simple calculations of drag on instruments and cables using the equation Drag $= \frac{1}{2}C_D \rho U^2 A$, where C_D is a coefficient of drag, ρ is density of fluid, U is the speed of flow and A is the projected area in the flow direction. Suppose that a heavy, streamlined weight is used to keep a taut wire as near vertical as possible. A well-streamlined cast-iron weight of cylindrical cross-section, diameter d, could weigh about 110,000 d^3 N. Its drag coefficient would depend on Reynolds number but would be around $C_D = 0{\cdot}07$ for speeds of flow in estuaries[8]. Its drag would then be 27·5 d^2U^2 N. A 50 kg weight could have diameter 0·16 m. At a flow speed of 3 m/s, the drag would be 6·6 N, which is negligible compared with its weight of 490 N. However, its suspension cable may have diameter of at least 0·01 m, particularly if it is designed to carry signals as well as to support weight. If its drag coefficient is 1.3[9], its drag at 3 m/s would be 76·5 N per metre length. (A full analysis, as described by Wilson, would take into account the inclination of the cable to the flow direction.)

Consider now an instrument package consisting of a cylinder 1 m long and 0·1 m diameter with vertical axis, typical of many oceanographic instruments. The drag coefficient, due to the larger diameter, would be about 1·0 and its drag at 3 m/s would be about 450 N, nearly equal to the vertical force on the 50 kg weight considered above. Such equipment is quite unsuitable for use in high-speed flow as found in many estuaries. Some current meters are very well streamlined, but it is common to mount several instruments of different kinds on the same cable. It is then difficult to ensure low drag. The total drag on the suspension cable alone can be of order 800 N, in a stream 10 m deep and flowing at 3 m/s. The drag of a complete assembly of instruments can easily exceed 1200 N if a single suspension cable is used; it could exceed 2000 N if multiple cables or tubes are used. If the angle of the suspension cable to the vertical is to be kept less than 15°, the total effective weight on the cable at the surface must be nearly 4500 N; equivalent to 0·44 tonnes weight.

An alternative that has been used successfully is to mount current meters on a framework which can be rested on the bed. This is often necessary when

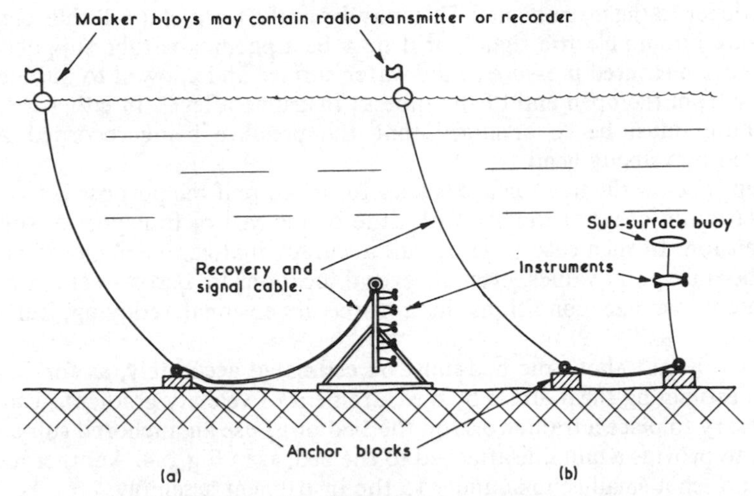

Figure 5.4. Location above bed by sub-surface buoy

shear stress is estimated from velocity measurements close to the bed (Nece and Smith[10]). The problem of measuring shear stress is discussed further in section 5.4.

5.3.2 Determination of location of instruments

Location in plan must usually be referred to the location of the point of suspension, be it vessel, buoy or stand. The greatest problem that arises is the determination of the distance of instruments downstream of the point of suspension. In the case of wire-suspended instruments the location in plan depends on the length of wire which, in turn, affects the depth of the apparatus below the water surface. In many cases, location in plan within 10 m or so is not critical, but determination of shear stress requires accurate knowledge of the height of velocity measurement above the bed. For this purpose location above bed forms must be known. Sand dunes in estuaries can have heights exceeding 1 m and slopes on the downstream face in excess of 1 in 5. In such cases it is essential to measure the height of the instrument above the bed at the instrument location. This is not at all easy if the instrument is streaming downstream of a vessel in a strong current. Modern techniques of position location and bed observation can help considerably. Exploration for oil has led to the development of devices such as transponders, which emit a pressure pulse in response to an acoustic signal from a ship on the surface. Transponders attached to an instrument package can be 'interrogated' by range-finding devices used by divers, from submersibles or from ships. They are capable of use over a range of up to 700 m with a resolution of 0·1 m. Three transponders on the bed have been claimed to give location of a receiving device to within 0·5 m over a wide area 11].

Elevation of instruments can be measured most easily by reference to the water surface as local datum. This can be done by attaching a pressure transducer to the instruments. This may be in the form of a cell that converts pressure into an electric signal, or it may be a pneumatic tube supplied with air under measured pressure at the water surface and allowed to bubble very slowly from the open end of the tube at instrument level. In either case, the mounting must be so arranged that the pressure being recorded is not affected by velocity head.

Depth below the free surface is a useful measure if the purpose is to obtain the variation of water quality with time or the rate of transport of solids in suspension. In such cases, the overall accuracy that can be achieved is quite low because point values, even at several locations in a cross-section, cannot represent average conditions in a three-dimensional, eddying, turbulent stream.

When height above the bed must be estimated accurately, as for example when estimating shear stress by measurement of velocity gradient, it may be necessary to place a framework on the bed or to use an anchored sub-surface buoy to provide a taut line attached to the bed, as in Fig. 5.4. Another method is to attach a small echo-sounder to the instrument assembly.

5.3.3 Variation of water state or quality with time

Variation of water state or quality with time has been discussed in Chapters 2 and 3 in connection with turbulence (see Fig. 3.1). Turbulent fluctuations in velocity give rise to an effective shear stress that is much greater than viscous shear. There are two ways of dealing with such turbulent fluctuations in water. One is to attempt to measure them and to analyse the resulting records; the shear stress being proportional to the product of the instantaneous fluctuations in two perpendicular components of velocity (eqn. 3.9). The problems of practical measurement are formidable and this has only been done as part of basic research into tidal behaviour. The other method is to record time-averaged velocities over a period long enough to smooth out the effects of turbulence but short compared with the tidal period. The aim is to ensure that the temporally averaged velocity does not change significantly during the observation. A period of five minutes' continuous measurement should be sufficient in most cases. In estuaries, however, there are masses of water having different properties that intermingle slowly, causing fluctuations in velocity over periods longer than ten minutes. For example, water held in an eddy down-current of a headland can have a different density from the main stream. Masses of this water may detach from the edge of the eddy and be carried into the stream from time to time, causing fluctuations of velocity. Similarly, the faster-moving water in the main stream may have different properties from the slower-moving water along the margins of an estuary. There is frequently a sharp demarcation between these bodies of water, with a visible line of small eddies and foam along the interface and a marked difference in turbidity and wave action on the surface. The location of the interface varies continuously. A vessel stationed close to the interface will detect abrupt changes of velocity as the interface moves past it. In such a situation, time-averaging over 5 minutes can be inaccurate, and the result depends strongly on the location chosen for the observations. Unfortunately it is not always easy to forecast the times and positions of interfaces at all positions in an estuary. The only way to reduce errors of this kind is to have several measuring points in a cross-section; but this involves much labour and the simultaneous use of several vessels and sets of measuring equipment.

5.4 MEASUREMENT OF SEDIMENT TRANSPORT[12]

Knowledge of the physics of sediment transport has been built up gradually over a very long period, yet no satisfactory unified theory has been evolved. Shear stress at the bed has been recognised to be the most important parameter in initiating particle movement and in sustaining the motion of coarse material along the bed. The role of turbulence in keeping solids in suspension in the stream has also been studied. Unfortunately, the link between shear stress, turbulence and the velocity field in time-varying three-dimensional flow is so complex that analysis is not yet possible.

Suspended load

This is a major factor in estuaries, particularly in the reaches which have appreciable salinity gradients. Any disturbance of the bed, by dredging or construction work, causes material to be brought into suspension. Coarse sand and silt settles back to the bed within a short distance, but fine material may be carried a long way by tidal currents. Monitoring of suspended load is comparatively straightforward; velocity and sediment concentration are measured at several points in a vertical using the techniques described in the previous section. There is one precaution that should be taken, however. Sediment concentrations increase rapidly towards the bed, as described in Chapter 4 and shown in figure 4.7. The highest concentrations occur in the region of lowest speed, but also in a region in which the accurate measurement of elevation from a vessel on the surface is most difficult. Layers of very high silt concentration can occur as in figure 1.12, usually with a thickness of 1 or 2 metres. Care is therefore needed to ensure the accuracy and adequacy of measurements close to the bed and a pragmatic approach must be adopted by surveyors. They must explore any zones of high concentration or high velocity gradients with particular care.

Bed load measurement

Different methods must be used when measuring the rate of flow of material close to the bed. In practice, direct measurement of bed load movement is very difficult. Several devices have been tried, but all interfere with the bed in some way, or are difficult to place so that they sample the desired layer of moving sand.

The rate of bed load transport varies over the length of a moving sand dune. When dunes are present, it is necessary to obtain samples at several positions relative to the dune and, if possible, to record the rate of movement of the dune itself; this is very difficult. Moreover, de Vries[13] has shown that there is a minimum size of sample, dependent on grain size, necessary for accuracy of bed-material sampling. Because of these problems, indirect methods of measuring the rate of bed load transport have been evolved. The main ones are:

(i) measurement of shear stress at the bed and use of one of the bed load transport equations, such as equations 8.15, 8.17, 8.18 or 8.19;
(ii) observation of movement of sand marked with suitable tracers;
(iii) measurement of total solid transport by means of hydrographic surveys of the bed at intervals of time.

The first method can indicate the rate of flux of coarse solids at a point in an estuary. If combined with measurement of solids moving in suspension, the total flux of solids past a point in plan can be estimated. Several such points in a cross-section are necessary to determine the total rate of flux. The second method can be used for fine solids moving in suspension as well as for bed load. The third method is used to examine the long-term consequences of solid transport including bed load and can lead to a measure of the net rate of solid transport through a cross-section. It cannot be used to estimate the total amount of solid in movement or the maximum rate of flux, yet these will be needed if the consequences of channel changes are to be predicted.

In practice, none of these methods is satisfactory and it is usual to use more than one of them in an attempt to give more confidence in the results.

Estimation of shear stress

This can be done by measurement of the velocity gradient close to the bed, either by direct measurement of velocity at known heights above the bed or by use of a device such as the Preston tube that records total head in a layer adjacent to the boundary. Nece and Smith[10] have described an experiment in which both methods were compared. The estimation is based on the 'law of the wall' of turbulent shear flow in which the time-averaged velocity profile is given by:

$$\frac{u}{U_*} = \frac{1}{K}\ln z + C \tag{5.2}$$

where u is the velocity averaged over a time long compared with turbulent fluctuations; K is the Von Karman constant; C is a constant dependent on boundary roughness Reynolds number $U_* k_b/\nu$; U_* is the shear velocity $(\tau_{z_0}/\rho)^{\frac{1}{2}}$; z is the height above a datum close to the solid boundary and k_b is the effective height of boundary roughness.

In practice, determination of boundary roughness and location of the datum level for an irregular surface is difficult. Von Karman's constant K is affected by sediment concentration in the water (Chapter 4). Moreover the equation was established semi-empirically for steady, uniform flow. However, close to the boundary, inertia effects are small compared with friction and it is found that in gradually-varied tidal flows, equation 5.2 applies quite well. Determination of the effective datum is best done by measuring velocity at several levels close to the boundary. Selection of a datum that will give the best fit to equation 5.2 is then possible, for example by plotting u versus z on semi-log paper or by numerical analysis.

Practical measurements of velocity gradients in tidal streams have been made by using at least five small current meters spaced over a depth of about 1·5 m. It is essential to keep the drag of this array as small as possible so that the supporting stand will not be moved by the strongest currents. Provided that the bed is reasonably plane and firm, the assembly can be lowered into place from the surface, but it is a great help if a diver can check the orientation before it is put into use. A difficulty arises if sand dunes are in motion. The base of the stand may be partially buried or sand scoured from it during its operation. Under these conditions, five or more current meters are essential, so that changes in bed level or local flow anomalies due to curvature of dune crests can be detected.

An alternative to the use of an array of current meters is to arrange for a single meter to be traversed mechanically over a short vertical distance. This method has been successfully used by the Hydraulics Research Station, Wallingford, England.[14]

Under some circumstances, it is possible to estimate mean shear stress at the bed without measuring velocity gradients close to the bed. The equations of motion integrated over the vertical can be written

$$\frac{1}{g}\frac{\partial U}{\partial t}+\frac{\partial}{\partial x}\left(\frac{U^2}{2g}\right)+D-\frac{\tau_{z_0}}{\rho g H}+i_x = 0 \tag{5.3}$$

where U is the velocity averaged over depth H, D is $(H/2\rho)\ \partial\rho/\partial x$, ρ is density, i_x is water surface slope and τ_{z_0} is shear stress at the bed. This equation includes the gradients of kinetic and potential energy and takes into account the effect of density gradients on the latter. It does not include effects due to curvature in plan.

The individual terms of equation 5.3 expressed in finite difference form, can be estimated from field measurements taken at frequent intervals and at cross-sections separated by distance δx, as described in Chapter 2 and illustrated in figure 2.8. Halliwell and O'Connor[15] have used this method to estimate shear stress in a straight reach of the Mersey Estuary and found that the result compared well with estimates based on velocity measurements at about 1 m above the bed. It also compared well with estimates based on the cross-sectional and depth-mean velocities, on the assumption that the logarithmic distribution of velocity expressed by equation 5.2 was applicable. However, the Mersey is a well-mixed estuary and the latter finding would only hold in similar cases. Equation 5.3 applies to all cases but only gives estimates of shear stress of the same order as those found by measuring velocity gradients close to the bed when applied to straight or nearly straight reaches of an estuary. This is an expensive method of estimating shear stress, but it can be used if a comprehensive programme of measurements is being undertaken for other purposes.

Tracer techniques [16]

These can give quantitative results in steady unidirectional flows in confined channels but have also been used qualitatively in tidal situations. Accuracy in unidirectional flows can be high, as for example, in experiments described by Rathbun and Nordin[17] where transport rates were predicted to within 14 % of observed values. Such accuracy could not be expected in tidal situations because rapid dispersal of the tracer makes accurate surveying of the tracer patch impossible during the oscillating flow. In the outer zone of estuaries, storms can bury tracer material beneath thick layers of unmarked sediment. Nevertheless, quantitative estimates are possible in particular cases such as estimation of infill rates in projected dredged channels[18].

Generally, three methods of study are available and may be used for both bed and suspended load movements.

(i) The continuous dilution method, in which a tracer is injected continuously onto the channel bed at a fixed point and detected continuously as a position down-drift from it.

(ii) The time-integrated method, in which a tracer is injected instantaneously at a point or as a line source spread across the channel width and detected continuously at a few points.

(iii) The space-integrated method, which is similar to (ii) with the exception that the tracer patch is tracked spatially rather than temporally at a fixed point.

All the methods entail the use of either a natural or artificial sediment which has been labelled with a suitable tracer. Sand, gravel and ground glass particles are commonly used and may be labelled by: (*a*) painting the sediment grains; this is usually used for gravel; (*b*) mixing the sediment with a binder and a fluorescent dye; this is usually used for sands, particularly in shallow water areas where radioactive methods could constitute a health hazard; (*c*) radioactive irridation, which is usually used for sand or glass particles containing an inactive isotope; (*d*) adsorption of radioactive materials on to particles from solutions such as radioactive gold dissolved in strong acids.

Method (i), continuous dilution, provides the most reliable results if rapid dispersal is likely to occur. The gradual accumulation of tracer at several points can be monitored. In steady flows, a steady state will eventually be reached; but in tidal flows the rate of arrival at a point will vary during the spring–neap tidal cycle. Method (iii) consists of instantaneously injecting tracer onto the estuary bed, followed by monitoring of spatial spread of the tracer patch. This is achieved, in the case of bed load transport, by either lowering a radioactive detection probe on to the estuary bed at a series of points covering the patch or by repeatedly towing it across the bed so that the whole patch is explored. Alternatively, if fluorescent material is used, a series of vertical cores are taken on a grid which must be large enough to cover the moving patch. Such cores are also required for quantitative radioactive tests so that the depth of tracer mixing can be found and probe readings corrected for absorption by the bed sediment. Method (iii) has been used many times in tidal streams, but is difficult to interpret because of the rapidity of change of flow. For two or three tidal cycles after the moment of injection, results depend strongly on the tide-time of injection. The time required to identify the whole area of a patch is long. Errors can be reduced by surveying at the same tide-time on successive tides and by using a tracer with a long useful life—such as Scandium 46 with a half-life of 85 days. This means, however, that many weeks must elapse between injections at one location. Volume of the sediment transported per second is given (figure 5.5, [17])

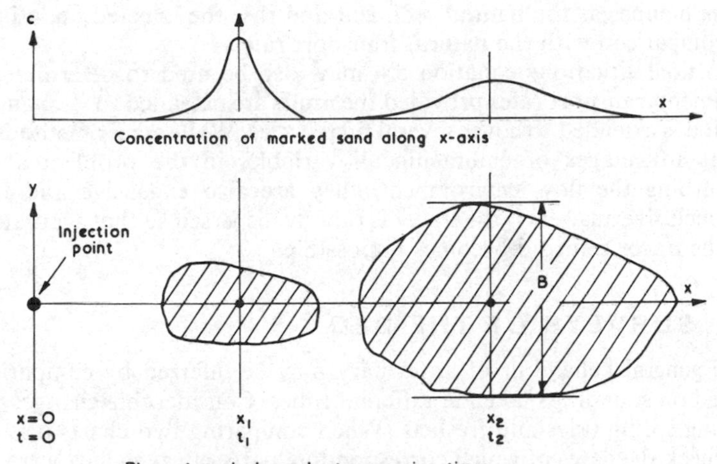

Figure 5.5. Area-integration of tracer-marked sand in uni-directional flow

by the equations:

$$Q_s = \gamma_b(1-m)\ Vd_0B \tag{5.4a}$$

where

$$V = (\bar{x}_2-\bar{x}_1)\ \Delta t \tag{5.4b}$$

and

$$\bar{x} = \int_0^\infty Cx\mathrm{d}x/\int_0^\infty C\mathrm{d}x \tag{5.4c}$$

γ_b = specific weight of the solid
d_0 = mean depth of tracer mixing over the time interval $\Delta t = t_2-t_1$
m = porosity of the bed
B = width of the patch of marked material at distance $\bar{x}_2$ from the injection point
$\bar{x}$ = position of the centroid of the longitudinal tracer concentration curve while subscripts 1, 2 refer to longitudinal positions, x_1 and x_2
C = mean cross-sectional tracer concentration. In the case of radio-active tracers, this must be referred to the level of natural background radiation as datum

In flat bed flows, the depth of mixing is very small, the sediment moving in a thin layer on the bed. In dune bed flows tracers may be buried in layers over the whole dune height. It is then impossible to determine C except by extensive core sampling. Notice that the area under the concentration – distance graph is proportional to the total amount of tracer injected. Reliable results can only be expected if more than 80 % of the tracer is detected. While this may be possible in some rivers, it is impossible in estuaries in which the pattern is rapidly changed by tidal flow reversals. In using equation 5.4 it is also assumed that the tracer is deposited initially at a point, that the tracer behaves in the same manner as the natural sediment and that the injected quantity is small in comparison with the natural transport rate.

In tidal situations equation 5.4 may also be used to determine net tidal sediment transport rates provided the limits are extended to $\pm\infty$ and the time period is extended to cover several tidal cycles. While tracer methods have the great advantages of embracing all variables in the problem and of not disturbing the flow environment, they are also expensive and may yield inconclusive answers if the tracer is rapidly dispersed so that accurate location of the tracer centroid becomes impossible.

5.5 SURVEYS OF THE BED

The general behaviour of an estuary may be inferred by comparing charts based on soundings taken at different times. Considerable changes can occur during spring tides and freshets. When comparing two charts it is necessary to check the dates at which corresponding parts of each chart were surveyed. Useful comparisons between two charts are only possible if surveys were done in the same season and after similar sequences of tides and river flows. In practice this is highly unlikely and is almost impossible to arrange. In some regions, however, there is a seasonal pattern of rainfall and dry weather which is so well defined that surveys can be arranged at comparable times, such as the end of a dry season during neap tides. However, a full survey of an estuary may take several weeks to complete, during which spring–neap cycles can cause significant changes of bed level. In such a case results may be misleading if a survey which was begun at the seaward end of an estuary at the start of a neap tide cycle in one year, is compared with one that began at the landward end in another year.

Two examples illustrate the problem. Figure 10.2 shows the cubic capacity of the Mersey Estuary as shown by quinquennial surveys. This is based on measurements taken along fixed cross-sections of the estuary spaced approximately 300 m apart. Soundings were by hand lead up to 1946 and by echo soundings from 1946. It will be seen that, between 1861 and 1911, there was considerable variation in estimates of cubic capacity but that no long-term trend was apparent. From 1911 onwards, progressive changes have occurred. The bed of parts of the Mersey and parts of Liverpool Bay is covered, at times, by sand dunes with crests arranged roughly normal to the dominant flow directions. These dunes may be up to 2 m in height, though their mean height is usually much less. Crest lengths are typically at least 10 times the height. It is quite possible for sounding lines to be parallel to dune crests and to follow them for a considerable distance. Adjacent lines might fall, by chance, near the crest or trough of adjacent dunes, causing the mean bed level to be under- or over-estimated. An error of only 0·1 m in surveying a spoil dumping ground covering 25 km^2 would lead to an error of $2{\cdot}5\times10^6$ m^3 – a greater volume than the estimated annual accretion in the Mersey due to natural causes. This kind of error can be anticipated and reduced by using side-scan sonar simultaneously with echo sounding, or by taking a few lines of soundings at right angles to sand dunes so that their size and location relative to other lines of soundings can be measured.

Table 5.1. *Volume changes in the River Hooghly shown by successive surveys*

1954	Hooghly Point to Calcutta	Seaward reach: Hooghly Point Royapur	Landward reach: Royapur to Calcutta
May/June	580	294	286
Nov./Dec.	583	280	303
	+3	−14	+17
Total volumes in million cubic metres below (*Datum*+5·5 m)			
May/June	232	107	125
Nov./Dec.	253	110	143
	+21	+3	+18
Channel volumes in million cubic metres below Datum			
May/June	348	187	161
Nov./Dec.	330	170	160
	−18	−17	−1
Inter-tidal volumes in million cubic metres between Datum and +5·5 m			

+ indicates erosion — indicates accretion

The other example is taken from the River Hooghly (see figure 10.13) the behaviour of which is markedly seasonal. Little rain falls on its catchment between October and mid-June. It is surveyed bi-annually over much of its length, one survey being done in November and December, just after fresh water discharge has fallen to a low level, the other being done in May and June, at the end of the dry season. Table 5.1 shows the results of two successive surveys of a section of tidal river in the middle reaches. Hooghly Point is largely governed by tidal conditions; the tidal rate of influx during spring tides may exceed 30000 m^3/s compared with fresh water flow of less than 6000 m^3/s. At Calcutta, on the other hand, the highest rate of tidal influx is of the order of 7600 m^3/s and during neap tides tidal flow may be less than fresh water flow.

The table shows that there is seaward movement of bed material during the wet season (June–October) and that most of the siltation occurs on the inter-tidal shoals in the seaward reaches. The converse occurs during the dry season; the channels in the landward sections accrete while material is eroded from the shoals near Hooghly Point. It is evident that the timing of surveys of the River Hooghly is critical if estimates of long-term sand movement are to be based on them.

5.6 SUMMARY

The behaviour of a system as it exists can be studied by field measurements, but the size of estuaries makes comprehensive study difficult. Prediction of future behaviour can only be made by use of some kind of simulator, such as mathematical equations, analogue or digital representation, or by physical modelling of an estuary. In each case, field measurements are needed to provide boundary conditions and data for calibration.

Programmes of field measurements must be planned to suit the requirements of each new situation. These are not, therefore, discussed in this book. However, there are several factors that affect the economy and accuracy of all field measurements. These are as follows.

(1) Overall behaviour can only be determined by field measurements at a large number of points.

(2) River flows, wind, wave action and influx of pollutants, unlike tides in the sea, cannot be forecast in advance. As a result, it is difficult to correlate measurements taken at different places and times.

(3) The correlation of different observations is easier if a reference site for monitoring typical variables is operated throughout a programme of measurements. This is in addition to measurement of the main independent variables such as tidal rise-and-fall and fresh water influx.

(4) Care must be taken to ensure that water surface level is known at a particular place and time. The installation and maintenance of tide gauges can affect measurement. Transfer of level from a tide gauge to a floating vessel needs particular care.

(5) Echo sounders can only provide a line of soundings, while sounding leads, used in the preparation of nearly all charts prior to 1940, only give point soundings. Use of side-scan sonar can help in identification of features between soundings, thus reducing the risk of systematic error.

(6) The location of measuring instruments in plan and elevation presents problems due to the strength of currents and the irregularity of the bed. For some purposes, particularly the measurement of shear stress, instruments should not be suspended beneath boats but should be mounted on the bed.

(7) Rate of flux of sediment in suspension can be estimated from field measurements but only to a low degree of accuracy. Sediment discharge including material moving along the bed cannot be measured directly, but can be inferred from knowledge of local shear stress or measured indirectly by use of a suitable radioactive or fluorescent tracer material. The accuracy of estimation is low and it is usual to use more than one method to reduce the chances of error.

(8) In suitable cases, volumetric changes can be estimated from surveys of the bed, but gross errors can arise if surveys are made at differing seasons or under different tidal conditions.

REFERENCES

[1] Ishiguro, S., Tidal analogues, *Encyclopaedic Dictionary of Physics*, Pergamon Press (1963)

[2] Harder, J. A. and Nelson, J. O. Analog modelling of the California Delta system, *Trans. ASCE*, HY4, July (1966)

[3] Inglis, Sir C. C. and Allen, F. H. Regimen of the Thames estuary: currents, salinities and river flow, *Proc. Inst. Civil Eng.*, **7**, 827–878 (1957)

[4] Halliwell, A. R. and Perry, J. G., Hydrodynamic studies of tide gauges, *J. Hyd. Res.*, **7**, 4, 485 (1969)
[5] Guliev, U. M., On squat calculations for vessels going in shallow water and through channels, *PIANC Bulletin*, **7**, 1 (1971)
[6] Haslett, R. W. G. and Honnor, D., Simultaneous use of sideways-looking sonar strata recorder and echo sounder, *The Radio and Electronic Engineer*, **33**, No. 6, June (1967)
[7] Wilson, Basil W., Characteristics of deep-sea anchorages in strong ocean currents, *PIANC Bulletin (1965)*, nos. **17** and **18**
[8] Goldstein, S., *Modern developments in fluid mechanics*, vol. II, p. 425, Dover Press (1965)
[9] Goldstein, S., *Modern developments in fluid mechanics*, vol. II, p. 508, Dover Press (1965)
[10] Nece, R. E. and Smith, J. D., Boundary shear stress in rivers and estuaries, *Am. Soc. Civil Eng.*, **60**, WW2, 335–357, May (1970).
[11] *Offshore Services*, p. 45, Spearhead Publications, September (1973)
[12] Graf, W. H., *Hydraulics of sediment transport*, Ch. 13, Sediment measuring devices, McGraw-Hill (1971)
[13] de Vries, M., On measuring discharges and sediment transport in rivers, *Int. Seminar on hydraulics of alluvial streams*, *IAHR* (New Delhi, India, 1973).
[14] Thorn, M. F. C., Deep tidal flow over a fine sand bed, *Proc. XVI Congress, IAHR*, Paper A 27 (1975).
[15] Halliwell, A. R. and O'Connor, B. A., Shear velocity in a tidal estuary, *Proc. 11th Conference on Coastal Engineering*, Vol. II, Ch. 88 (1968)
[16] Gasper, E. and Oncescu, M., *Radioactive tracers in hydrology*, Ch. 11, Elsevier (1972)
[17] Rathbun, R. E. and Nordin, C. F. Tracer studies of sediment transport processes, *Am. Soc. Civil Eng.*, **97**, HY9, 1305–1316, September (1971)
[18] Crickmore, M. J., The use of tracers to determine infill rates in projected dredged channels, *Dredging*, Institution of Civil Engineers, London (1968).

6

The study of tidal systems: mathematical tidal models

Various analytical, numerical and simulation techniques are available for use in predicting the effects of engineering works on estuarine regimen. They may generally be divided into two types.

(i) Those in which an attempt is made to solve the mathematical equations which govern the physical process: these include closed-form analytical solutions[9, 24] as well as numerical solutions using computers.

(ii) Those in which an attempt is made to simulate the physical process by indirect mathematical equations. For example, statistical functions may be used to simulate the movement of pollutants in an estuary.

Both types are applicable to estuarine problems but may be incapable of producing a complete answer to a particular engineering situation because of the complexity of the governing equations and boundary conditions, or an inadequate knowledge of the physical processes involved. Thus, although mathematical models are used on a wide variety of problems, they tend to be limited, particularly in problems involving estuarine regimen, to rather simplified situations. Often they are used in conjunction with other methods. For example, the effect of a proposed tidal barrier on the longitudinal distribution of pollutants in a well-mixed estuary might be investigated by using a one-dimensional mathematical model which is in turn provided with information on salinity and flow from a physical model of the estuary. Field data would also be required for both models.

Mathematical–physical model combinations are recommended for engineering problems involving large capital investment, since a relatively inexpensive mathematical investigation can often provide valuable insight into a problem as well as providing a guide to the best testing programme for any subsequent physical model investigations. Certainly the mathematical model once developed has many advantages over its physical counterpart. For example, an established digital computer model can be used easily and quickly, often on standard computing equipment. This enables model parameters to be varied over a wide range so that their importance can be assessed. It takes up little storage space, is readily adapted to new situations and is relatively inexpensive.

A particular mathematical model also has many features in common with the physical model (see Chapter 8 for details of the latter). For example, once the model has been 'constructed', it must be 'calibrated' to reproduce known conditions in the estuary with an acceptable accuracy. This process consists of optimisation of 'scale effects' which result from equation simplification or numerical approximations inherent in the mathematical techniques used for solving the equations or computer round-off errors. Calibration can be a tedious process and requires considerable experience of the model and physical situation in order to achieve the best results. Exact agreement with nature is not possible due to scale effects but the model may be optimised to reproduce the more important aspects of the problem. For example, a one-dimensional tidal mathematical model used for storm surge investigations would be optimised so as to obtain the best agreement with observed water surface profiles rather than flow velocities.

6.1 PARTICULAR MATHEMATICAL MODELS

Many estuarine engineering works produce changes in tidal characteristics and consequential changes in mixing and sedimentation processes. The immediate effects of the works is to alter tidal velocities and water surface elevations. As a result of these changes, internal water circulation patterns are altered and may lead eventually to changes in the concentration of substances contained in the water and on the estuary bed. The areas in which mathematical models are of most use are as follows.

(i) Prediction of tidal quantities

The estimation of water surface elevations and flow velocities in rivers, estuaries and shallow seas by mathematical models is well established. Consequently such models are widely used for the solution of engineering problems and have comparable accuracy with the more traditional physical models.

(ii) Prediction of water quality

Water quality parameters include the temperature of the water, its biochemical oxygen demand (BOD), its dissolved oxygen content (DO), the concentration of salt, nitrates, phosphates, ammonia, bacteria and toxic chemicals. They can be studied with the aid of models based on analytical and numerical solutions of the mass-continuity equations (see Chapter 3), and on numerical simulation techniques which are often loosely based on mass-continuity principles. The susceptibility of mixing processes to random influences, such as transient wind and wave action, makes models in this field less reliable than those described under 'Prediction of Tidal Quantities'. However, the mathematical approach is still to be preferred since comprehensive field measurements cannot, by themselves, be used for prediction of the effects of change and are expensive and time consuming, while physical models suffer serious scale effects.

(iii) Prediction of sedimentation parameters

The complicated nature of sediment transport in real tidal situations has limited the development of adequate mathematical models. Engineering problems are solved with the aid of physical models, in the case of bed load transport, and by field observations in the case of fine suspended sediment movement. Recently, however, attempts have been made to produce one-dimensional mathematical models for suspended and bed load sediment transport[1, 2] which are based on the equations of fluid motion and fluid and sediment continuity. These represent a step in the right direction but a better understanding and mathematical description of the physical process is required, to improve the accuracy of the models. Attempts have also been made to predict changes in estuarine nodal shoaling zones due to changes in flow characteristics[3, 4, 5]. These methods are useful but can only provide an indication of likely changes in a real estuary due to simplifications in their derivation.

Mathematical models are, therefore, of most use in areas (i) and (ii), and have the widest application in area (i). In this chapter, the various solution methods which form the basis of mathematical models in area (i) will be described. Details of mathematical models used in area (ii) are deferred until Chapter 7.

6.2 TIDAL MODELS

Mathematical tidal models are based upon the equations of fluid motion and continuity (see equations 2.1–2.4). Unfortunately, most three-dimensional situations are far too complicated for solution by these equations and use is made of simulation techniques (see Chapter 8). Simplified forms of the equations may, however, form the basis of three-dimensional models in shallow seas[6]. Fortunately many tidal flows can be considered as one or two-dimensional in form. For example, density and Coriolis effects can be ignored in narrow estuaries with little fresh water discharge or with large tidal prisms. Often the major interest is the change in surface elevation produced by engineering works and consequently details of the flow structure assume less importance. In such cases, equations 2.16 and 2.17, based on cross-sectional average values, form the basis of mathematical models:

$$\frac{1}{g}\frac{\partial U}{\partial t}+\beta'\frac{U}{g}\frac{\partial U}{\partial x}-i_s-i_f=0 \tag{6.1}$$

$$\frac{\partial Q}{\partial x}+B\frac{\partial z_1}{\partial t}=0 \tag{6.2}$$

where $$i_s=-\frac{\partial z_1}{\partial x};\quad i_f=-\frac{U|U|}{C^2R}$$

If details of the internal flow structure of narrow stratified estuaries is required, it is necessary to solve momentum and continuity equations for both upper and lower layers. In many problems, average flows provide sufficient information and consequently one-dimensional equations are used in each layer connected by an inter-facial shear stress or inter-layer exchange flows[29].

In wider estuaries, where flow directions deviate considerably ($> 20°$) from the mean axial direction of the estuary due to the presence of side-channels, Coriolis force or wind action, a one-dimensional description of the flow field is inadequate. Two-dimensional plan models are used in such situations, as for example, in the coastal zone at the entrance to an estuary, and are based upon the depth-integrated forms of the equations 2.13, 2.14 and 2.15.

The particular solution method to be used for equations 6.1 and 6.2 depends on the spatial and temporal variation in estuarine geometry, the magnitude of fresh water discharges and the characteristics of the ocean tide at the estuary entrance. Analytical methods can be used in simple shaped estuaries, as, for example, those which can be regarded as having a constant bed level and parallel sides, and with small fresh water discharges and a dominant tidal component (e.g. M_2 or O_1). Numerical methods are preferred for problems involving mixed tides where diurnal and semi-diurnal components are both significant or where estuary geometry changes rapidly in space and time. The numerical approach also allows a better representation of the frictional term as well as the effects of wind-induced surface shear forces, Coriolis force, irregular channel shapes and varying fresh water flow rates.

6.3 ANALYTICAL METHODS

In deep estuaries, frictional effects are small and equations (6.1) and (6.2) reduce to the classical wave equation for estuaries of constant mean depth (H) and cross-section subject to tides of small vertical amplitude.

i.e.
$$\frac{\partial^2 U}{\partial t^2} = c_0^2 \frac{\partial^2 U}{\partial x^2} \tag{6.3}$$

$$\frac{\partial^2 \eta}{\partial t^2} = c_0^2 \frac{\partial^2 \eta}{\partial x^2} \tag{6.4}$$

where $c_0 = (gH)^{\frac{1}{2}}$; η is the water surface elevation relative to mean water level and the origin of x is taken at the estuary entrance. These equations, (6.3) and (6.4), are satisfied by a single harmonic function for infinitely long estuaries of uniform depth and section, i.e.

$$\eta = \eta_m \cos(\sigma t - kx) \tag{6.5a}$$

$$U = \frac{\eta_m c_0}{H} \cos(\sigma t - kx) \tag{6.5b}$$

where $\sigma = 2\pi/T = kc_0$, $k = 2\pi/L_0$, η_m, T and L_0 are the maximum amplitude and the period and wavelength of the tidal wave respectively.

Equation 6.5*a* represents wave motion of amplitude η_m which is harmonic in space and time. High water occurs when $t_{\text{hw}} = kx_{\text{hw}}/\sigma = x_{\text{hw}}/c_0$ and its rate of translation through space $(\mathrm{d}x/\mathrm{d}t)_{\text{hw}}$ is c_0. (Suffix $_{\text{hw}}$ indicates a value occurring at high water.) The wave is thus referred to as a progressive wave of phase velocity c_0. Times of maximum velocity are seen from equation 6.5 to coincide with times of high and low water level. Such conditions are found in the open sea or near the entrance to a long estuary (see also figure 4.13). In short friction-less estuaries the tidal wave will be reflected from the head of the estuary causing a wave to propagate back out of the estuary, i.e.

$$\eta = \eta_m \cos(\sigma t + kx).$$

Superposition of the incident and reflected waves gives rise to the equations:

$$\eta = 2\eta_m \cos \sigma t \cos kx \tag{6.6a}$$

$$U = \frac{2\eta_m c_0}{H} \sin \sigma t \sin kx \tag{6.6b}$$

which represents a standing wave system $(\mathrm{d}x/\mathrm{d}t))_{\mathrm{hw}} = 0$ of amplitude $2\eta_m$ with mean water level occurring continuously at $x = L_0/4$, $3L_0/4$, etc. and high/low water at $x = 0, L_0/2, L_0$, etc. Notice that maximum velocities occur at times of mean water level.

Water surface elevations and tidal velocities in real systems can be considered to be the combination of a large number of such incident and reflected waves, each having a distinct amplitude, frequency and wavelength.

Equations 6.1 and 6.2 may be solved analytically for some simple shaped rectangular estuaries[7]. If the friction term is included in the equations it must first be linearised before a solution can be obtained. Thus i_f in equation 6.1 can be replaced by a term of the form

$$i_f = f_0 U \tag{6.9}$$

where f_0 is a constant for the estuary and may be determined by equating the work done over a tidal cycle by the general frictional term $\tau_{z_0}/\rho g H$ to that done by the linearised term[8]. For example, a particular solution to equations 6.1–6.3 for the case of an exponentially shaped estuary of width $(B = B_0\,\mathrm{e}^{2ax})$ and constant tidal mean depth (H) and with a tidal barrier $(U = 0)$ at its head $(x = 0)$ has been given by Hunt[9]:

$$\eta = a_1 \mathrm{e}^{-ax} b_1(x) \cos(\sigma t + g(x) - \tfrac{1}{2}(\phi + \psi)) \tag{6.10}$$

$$U = \frac{2\sigma a_1 \mathrm{e}^{-ax}}{H} (\sinh^2(\alpha_0 x) + \sin^2(\beta_0 x))^{\frac{1}{2}} \sin(\sigma t - \Omega) \tag{6.11a}$$

where

$$\Omega = \tan^{-1}\{\cot(\beta_0 x) \tanh(\alpha_0 x)\} \tag{6.11b}$$

$$\alpha_0^2 - \beta_0^2 + \sigma^2/gH = a^2 \tag{6.11c}$$

$$2\alpha_0\beta_0 - \sigma f_0/gH = 0 \tag{6.11d}$$

$$\beta_0 \tan(\phi) = \alpha_0 - a; \quad \beta_0 \tan(\psi) = \alpha_0 + a \tag{6.11e}$$

$$g(x) = \tan^{-1}\left\{\left[\frac{\sec(\psi)\,\mathrm{e}^{2\alpha_0 x} - \sec(\phi)}{\sec(\psi)\,\mathrm{e}^{2\alpha_0 x} + \sec(\phi)}\right] \tan(\beta_0 x - \tfrac{1}{2}(\psi - \phi))\right\} \tag{6.11f}$$

$$b_1(x) = \beta_0\{\sec^2(\psi)\,\mathrm{e}^{2\alpha_0 x} + \sec^2(\phi)\,\mathrm{e}^{-2\alpha_0 x} + 2\sec(\psi)\sec(\phi)\cos(2\beta_0 x - \psi + \phi)^{\frac{1}{2}} \tag{6.11g}$$

$$\sigma = 2\pi/T; \quad a > \sigma/(gH)^{\frac{1}{2}}; \quad f_0 = 8g|U_0|/(3\pi C^2 H) \tag{6.11h}$$

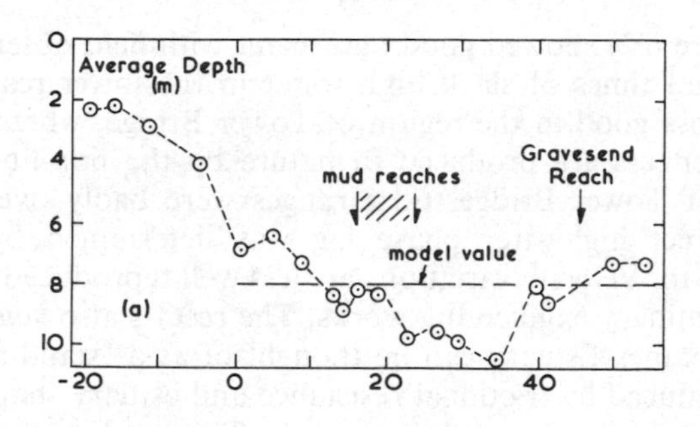

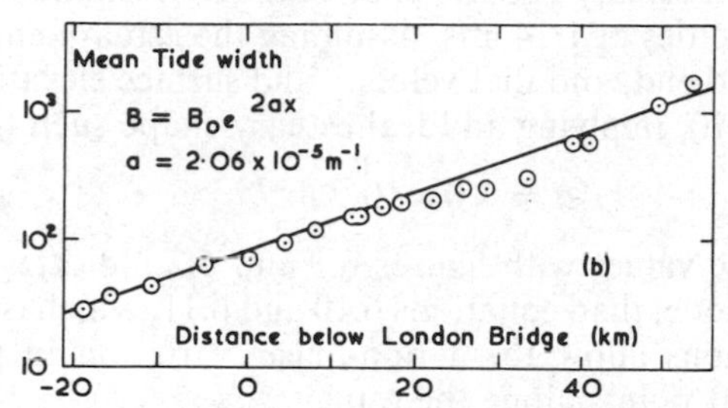

Figure 6.1. Average depths (*a*) and widths (*b*) along the Thames Estuary at mean tide level (based on figures 1 and 2 of reference[9], courtesy the Royal Astronomical Society)

U_0 is a spatially averaged maximum tidal velocity and $|\ |$ is its modulus; a_1 is a constant equal to $\eta_0/b_1(0)$; η_0, $b_1(0)$ being the maximum value of η and the value of $b_1(x)$ at the tidal barrier ($x = 0$) respectively; C is Chezy's coefficient.

These equations (6.10 and 6.11) require that:

(i) $\eta < H < B < L_0$;

(ii) the estuary has a rectangular section of slowly varying width;

(iii) fresh water in-flows to the estuary are negligibly small;

(iv) convective accelerations are negligible ($U\,\partial U/\partial x = 0$);

(v) the dominant tidal constituent (e.g. M_2) is substantially greater in magnitude than any other so that ocean tide can be regarded as simple harmonic.

Equations 6.10 and 6.11 have been applied by Hunt[9] to the Thames Estuary in the UK. This has a nearly exponential shape ($a = 2{\cdot}06 \times 10^{-5}\ \text{m}^{-1}$) but a varying mean depth shown in figure 6.1a, b: a mean value of 7.93 m was used in equations 6.10 and 6.11. Figure 6.1a suggests that the analytical solutions are only applicable to the estuary downstream of London Bridge. This is confirmed by the results shown in figure 6.2, derived from equations 6.10 and 6.11 with $\sigma = 1.45 \times 10^{-4}\ \text{s}^{-1}$, $U_0 = 0{\cdot}915$ m/s, and $x = 0$ located at Teddington Weir, 30 km upstream of London Bridge. The factor f_0 was calculated by Hunt by fitting equation 6.11 to observed data. A Chezy coefficient of $64 \pm 1\ \text{m}^{\frac{1}{2}}/\text{s}$ was found, which is very similar to values for the English Channel, the Bristol Channel and the Irish Sea[9].

The model (figure 6.2) showed good agreement with field observations for tidal elevations, and times of slack high water in the lower reaches of the estuary, but was less good in the region of Tower Bridge, where significant wave interaction effects are produced in nature by the non-linear friction term. Upstream of Tower Bridge, tidal ranges were badly over-estimated although the correct highwater phase lag was determined by Hunt by reducing H to 4.58 m. Velocity variations are less well reproduced but may be adequate for preliminary engineering works. The results also suggest[9] that the tides in the Thames Estuary can be thought of as a 'standing wave' of variable phase produced by frictional resistance and estuary shape.

It is of interest to compare conditions in the Thames Estuary with those predicted by the ideal estuary equations of Chapter 4. Equation 4.16 predicts maximum tidal velocities of 1·14 m/s, assuming the estuary entrance ($x = 0$) to be located at Southend, and that velocity and surface elevations are out of phase by 112° (3·75 h), implying an ideal estuary shape such that

$$a = 2{\cdot}0 \times 10^{-5}\ \text{m}^{-1}$$

Comparison of these values with figures 6.1 and 6.2 indicates that equation 4.16 is a less exact model than equations 6.10 and 6.11, which is not surprising since Hunt's equations allow for a non-linear variation in phase lag and maximum velocity at points along the estuary.

The analytical equations may also be applied to sections of an estuary in order to allow for variations in α_0 and β_0. For example, the analytical solution of equations 6.1 and 6.2 for a rectangular estuary of constant depth and parallel sides ($a = 0$) with a tidal barrier at the estuary head is given by Hunt[9] as:

$$\eta = \eta_0/2\{e^{\beta_0 x}.\cos(\alpha_0 x+\sigma t)+e^{-\beta_0 x}.\cos(\alpha_0 x-\sigma t)\} \tag{6.12}$$

$$U = \frac{\sigma\eta_0}{2H\sqrt{(\alpha_0^2+\beta_0^2)}}\{e^{-\beta_0 x}.\cos(\alpha_0 x-\sigma t-\theta)-e^{\beta_0 x}.\cos(\alpha_0 x+\sigma t+\theta)\} \tag{6.13a}$$

where

$$\beta_0^2-\alpha_0^2+\sigma^2/gH = 0 \tag{6.13b}$$

$$2\alpha_0.\beta_0 = \sigma f_0/gH \tag{6.13c}$$

$$\tan\theta = \beta_0/\alpha_0 \tag{6.13d}$$

Notice that equation 6.12 represents the addition of two damped progressive waves of equal amplitude at $x = 0$; one travelling into the estuary and the other being reflected from the estuary head ($x = 0$).

Differentiation of equation 6.12 with respect to time relates α_0, β_0 to times of high water (t_{hw}) since $\partial\eta/\partial t = 0$ when $t = t_{\text{hw}}$:

$$\tan(\sigma t_{\text{hw}}) = -\tan(\alpha_0 x).\tanh(\beta_0 x) \tag{6.14}$$

Substitution of equation 6.14 into 6.12 relates high water levels to high water level at $x = 0$, i.e.

$$\frac{\eta_{\text{hw}}}{\eta_0} = \{(\cos(2\alpha_0 x)+\cosh(2\beta_0 x))/2\}^{\frac{1}{2}} \tag{6.15}$$

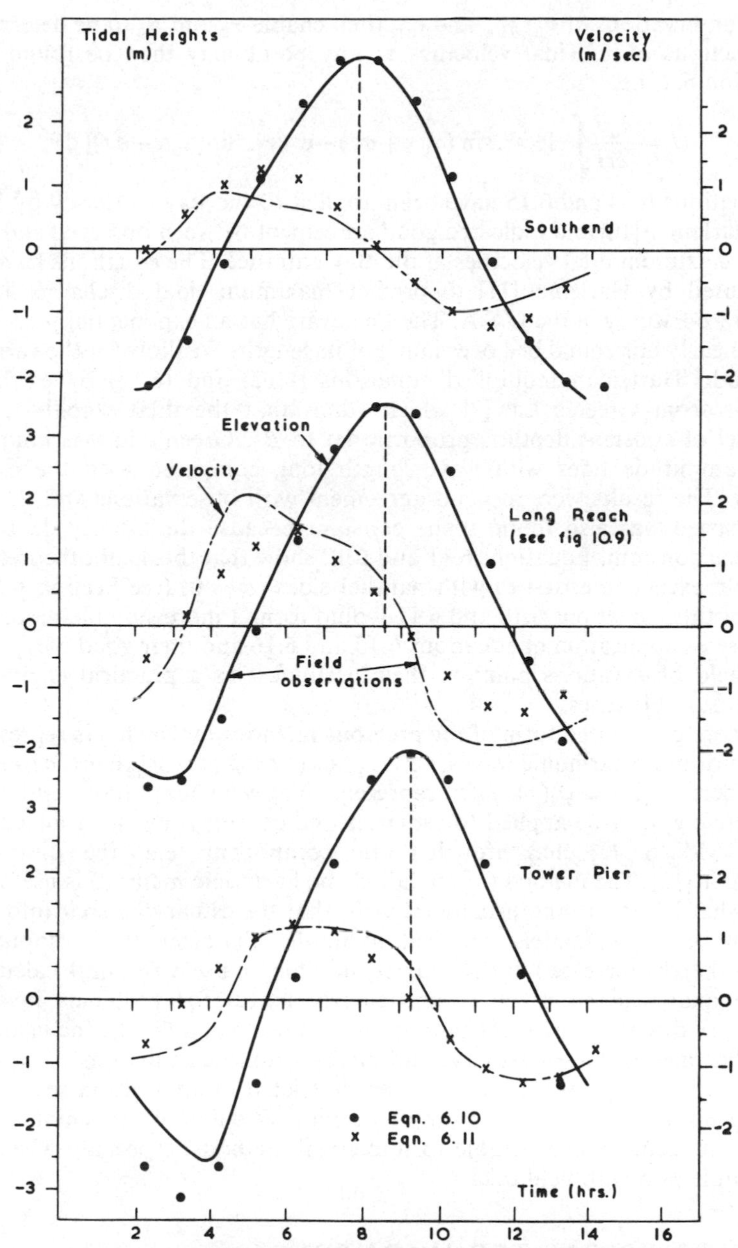

Figure 6.2. Observed and calculated tidal heights and mean velocities for the Thames Estuary (based on figure 3 of reference[9], courtesy the Royal Astronomical Society)

Field observations of η_0, η_{hw} and t_{hw} then enable α_0 and β_0 to be determined as functions of x. Tidal velocities at any point may then be found from equation 6.2, i.e.

$$U = \frac{\eta_0 \sigma}{2H} \int_0^x [e^{\beta_0 x} . \sin(\alpha_0 x + \sigma t) - e^{-\beta_0 x} . \sin(\alpha_0 x - \sigma t)] \, dx \qquad (6.16)$$

Equations 6.14 and 6.15 have been applied to the Bay of Fundy by Ippen and Harleman[10] and indicated good agreement between observed and computed maximum tidal velocities at the Bay entrance. The equations have also been used by Harleman[11] to predict maximum tidal discharges in the Delaware Estuary in the U.S.A. The Delaware has an exponential plan shape and a nearly horizontal bed over much of its length. To allow for the variation in width, Harleman multiplied equations (6.12) and (6.15) by e^{-ax}; this follows from Green's Law[7] which states that the tidal amplitude in a channel of constant depth is proportional to $B^{\frac{1}{2}}$. Green's Law is limited to small-amplitude tides with wave length long compared with the estuary length. The results were in good agreement with observations and the tidal wave speed was also found to be constant because the estuary depth was assumed constant. Equations 6.11 and 6.12 show that this is not theoretically possible except in estuaries with parallel sides ($a = 0$) (see Section 6.5). In view of this, equations 6.10 and 6.11 would form a more suitable model, but the ease of application of equations 6.13 and 6.16 and their good comparison with field observations point to their usefulness as a practical engineering tool in suitable cases.

A more generalised form of the previous methods in which η is represented by two or more harmonic waves, i.e. $\eta_{m1} \cos(\sigma t + \phi_1(x))$, $\eta_{m2} \cos(2\sigma t + \phi_2(x))$ etc. where η_{m1}, η_{m2}, $\phi_1(x)$, $\phi_2(x)$ represent tidal wave amplitudes and phases respectively, may be applied to estuaries and coastal zones with mixed tides (O_1, K_1, M_2, S_2, K_2 etc.) provided one component (e.g. the diurnal) is dominant[12]. The method (often called the harmonic method) is useful, but somewhat laborious and usually requires that the estuary be split into small lengths of approximately constant hydraulic characteristics: computation time is thereby increased. It is, however, suitable for use with a desk calculator.

The analytical approach is particularly useful for problems involving boundary discharge or level changes, such as, for example, the inclusion of a tidal barrage across an estuary, and enables rapid calculations to be made with simple equipment. It is, however, restricted in application to estuaries having simple geometry and a simple-harmonic tidal input. In suitable cases, it has an accuracy comparable to numerical methods, especially when used in conjunction with field data.

6.4 NUMERICAL TECHNIQUES

The numerical techniques are based on three main approaches: (i) method of characteristics; (ii) finite difference method; (iii) finite element method.

The finite difference method has been used most widely, but the method of finite elements has been found to have advantages, particularly in ease of application to new situations.

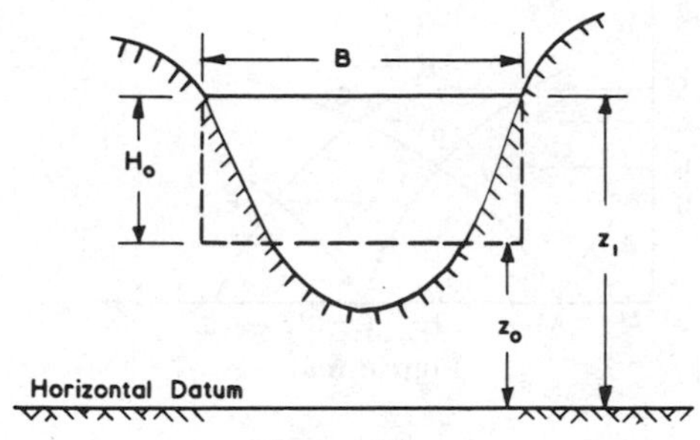

Figure 6.3

6.5 CHARACTERISTICS

Suppose that a simple shaped cross-section of an estuary can be replaced (i.e. schematised) by an equivalent rectangular section of comparable hydraulic properties as in figure 6.3. Define a characteristic velocity (c'') by the expression

$$c = (gH_0)^{\frac{1}{2}} = [g(z_1 - z_0)]^{\frac{1}{2}} \tag{6.17}$$

which represents the phase velocity of a small-amplitude water wave in the channel. Equations 6.1 and 6.2 may now be re-written in terms of c'' and H_0, i.e.

$$\frac{\partial U}{\partial t} + U\frac{\partial U}{\partial x} + 2c''\frac{\partial c''}{\partial x} + g\frac{\partial z_0}{\partial x} + g\frac{U|U|}{C^2 H_0} = 0 \tag{6.18}$$

$$2\frac{\partial c''}{\partial t} + 2U\frac{\partial c''}{\partial x} + c''\frac{\partial U}{\partial x} + \frac{Uc''}{B}\frac{\partial B}{\partial x} = 0 \tag{6.19}$$

Addition and subtraction of these equations gives the two equations

$$\frac{\mathrm{d}}{\mathrm{d}t}(U + 2c'') = -g\frac{\partial z_0}{\partial x} - \frac{gU|U|}{C^2 H_0} - \frac{Uc''}{B}\frac{\partial B}{\partial x} \tag{6.20}$$

and

$$\frac{\mathrm{d}}{\mathrm{d}t}(U - 2c'') = -g\frac{\partial z_0}{\partial x} - \frac{gU|U|}{C^2 H_0} + \frac{Uc''}{B}\frac{\partial B}{\partial x} \tag{6.21}$$

where

$$\frac{\mathrm{d}(\ )}{\mathrm{d}t} = \frac{\partial(\ )}{\partial t} + \frac{\mathrm{d}x}{\mathrm{d}t}\cdot\frac{\partial(\ )}{\partial x} \tag{6.22a}$$

and

$$\frac{\mathrm{d}x}{\mathrm{d}t} = U \pm c'' \tag{6.22b}$$

It is assumed that $\beta' = 1$ in equation 6.1 and that the channel is wide enough for R to be replaced by H_0.

Equations 6.20–6.22 are equivalent to equations 6.1 and 6.2 and have a simple graphical interpretation. Consider the space/time plane $(x-t)$ of figure 6.4. Equation 6.22b implies that a disturbance in the flow propagates downstream at speed $(U+c'')$ along characteristic lines such as AB, and upstream at speed $(U-c'')$ along lines such as AD.

Equations 6.20 and 6.21 can be integrated with respect to time along lines

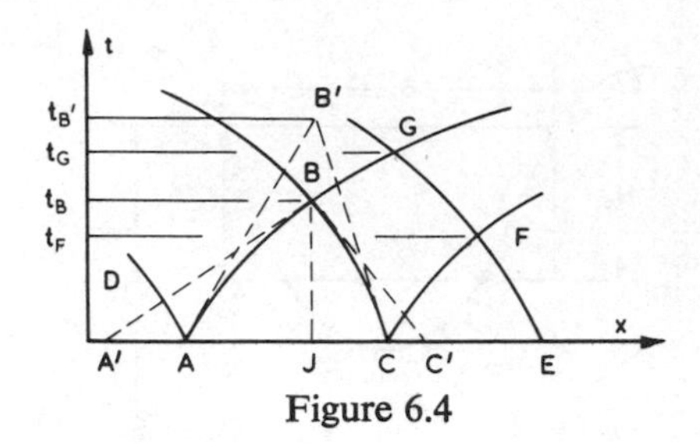

Figure 6.4

AB and AD to give relationships between U and c'' over the space-time plane. At a point such as B, the value of U and c'' can be determined by integrating equation 6.20 along the 'forward characteristic' AB and equation 6.21 along the 'backward characteristic' CB:

$$U_B+2c''_B = U_A+2c''_A-\int_0^{t_B}\left[g\frac{\partial z_0}{\partial x}+g\frac{U|U|}{C^2H_0}+\frac{Uc''}{B}\cdot\frac{\partial B}{\partial x}\right]\mathrm{d}t \qquad (6.23)$$

$$U_B-2c''_B = U_c-2c''_c-\int_0^{t_B}\left[g\frac{\partial z_0}{\partial x}+g\frac{U|U|}{C^2H_0}-\frac{Uc''}{B}\cdot\frac{\partial B}{\partial x}\right]\mathrm{d}t \qquad (6.24)$$

Addition and subtraction of the equations determines U_B and c''_B, provided that the terms in the integrals are known functions of space and time. Because some of these terms depend on the solutions for U and c'', an iterative procedure must be used. Given known initial conditions (U, H_0 for all x at $t = 0$) and boundary conditions (e.g., H_0 for all t at $x = 0$ and the fresh water discharge and level at the other end of the channel), equation 6.22b enables B' and $t_{B'}$ to be found using initial values of U and c'' at points A and C. Once $U_{B'}$ and $c''_{B'}$ have been estimated from equations 6.23 and 6.24 with initial values in the integral terms, new slopes can be computed from equation 6.22b and a new estimate of $t_{B'}$ made, together with a revised estimate of $U_{B'}$ and $c''_{B'}$.The process is repeated until successive values of $U_{B'}$, $c''_{B'}$ differ by less than a specified amount, that is, when B′ converges to a point very close to B. Clearly convergence occurs more rapidly if A is close to C. The process is continued from other points along the x axis to determine points such as F. Once B and F are known, points such as G can be located. The computation thus enables values of U and c'' and hence η to be determined at all points in the $x-t$ plane. Notice that only one characteristic need be considered at the upstream and downstream boundaries to obtain the solution.

The method provides a valuable graphical indication of the role of the various physical processes and can be adapted for computer use by the adoption of regular space and time intervals (i.e., AC = CE and $t_B = t_F$, etc.). In this case, the iterative procedure starts initially from B by locating points A′ and C′ using U, c'' data corresponding to point J. Once A′ and C′ are found, revised slopes can be drawn from B and the process repeated until some limiting error bound is reached.

The characteristic method has the advantage of clearly demonstrating transformations in the tidal wave as it proceeds through the estuary. Con-

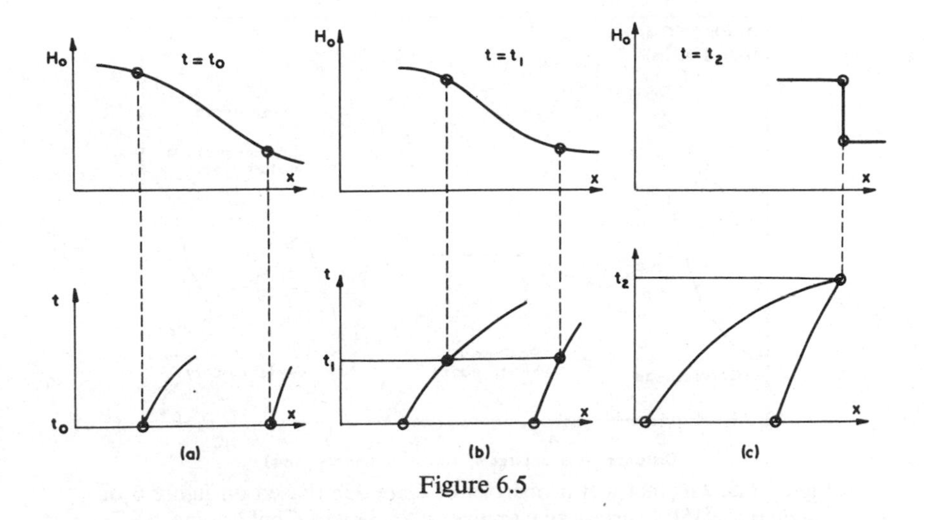

Figure 6.5

vergence of the forward characteristics indicates the steepening of the wave form, while their crossing indicates the formation of a tidal bore (see figure 6.5).

Once a steep-fronted wave is formed, vertical accelerations become large and may result in a spread of energy over shoals in wide, shallow reaches, or in the formation of a highly turbulent travelling surge wave or bore in steep-sided reaches. Equation 6.1 applies quite well to conditions on either side of the bore but is not applicable in the bore itself. However, corrections can be made to account for bore formation[13].

Equation 6.23 and figure 6.5c show that the conditions most suited to bore formation in an estuary are for the incoming flood tide to meet a steep bed slope or a rapid reduction in cross-sectional width at about mean tide level when velocities are highest.

Equations 6.23 and 6.24 also confirm the effect of estuary shape on wave speed. This is small if $Uc'' \partial(\ln(B))/\partial x$ is less than $gU|U|/(C^2H_0)$; the bed slope term being neglected because many estuaries have very flat bed slopes. In the case of the Delaware Estuary referred to earlier, the friction term was three to four times as large as the slope term ($\partial z_0/\partial x$ was zero) and would help to explain the good results obtained from the analytical solution based on the incident and reflected wave method.

The characteristic method is considered by Liggett and Woolhiser[14] to be the most accurate method available for problems involving flood routing on rivers and has distinct advantages over other numerical methods. For example, equations 6.20 and 6.21 are total differentials and are solved by numerical integration of the right-hand side. Consequently there are no truncation errors caused by neglect of higher-order terms of series expansions (see equation 6.27) and the necessary extrapolations can be improved by iteration to give any desired degree of fit. The method is also inherently stable, although the accuracy of solution that is finally obtained must depend on the accuracy of schematisation and of the numerical integration over finite

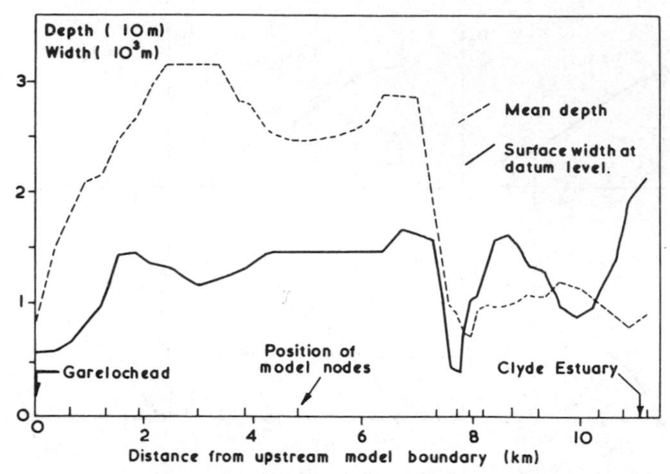

Figure 6.6. Geometrical properties of Gareloch (based on figure 6 of reference[15], courtesy the American Society of Civil Engineers)

steps. The characteristic method is also useful for stratified flow situations since the movement of the interface between layers can be readily tracked through space and time. Other solution methods, such as finite-difference techniques, have difficulties in dealing with problem parameters which show large spatial or temporal gradients (see section 6.9).

The characteristic method is of most use in the estuarine zone for problems involving little friction, or shape effects, such as occur in deep or short estuaries. It has, for example, been used by Ellis[15] to study velocities and tides in a deep fiord-type inlet, the Gareloch, situated on the north side of the Clyde Estuary in Scotland. This deep elongated inlet has a U-shaped cross-section, with a step across its width at one third of its length from its junction with the Clyde (see figure 6.6). It has a small fresh water flow in comparison with the inter-tidal volume. The prediction of surface elevations was found to be good[15], as might be expected in a short system. The longitudinal variation of velocity and surface elevation (figure 6.7) shows the effect of the contraction at its mouth, the Rhu Narrows.

The characteristic method also has several disadvantages which have resulted in engineers prefering alternative methods of solution. It becomes unsuitable for machine computation once estuarine geometry changes produce large variations in the friction term, and corrections need to be applied to the basic equation to allow for momentum exchange between tidal flats and the main flow channels, or for the formation of a bore tide. Characteristic intersections can occur at any point within a channel and therefore boundary data must be provided for all points. In practice they may be obtained by interpolation between values known at fixed cross-sections. An alternative is to calculate characteristic slopes at intersections and then to interpolate from the nearest fixed cross-sections. These methods work quite well when used for calculation of levels and mean axial flows along a channel, but become cumbersome when applied to flows that are two-dimensional in plan.

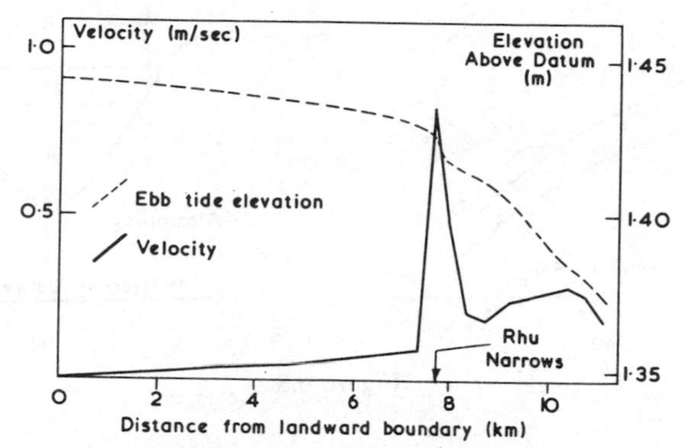

Figure 6.7. Velocities and water levels in Gareloch – characteristic method (based on figure 11 of reference[15], courtesy the American Society of Civil Engineers)

6.6 FINITE-DIFFERENCE METHODS

These methods consist of replacing the differential terms in equations 6.1 and 6.2 by discrete spatial or temporal differences. The simplest application of the method is to equation 6.2. Consider a length of estuary in which the water surface level (z_1) changes by an amount dz_1 at point x_2 during the time interval $\Delta t = (t_3 - t_1)$, while the cross-sectional area (A) changes by dA (figure 6.8).

Equation 6.2 applies at all points in space and time and consequently is applicable to conditions within the small area Δt, Δx in the $x-t$ plane (figure 6.9).

Conditions at the centre of the area (co-ordinates x_2, t_2) can be related to conditions at the edges of the box using Taylor series expansions for the various terms in equation 6.2. For any general function (f) the Taylor series expansion about point x_2 at time t_2 are:

$$f_{x_3t_2} = f_{x_2t_2} + (\Delta x/2)\left.\frac{\partial f}{\partial x}\right)_{x_2t_2} + \frac{(\Delta x/2)^2}{2!}\left.\frac{\partial^2 f}{\partial x^2}\right)_{x_2t_2} + \frac{(\Delta x/2)^3}{3!}\left.\frac{\partial^3 f}{\partial x^3}\right)_{x_2t_2} + \ldots \quad (6.25)$$

$$f_{x_1t_2} = f_{x_2t_2} - (\Delta x/2)\left.\frac{\partial f}{\partial x}\right)_{x_2t_2} + \frac{(\Delta x/2)^2}{2!}\left.\frac{\partial f}{\partial x}\right)_{x_2t_2} - \frac{(\Delta x/2)^3}{3!}\left.\frac{\partial^3 f}{\partial x^3}\right)_{x_2t_2} + \ldots \quad (6.26)$$

where subscripts refer to values of the functions in space and time.

Equation 6.2 may thus be replaced by the equation

$$(Q_{x_3t_2} - Q_{x_1t_2})/\Delta x = -B_{x_2t_2}(\partial z_1/\partial t)_{x_2t_2} + O(\Delta x^2) \quad (6.27a)$$

$$= -(\partial A/\partial z_1)_{x_2t_2} \cdot (\partial z_1/\partial t)_{x_2t_2} + O(\Delta x^2) \quad (6.27b)$$

where $Q_{x_3t_2}$ refers to the discharge at point x_3 at time t_2 and the quantity $O(\Delta x^2)$ is a truncation term and includes such quantities as $\Delta x^2/4.(\partial^2 Q/\partial x^2)_{x_2t_2}$,

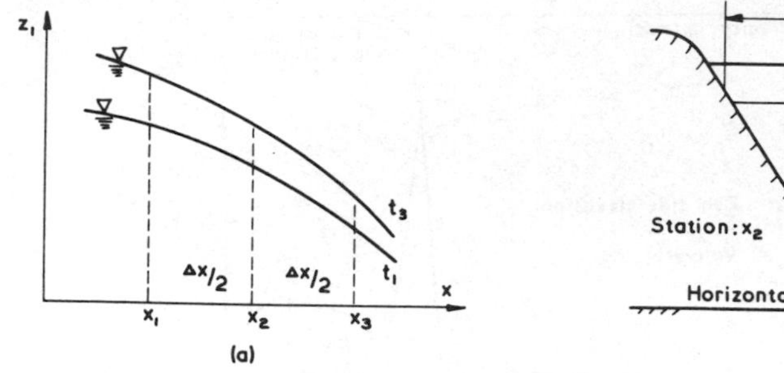

Figure 6.8

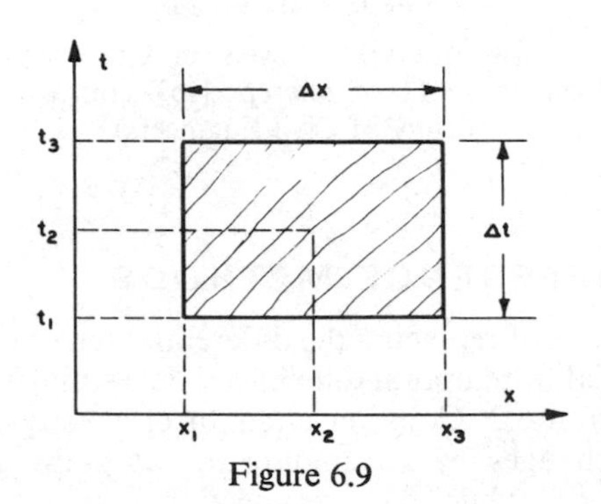

Figure 6.9

$\Delta x^4/192.(\partial^4 Q/\partial x^4)_{x_2 t_2}$, etc. from the Taylor series. It is usually assumed that the residual quantity is negligible for sufficiently small values of Δx.

Finite-difference equations are often classified by the lowest power occurring in the residual error. For example the central-difference scheme leading to equation 6.27 produces a second order correct $O(\Delta x^2)$ equation. Clearly third, fourth, or higher order finite-difference equations represent better approximations to the original differential equations.

The terms on the right-hand side of equation 6.27 may also be estimated from their values at x_3 and x_1 by use of equations 6.25 and 6.26. If the rate of change of these quantities is small, they are simply replaced by their average values over distance Δx. If Δx is sufficiently small and cross-sectional properties change slowly between x_1 and x_3, then values of z_1 and A at x_2 may be used for the right-hand side of equation 6.27. Equation 6.27 thus enables cross-sectional average tidal discharges to be computed at position x_3 if discharges are known at position x_1 and water surface elevations and estuary geometry are known at points x_1, x_2 and x_3. Discharge at any other section can be found by repeated use of equation 6.27 once Q_3 has been determined.

This method, known as cubature, presents a good means of determining tidal discharges provided that water surface levels are known at close longitudinal intervals (Δx) and are accurately determined. In practice this requires Δx values of some 1–10 km and an accuracy of level better than 10 mm. Tide gauges must, therefore, be sited free from any disturbance and the gauge itself

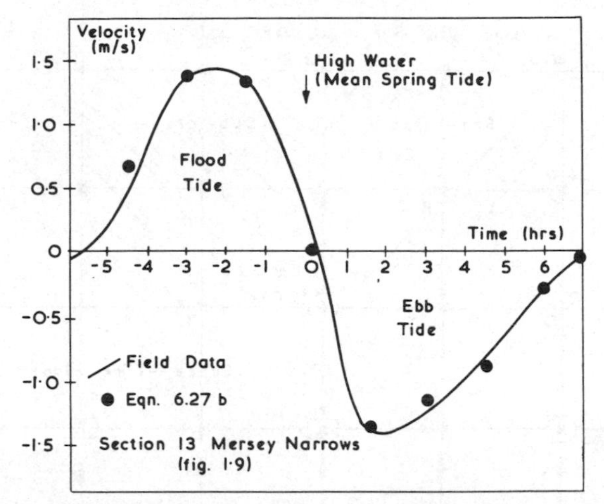

Figure 6.10. Comparison of observed and computed cross-sectional mean velocities — Mersey Narrows (after Burke[16])

must be carefully maintained. Solutions of equation (6.27) should only be accepted if field checks on discharges can be made.

The cubature method has been applied to many estuaries for determining tidal flow rates, leading in turn to evaluation of equations of motion such as equation 6.1. It was used in the analysis of data from the River Hooghly shown in figure 2.8 and by Burke[16] to determine tidal discharges in the Mersey Estuary in the UK (see figure 6.10).

The finite-difference approach may also be applied to both equations 6.1 and 6.2 to determine simultaneous tidal discharges and water surface levels.

In this case the finite difference equations can be usedin two distinct ways.

(1) The equation of continuity 6·2 may be solved using initial values to give solutions for water level at a later time step. The levels thus found may be used with previously-calculated flow data to solve equation 6.1 at the next, later time step. *Explicit* solutions of the equation are thus obtained at each mesh point, depending only on previously calculated or boundary values.

(2) Equations 6.1 and 6.2 may be solved simultaneously at each time step, the solutions for level changes at any point from equation 6.2 being used to solve equation 6.1, while solutions for velocity from equation 6.1 are used to solve equation 6.2. The result is a set of simultaneous equations that can only be completed by boundary values from the two ends of the system. The equations thus give *implicit* solutions for level and flow at each mesh point.

Both schemes are used in tidal computations and are therefore considered here.

6.7 EXPLICIT SCHEME

It is proposed to determine tidal velocities and water surface elevations at fixed time (Δt) and space (Δx) intervals. The x–t plane is therefore divided

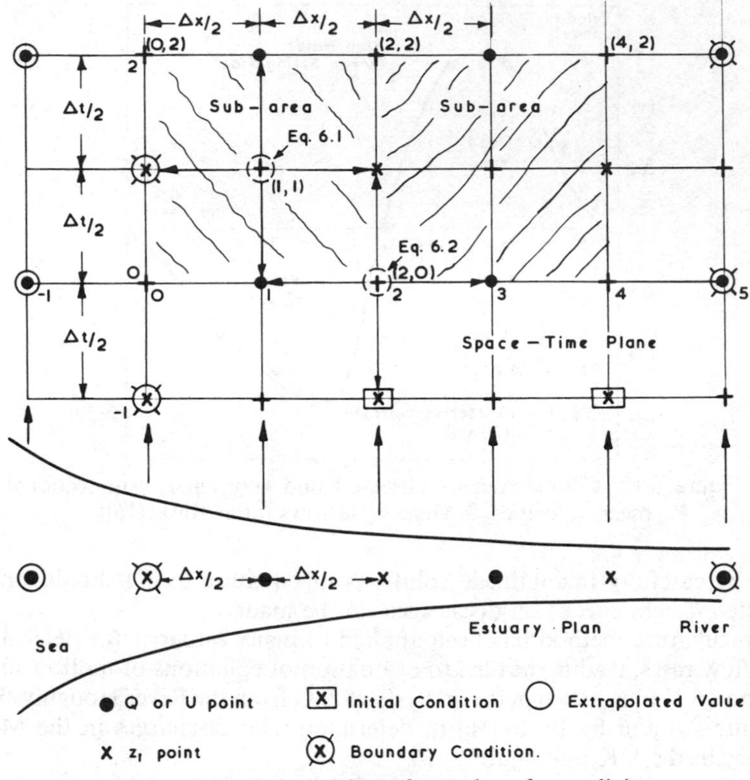

Figure 6.11. Sub-division of x–t plane for explicit finite-difference scheme

into a number of small sub-areas, e.g. 0,0; 2,0; 2,2; 0,2 in figure 6.11. Equation 6.1 is then applied to the centre of each sub-area, i.e. point 1,1 in figure 6.11, while the continuity equation 6.2 is used to connect the various sub-areas together. Application at point 2,0 links four sub-areas within the larger area defined by co-ordinates 0, – 2; 0,2; 4,2 and 4, – 2.

The momentum and continuity equations are now written in finite difference form by relating the terms at positions 1,1 and 2,0 to neighbouring positions using equations 6.25 and 6.26. The momentum equation may be written as:

$$\frac{U_{1,2}-U_{1,0}}{g\Delta t}+\frac{U_{1,1}(U_{2,1}-U_{0,1})}{g\Delta x}+\frac{(z_1)_{2,1}-(z_1)_{0,1}}{\Delta x}+\frac{U_{1,1}|U_{1,1}|}{C_{1,1}^2 H_{1,1}}+O(\Delta t^2, \Delta x^2)=0 \quad (6.28)$$

where subscripts refer to position in the x–t plane, i.e. $U_{1,2}$ is the velocity at position $x=1$, $t=2$. The finite difference equation is seen to be correct to the second order in space and time and to be a good model of the difference equation since the truncation error term tends to zero as Δt and Δx tend to zero. Again β' has been taken as unity in equation 6.1, and R has been replaced by H for large width:depth ratios; both quantities could easily be retained in equation 6.28.

The non-linear terms in equation 6.28 can be linearised and approximated in several ways. For example, the velocities $U_{1,1}$, $U_{2,1}$ and $U_{0,1}$ in the convective acceleration term can be replaced by the relationships[17]:

$$U_{1,1} = \tfrac{1}{2}(U_{1,2}+U_{1,0})+O(\Delta t^2) \tag{6.29a}$$

$$U_{2,1} = \tfrac{1}{2}(U_{3,0}+U_{1,2})+O(\Delta t.\Delta x) \tag{6.29b}$$

$$U_{0,1} = \tfrac{1}{2}(U_{1,2}+U_{1,0})+O(\Delta t.\Delta x) \tag{6.29c}$$

The convective term may now be written as:

$$\frac{U_{1,1}(U_{2,1}-U_{0,1})}{g\Delta x} = \frac{U_{1,2}(U_{3,0}-U_{-1,0})}{4g\Delta x} + \frac{U_{1,0}(U_{3,0}-U_{-1,0})}{4g\Delta x} + O(\Delta t) \tag{6.30}$$

An alternative representation is to retain only the first term on the right-hand side of equation 6.30, multiplied by two. This formulation has the advantage of increasing numerical stability by placing more emphasis on future time levels. It should be noted, however, that high order approximations for the convective term are rarely required since its value is usually small compared to the friction term.

The resistance term may also be approximated, using equations 6.25 and 6.26, as

$$\frac{U_{1,1}|U_{1,1}|}{C_{1,1}^2 H_{1,1}} = \frac{U_{1,2}|U_{0,1}|}{C_{1,1}^2 H_{1,1}} + O(\Delta t) \tag{6.31}$$

The accuracy of this representation improves to second order ($O(\Delta t^2)$) as the temporal acceleration tends to zero. However, the use of equations 6.30 and 6.31 reduces the local accuracy of difference equation 6.28 to first order. The total model, as represented by equations 6.28–6.31, should therefore be operated with as small a time step as convenient for stability and cost considerations, so that local truncation errors are minimised.

The continuity equation is also written in difference form by application of equations 6.25 and 6.26 to 6.2, i.e.:

$$(z_1)_{2,1} = (z_1)_{2,-1} + \frac{[(UA)_{3,0}-(UA)_{1,0}]\,\Delta t}{B_{2,0}\Delta x} + O(\Delta t^2, \Delta x^2) \tag{6.32}$$

If initial and boundary conditions are known (see figure 6.11), equation 6.32 can be used at time level (0) to determine surface elevations at points 2,1 and 4,1. This requires an iterative process since the values of B_{20} and A_{30} depend upon the water surface elevation at point 2,1. Initially, the water surface elevation at point 3,0 can be taken as the mean value of points 4, −1 and 0,1 and thereafter as the mean of points 4, −1 and 2,1. The water surface elevation at point 1,0 is found as the average value of positions 2, −1 and 0,1 thus enabling A_{10} to be found. Application of equations 6.28–6.31 at time level (1) then enables velocities or discharges to be calculated at points 1,2 and 3,2. This also requires the velocity value at point −1,0 and Chezy

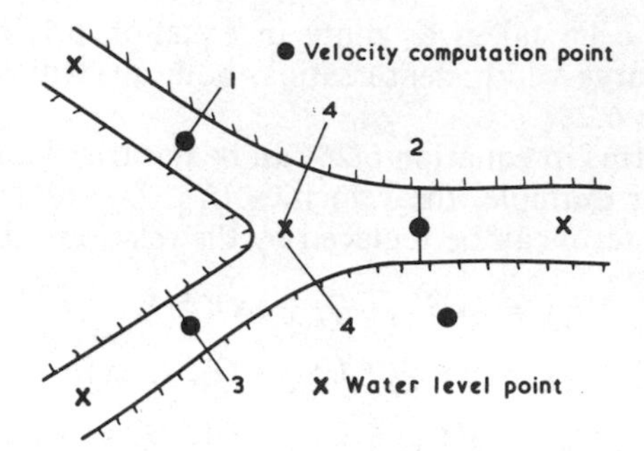

Figure 6.12.

coefficient and depth values at points 1,1 and 3,1. The velocity may be obtained by extrapolation using the known values along time line (0), while the depths are found as the mean value of points 4,1 and 2,1 and 2,1 and 0,1. Equations 6.32 and 6.28–6.31 are then applied on alternate time steps to determine z_1 and U at fixed points covering the x–t plane. This explicit difference scheme is thus often referred to as the *leap-frog* method and has first-order accuracy.

The method may also be extended to include the effects of a longitudinal salinity gradient. This was found by Rossiter and Lennon[17] to produce a change in mean water level of some 150 mm in the Thames Estuary.

Lateral inflows due to streams and sewers, the effects of islands, the momentum exchange from tidal flats, energy losses due to rapid changes in channel geometry and the formation of tidal bores[18] can all be included in the finite difference model. Bifurcations in estuary channels are readily included in the scheme by arranging for a level-computation point to be placed at the channel junction (see figure 6.12).

The computation proceeds upstream to the junction where the water level $(z_1)_{4,1}$ is calculated from a similar equation to equation 6.32, that is;

$$(z_1)_{4,1} = (z_1)_{4,-1} + (Q_m - Q_{2,0})\,\Delta t/(B_{4,0}\,\Delta x) \tag{6.33a}$$

$$Q_m = (Q_{1,0} + Q_{3,0})/2 \tag{6.33b}$$

and $B_{4,0}$ is a suitable average value between sections 1, 2 and 3. Once $(z_1)_{4,1}$ is known, the computation can proceed independently up each channel.

Energy losses due to the bifurcation, as well as bends and slowly changing boundary surfaces are assumed to be proportional to U^2 and are included in the friction coefficient.

6.8 IMPLICIT SCHEMES

The x–t plane is again divided into a series of sub-areas (see figure 6.13) to which the equations of momentum and continuity are applied. The momentum and continuity equations are both located at time level (1) in this schematisation, but the continuity equation is displaced in space by one Δx

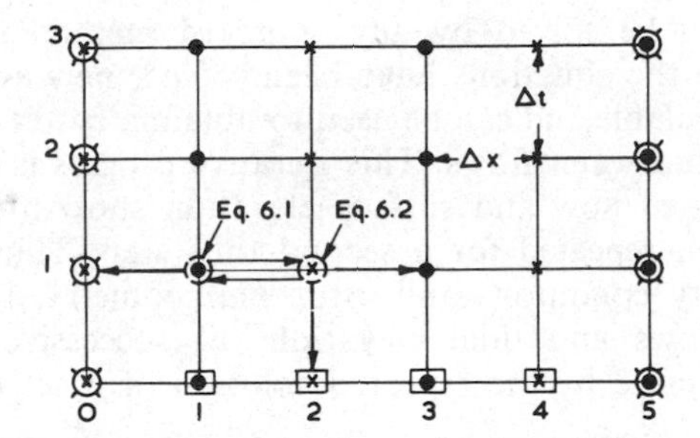

Figure 6.13. Implicit schematisation of the x–t plane

interval. Using equations 6.25 and 6.26, with $\Delta x/2$, $\Delta t/2$ replaced by Δx, Δt, the momentum equation becomes:

$$\frac{U_{1,1}-U_{1,0}}{g\Delta t}+\frac{U_{1,1}(U_{2,1}-U_{0,1})}{2g\Delta x}+\frac{(z_1)_{2,1}-(z_1)_{0,1}}{2\Delta x} + \frac{U_{1,1}|U_{1,1}|}{C^2_{1,1}H_{1,1}}+O(\Delta t, \Delta x^2) = 0 \quad (6.34)$$

The non-linear terms are again linearised. For example, the convective acceleration term may be written at time level (0) in terms of known velocity points by use of equations 6.25 and 6.26, i.e.

$$\frac{U_{1,1}(U_{2,1}-U_{0,1})}{2g\Delta x} = \frac{U_{1,1}(U_{3,0}-U_{-1,0})}{4g\Delta x}+O(\Delta x) \quad (6.35)$$

An alternative estimate is obtained by using the approximation[18]

$$\frac{U_{1,1}(U_{2,1}-U_{0,1})}{2g\Delta x} = \frac{U_{1,1}(U_{3,0}-U_{1,0})}{2g\Delta x}+O(\Delta x) \quad (6.36)$$

This approximation has the advantage of operating over a smaller space interval (see p. 171) and also avoids the use of extrapolated velocity values at the seaward boundary.

The friction term is also linearised to give

$$\frac{U_{1,1}|U_{1,1}|}{C^2_{1,1}H_{1,1}} = \frac{U_{1,1}|U_{1,0}|}{C^2_{1,1}H_{1,0}}+O(\Delta t) \quad (6.37)$$

The finite-difference form of the continuity equation follows in a similar manner to equation 6.34. The result is:

$$\frac{(AU)_{3,1}-(AU)_{1,1}}{2\Delta x}+\frac{B_{2,1}[(z_1)_{2,1}-(z_1)_{2,0}]}{\Delta t}+O(\Delta t, \Delta x^2) = 0 \quad (6.38)$$

Application of the momentum difference equations 6.34–6.37 to points 1,1 and 3,1, and the continuity difference equation 6.38 to points 2,1 and 4,1, produces four equations for the four unknown quantities at time level (1). These equations cannot be solved directly because the continuity equation requires data at time level (1). It is necessary, therefore, to approximate $A_{3,1}$, $A_{1,1}$ and $B_{2,1}$ by values at time level (0), i.e., $A_{3,0}$, $A_{1,0}$ and $B_{2,0}$. These equations may then be solved by any standard method such as pivotal condensation. Once the equations have been solved, new estimates of $A_{3,1}$, $A_{1,1}$ and $B_{2,1}$ are available and can be used to obtain a better estimate for the surface elevations and water flows. This iterative process is continued until successive estimates of flow and surface elevation show little change. The total process is then repeated for a second time step. Thus, given known initial and boundary conditions and estuarine geometry, it is possible to determine water flows and tidal elevations at successive time intervals throughout a tidal cycle by the repeated solution of a set of linear simultaneous equations.

6.9 COMMENTS ON THE FINITE DIFFERENCE SCHEMES

When using the explicit or implicit scheme, two main criteria must be satisfied.

(1) The scheme must be stable. This means that any errors introduced into the scheme due to the neglect of small order terms and by computer round-off errors should decay during the course of computation. Stability can be examined by expressing the error at any time and position in terms of a Fourier series, and then determining the change produced by the difference equations in one Fourier component (since the system is linear) over a time interval. Clearly the scheme is stable if the amplitude of the error component is less than unity. This is always found to be true for the implicit scheme but is only so for the explicit scheme if the criterion

$$\frac{\Delta x}{\Delta t} \geqslant \left(g\frac{A}{B}\right)^{\frac{1}{2}} \Big/ (1+\tfrac{1}{2}a_0)^{\frac{1}{2}} \tag{6.39}$$

is observed[18]. In this case, at point 1,1 of figure 6.11

$$a_0 = \frac{2g\Delta t U_{0,1}}{C_{1,1}^2 H_{1,1}} + \frac{\Delta t}{\Delta x}(U_{3,0} - U_{-1,0}) \tag{6.40}$$

Since, in general, the friction term is greater than the velocity term in equation 6.40, stability is usually obtained even if (a_0) is taken equal to zero in equation 6.39. The stability criterion simply implies that the speed of computation should be faster than the rate at which errors spread through the physical system. Equation 6.39 ensures global stability if the maximum depth of flow in the system is used. Because of the rapid damping of errors due to friction it is possible to assign arbitrary values to the initial velocities and surface elevations in both schemes. The equation of continuity should be satisfied by these initial values and this can be done easily if the water surface is assumed to be horizontal, with zero velocity at all points. During the first time step, fresh water inflow Q can be introduced at the head, and tidal rise or fall begun at the seaward boundary. Convergence is rapid[25]; the second tidal cycle is usually indistinguishable from the third (see figure 6.14).

(2) The scheme must be accurate. It is impossible to calculate the total accuracy of the difference schemes. Examination of the local truncation error,

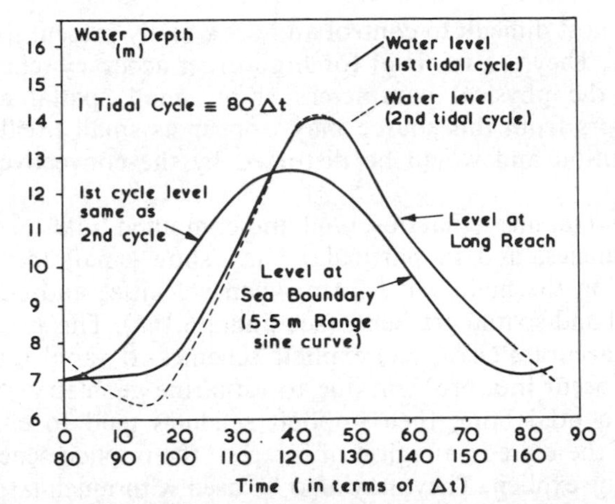

Figure 6.14. Convergence of a finite-difference tidal model of the Thames Estuary after two tidal cycles (based on figure 4 of reference[25], courtesy *The Engineer*, 1960)

i.e. the difference between the differential equation and the difference equation at any point (the term $O(\Delta t^2, \Delta x^2)$ in equation 6.28), enables the local accuracy of one scheme to be compared with another. Schemes with high-order error terms, as for example $O(\Delta t^4, \Delta x^4)$, are more accurate than first-order schemes $O(\Delta t, \Delta x)$, and schemes directly proportional to Δt, Δx or higher powers are preferable to those involving quotients, i.e. $\Delta t^2/\Delta x$, since the truncation error may not reduce to zero in the lattermost case as Δt and Δx tend to zero.

A deformation factor analysis may also be undertaken[19] to determine the change in amplitude and phase of the various wave components making up the real solution to equations 6.1 and 6.2 as they propagate over a distance equal to their wavelength. Again, this analysis does not enable the total accuracy of the scheme to be determined but will enable a choice to be made between the potentially most accurate of a number of different schemes. Ultimately, acceptable accuracy is determined by repeated numerical testing at successively smaller space and time intervals and by checks against known analytical solutions or good field observations.

Errors in finite difference schemes involving mass flow rates usually appear in three forms.

(*a*) A change in total mass flow: conservation errors. These can usually be checked by numerical summation of the total mass of the system, including all boundary losses and gains, at two widely different points in time.

(*b*) A damping out of short wavelength components making up the real solution: dispersion errors, also called numerical diffusion. These can be avoided by ensuring that difference schemes are properly centred and that the schemes are of a high order of accuracy, particularly if advective terms are dominant in the equations (see Chapter 7). Errors from this source will be found to produce extra spreading of mass within solutions.

(*c*) The Fourier components making up the real solution are advected at different velocities to that of the physical flow velocity: dissipation errors.

These are the most difficult to control and are usually present to some degree in all schemes. They are minimal for high-order accuracy schemes ($O(\Delta x^4)$) and/or when the physical parameters show small spatial and temporal gradients. Errors from this source may appear as small oscillations in the numerical solution and would be distorted by the convective acceleration term.

Errors (*b*) and (*c*) mean that observed and computed tidal curves may show slight discrepancies, and in particular may show small local oscillations superimposed on the main wave form when velocities and elevations have steep temporal and spatial gradients (see figure 6.16a). This is particularly so for first-order accurate $O(\Delta t, \Delta x)$ explicit schemes. If rapid rates of change are known to occur in a problem due to estuarine geometry changes or the formation of a tidal bore, then implicit schemes tend to eliminate small oscillations. If these are in a region of interest, the implicit scheme would be preferable to the explicit. They may also be used with much larger time steps than the explicit method since there are no stability criteria to satisfy and consequently it is also much easier to employ unequal spatial steps, which is useful in describing conditions in one part of the estuary in more detail than another, as for example, in the vicinity of islands. However, small space and time steps are usually required in shallow water estuaries with rapid change in tidal elevation, to prevent dissipation errors. Because the implicit method requires a more complicated computer program, it may have no advantage in such cases. In addition the explicit scheme requires less computing time and boundary data to advance the solution, since it has staggered space and time points. Increased accuracy can be obtained for estuaries with shallow depths and large tidal ranges by including higher order terms (Δt^2, Δx^2), etc. in the numerical solution, as for example, by using the Lax–Wendroff schematisation[14]. This method does, however, require considerably more computing effort and volume of data since velocities and surface elevations are also calculated at intermediate space and time levels. The explicit scheme would be expected to give good answers in the homogeneous tidal zone, because rapid rates of change of velocity and surface elevation rarely exist. In this zone the method of characteristics may give better accuracy than the finite-difference methods but if used with an interpolation routine so that velocities and surface elevations are produced at regular points in space and time, then its accuracy may be less than a finite-difference scheme using similar computer time.

6.10 GENERAL COMMENTS ON THE APPLICATION OF THE VARIOUS METHODS

The numerical methods described including the characteristic method, require boundary conditions at the landward and seaward ends of the estuary. At the landward end, these may be determined from river gauging records in the case of the river flows or from the condition that the tide is completely reflected ($U = 0$) in the case of zero river flow. At the seaward end, tide gauge records can be used. It is usually necessary to compute a few successive cycles with repeated identical tidal input. In reality, successive tides seldom

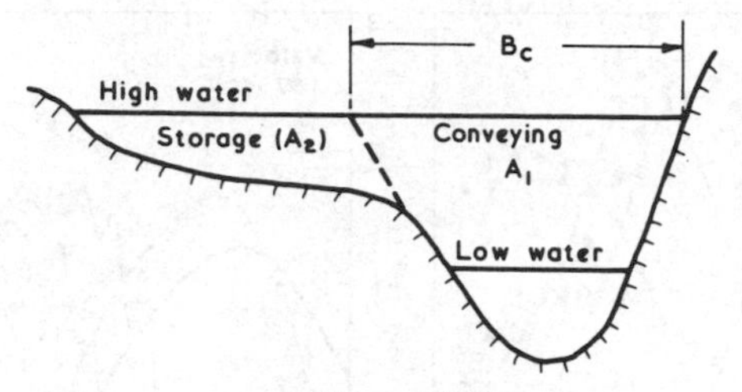

Figure 6.15

return to the same high- or low-water levels. When mixed diurnal and semi-diurnal effects are present, it may be convenient to use a diurnal period containing two high- and two low-waters. Some interpolation may be necessary to produce a repetitive cycle, and when this is done, exact agreement cannot be obtained between computed and observed tides throughout an estuary. To simplify comparisons in such cases, Rossiter and Lennon[17] decomposed the tide curves into their harmonic constituents, i.e. into mean tide level, diurnal, semi-diurnal, etc. components.

All models require geometric and frictional data, which depend upon estuary shape, the presence of bridge piers, jetties, dock entrances, etc., fresh water discharge, and bed sediment type and distribution. These data must also be representative of the finite lengths of estuary chosen as the space intervals (Δx). This is best achieved for the geometric data by determining the volume of water contained between model sections and lying below a given water level, together with the plan surface area between sections. A representative area and mean depth is then obtained by dividing the volume by the section length and by the surface area respectively. Determination of a representative hydraulic radius would require the calculation of a mean wetted perimeter from any individual cross-sectional surveys located between model sections. Usually mean depths, areas, surface widths and hydraulic radii are calculated at a series of levels covering the tidal range, and the information stored in the computer: typical vertical intervals might be 0·1 or 0·2 of the maximum tidal range. Alternatively, an equivalent rectangular or trapezoidal section might be chosen. This reduces computer storage and increases computation speed since tabular interpolation is a relatively slow machine process. This method may, however, require a change in schematization if extreme range tides are being investigated.

Allowance must also be made in irregular cross-sections for the small rate of flow over the shallow storage sections, compared with the deeper (conveying) sections, (see figure 6.15): most of the momentum flux is confined to the conveying section. Thus areas used in evaluating flow discharge and hydraulic radius might be reduced to allow for storage water as for example

$$A = A_1 + A_2 \frac{C_1}{C_2} \sqrt{\left(\frac{H_2}{H_1}\right)} \tag{6.41}$$

where H_1, C_1, H_2, C_2 are mean depths and Chezy resistance coefficients for the conveying and storage sections respectively. Often the same resistance

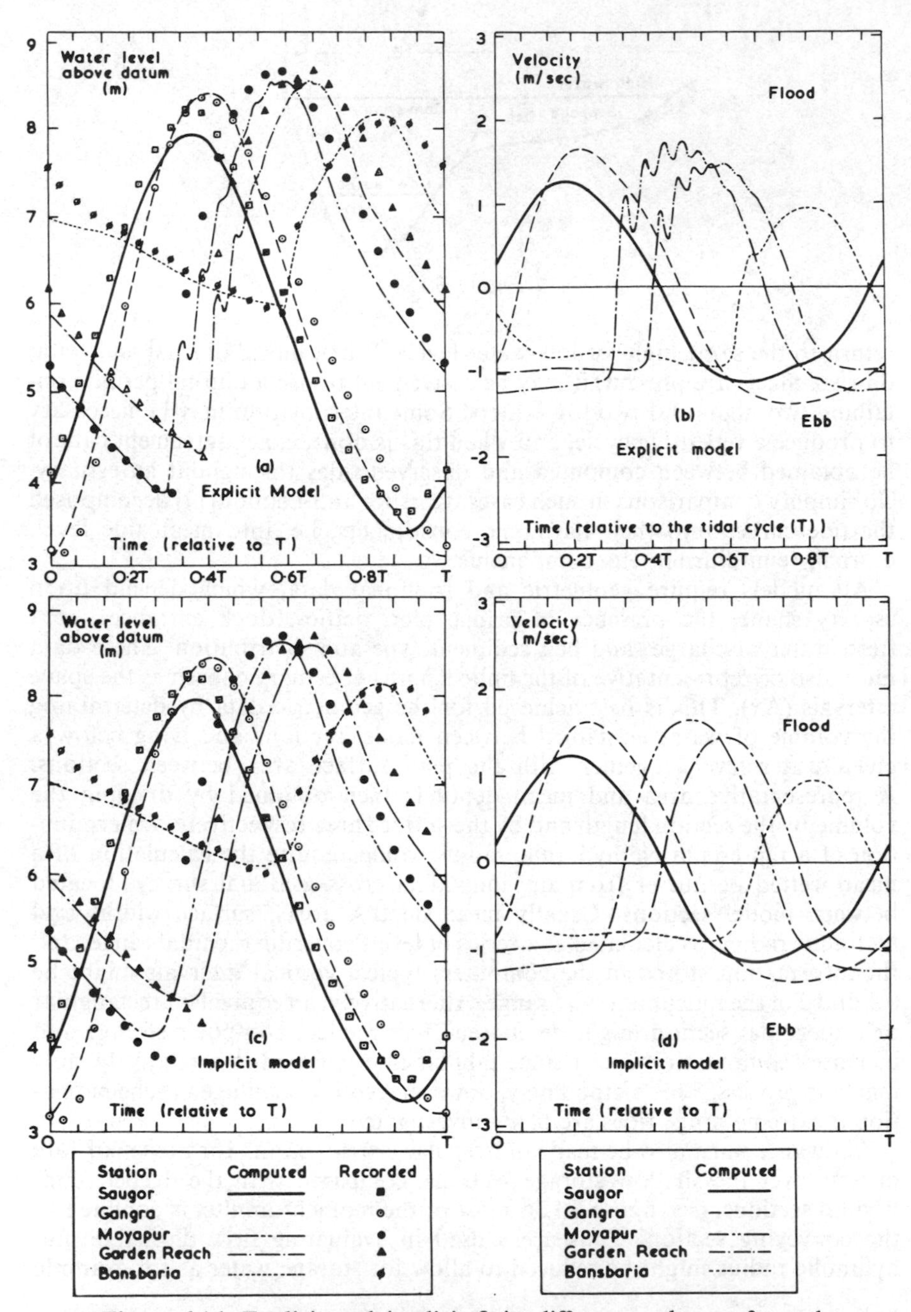

Figure 6.16. Explicit and implicit finite-difference schemes from the Hooghly estuary, India (based on figures 7, 8, 9 and 10 from reference[22], courtesy the American Society of Civil Engineers)

coefficient is used ($C_1 = C_2$) on the grounds that the Chezy term in the momentum equation already represents a lumped resistance parameter.

Friction parameters for use in a model can be obtained in several ways. The first choice may need to be modified during the process of calibration and will depend on the features of the channels. If they are straight and free from irregularities, bottom sediment may provide much of the friction and a frictional formula based on uni-directional flow might be used. For example, the equation

$$C = 16 \log (12H/k_b)\ m^{\frac{1}{2}}/s \tag{6.42}$$

where k_b is an equivalent sand grain, or boundary roughness, has been used by Kamphuis[20] for a study of the St Lawrence River in Canada. A composite friction factor might be used in shallow estuaries with tidal flats based on storage and main channel roughness as in uni-directional flow[21]. Corrections for the presence of bed forms may also be attempted[27]. Alternatively, a cubature analysis combined with equation 6.1 enables mean velocities and effective roughness coefficients to be found at points along the estuary.

Once initial roughness values have been selected, the model should be operated and friction coefficients adjusted to give the best agreement with prototype results. This calibration process can be rationalized by running the model for some three tidal cycles to ensure elimination of initial errors and then computing the effective friction coefficient from the finite-difference form of the momentum equation at half hourly intervals, using observed water levels. Any spurious results may then be eliminated and replaced by average values and the process repeated until a consistent set of coefficients is obtained. These are then used to predict prototype conditions; only slight changes should be required at this stage.

This optimization method was used with success in the application of a mathematical model to the River Hooghly in India[22], in which both explicit and implicit formulations were compared. Figures 6.16(a–d) show computed surface elevations and mean velocities for both formulations and their comparison with field observations. The occurrence of local instability in the explicit scheme is clearly seen and also its elimination by use of the implicit method. It should be noted that the results are identical except for the local oscillations and some rounding off of the low-water curve by the implicit method. However, while the greater smoothing ability of the implicit scheme may be desirable, it also eliminates the energy transfer process accounted for by the non-linear convective term. This term is responsible for the transfer of long period fluctuations into shorter ones (see Chapter 3), and its effect is apparent in prototype velocity records (see figure 6.17). The explicit scheme thus gives a better representation of the physical process than the implicit scheme but cannot model it precisely since wave components less than $2\Delta x$ were not represented in the model.

The finite difference method also allows the depth-integrated two-dimensional momentum and continuity equations to be solved in a somewhat similar manner to the one-dimensional solutions[19]. It is, therefore, possible to examine tidal motion in wide shallow seas subject to wind induced surface stresses and Coriolis force.

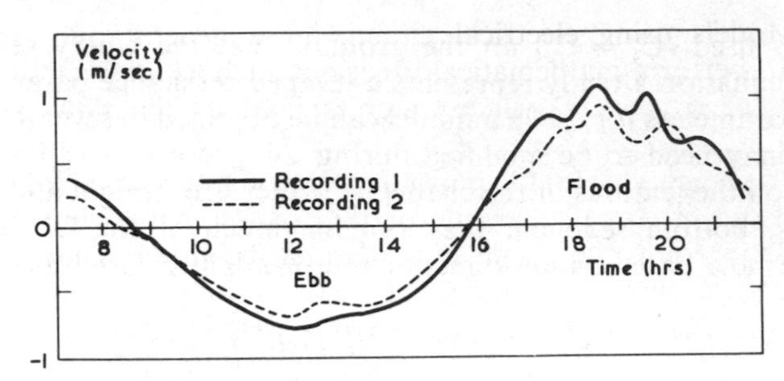

Figure 6.17. Observed velocities in the Hooghly at Moyaphur (see figure 10.13) (after McDowell and Prandle[22])

6.11 FINITE ELEMENT TECHNIQUES

In this method, the space domain is divided into elements having a convenient size. In two-dimensional problems, the elements can be triangular and can easily be adjusted in shape and size to fit complex boundary conditions[26]. Equations connecting the parameters are set up in a way that allows a residual error to be defined. This is then minimised; in practice, this necessitates solving a matrix. A method known as the Galerkin method has been used successfully by Taylor and Davis[28] to solve two-dimensional flow in wide shallow estuaries. It is usually applied only to the space domain, while conventional finite differences are used for the time domain. The estuary is divided into discrete lengths (one dimension), or areas (two dimensions), and the spatial variation of the unknown variables, for example surface elevations, specified using one or more terms of a power series involving the spatial co-ordinates. Substitution of these specified functions (shape functions) into the spatial terms of the differential equations 6.1 and 6.2 enables a residual to be defined which is then minimised using the Galerkin technique. Repetition of this process for all elements produces a set of first order simultaneous equations for the unknown quantities. These may be solved by any convenient numerical technique.

The method allows a greater degree of linking between the unknown parameters than in equivalent finite-difference schemes, and consequently it is usually more accurate for the same amount of computer work. The simplest forms of the method can also be reduced to implicit finite-difference schemes for one-dimensional problems. The method requires a complicated computer program which can increase the cost of using the method compared with finite-difference methods, but it has the twin advantages that new situations can be dealt with more easily than with the finite-difference method, while irregular boundary geometry is more easily simulated.

6.12 SIMULATION TECHNIQUES

These fall into two categories:

(i) (*a*) Models based upon the analogy between fluid and electric current flows: based largely on resistance–capacitance networks.

(*b*) Models using electrical circuits in a general purpose analogue computer to perform mathematical functions such as integration.

(ii) Physical models in which water is used as the flow medium (see Chapter 8).

Models in category (i) (*a*) are built for a special purpose, as for example the Rijkswaterstaat electrical analogue of the Rhine Delta, built to study the effect of the Deltaworks on river and estuarine flows. They usually have large physical dimensions and are difficult to adapt to different estuary systems.

Models in category (i) (*b*) enable the partial differential equations to be solved simultaneously. Electronic circuits representing the equations for each of several reaches of an estuary are inter-connected by means of a 'patch-board'. The constants appropriate to each reach are then pre-set on the computer. Tidal rise and fall is represented by a time-varying voltage, and the simultaneous response at any reach can be recorded. Once the patch-board has been set up it can be used for various estuarine problems, the only requirement being to adjust the pre-set parameters on the computer for each new situation.

Analogue computers can be used to represent estuaries as a small number of connected reaches in each of which parameters are 'lumped'[23]. However, integration proceeds continuously and instantaneously; the operator can see an immediate response to changes in model parameters.

6.13 SUMMARY

A mathematical model is the name given to any calculation procedure which produces the solution to a problem by the use of mathematical equations. Mathematical models are particularly useful for evaluating the effects of engineering works on tidal motion and the quality of tidal water. They are often used in conjunction with field and hydraulic model studies. The use of mathematical models is confined to simplified physical situations involving, in general, one or two space dimensions and time. However, they are often cheaper to use and can include effects which cannot be satisfactorily reproduced in reduced-scale physical models.

Many modern flow and water quality mathematical models use a digital computer and finite-difference or finite-element numerical techniques to solve basic hydrodynamic and continuity equations, but analytical methods, which were developed in pre-computer days, are still used today for particularly simple flow problems. The accuracy of tidal flow models is as good, if not better than the traditional physical model, provided care is taken to minimise numerical schematisation errors and friction parameters are determined from field results.

NOTATION

a, a_0	Numerical constant
a_1	Tidal constant $= [\eta_0/b_1(0)]$
$b_1(x)$	Tidal constant given by equation (6.11*f*)

c''	Phase velocity of a small amplitude gravity wave $= (gH_0)^{\frac{1}{2}}$
c_0	Value of c'' in water of depth $H = (gH)^{\frac{1}{2}}$
f_0	Friction function $(= 8\|U_0\|/3\pi C^2 H)$
g	Acceleration due to gravity
$g(x)$	Tidal function given by equation 6.11(e)
i_f	Friction slope
k	Tidal wave number $(= 2\pi/L_0)$
k_b	Equivalent bed roughness
t	Time co-ordinate
x	Longitudinal distance co-ordinate
z_1, z_0	Height of water surface and equivalent rectangular bed level above a horizontal datum respectively.
A	Flow cross-sectional area, or point in the x–t space-time plane
A'	Point in x–t space-time plane
A_1, A_2	Storage and conveying cross-sectional areas
B	Estuary water surface width, or point in x–t space-time plane
B_0	Water surface width at the estuary head ($x = 0$)
B'	Point in the x–t space-time plane
C	Chezy's coefficient
C_1, C_2	Chezy's coefficient for conveying and storage of parts of a cross-section respectively
C′, D, E, E′, F, G	Points in the x–t plane
H	Flow depth
H_1, H_2	Mean depth of storage and conveying cross-sections respectively
H_0	Depth of equivalent rectangular cross-sectional shape
J	Point in x–t space-time plane
L_0	Tidal wavelength
K_1, K_2	Diurnal and semi-diurnal tidal components
M_2	Moon's semi-diurnal tidal component
$O(\ )$	Indicates the order of error associated with a finite-difference equation
O_1	Diurnal tidal component
Q	Cross-sectional water discharge
Q_m	Mean cross-sectional flow downstream of a bifurcation
S_2	Semi-diurnal tidal component of sun
T	Tidal period
U	Cross-sectional flow velocity
U_0	Spatially-average maximum tidal velocity
$\alpha_0\beta_0$	Tidal constants in equation 6.11
η	Tidal elevation
η_0	Tidal elevation at $x = 0$
η_m	Maximum tidal elevation
η_{m_1}, η_{m_2}	Maximum tidal elevations of harmonic tidal components
θ	Tidal constant $(= \tan^{-1}(\beta_0/\alpha_0))$
ρ	Fluid density
σ	Tidal wave frequency $(= 2\pi/T)$
τ_{z_0}	Horizontal shear stress
ϕ	Tidal phase component given by equation 6.11
$\phi_1(x), \phi_2(x)$	Tidal phase angle of harmonic tidal components

Δt Time interval
Δx Space interval

REFERENCES

[1] Owen, M. W. and Odd, N. V. M., A mathematical model of the effect of a tidal barrier on siltation in an estuary, *Int. Conf. Utilization Tidal Power*, Atlantic Ind. Res. Inst., Halifax, Nova Scotia (1970)
[2] Chang, F. F. M. and Richards, D. L., Deposition of sediment in transient flow, *J. Hyd. Div. Proc. ASCE*, **97**, HY6, June (1971)
[3] Ippen, A. T. and Harleman, D. R. F., Two-dimensional aspects of salinity intrusion in estuaries: analysis and velocity distributions, Comm. on Tidal Hyd., *Corps of Eng. US Army Tech. Bull.*, no. 13 (1967)
[4] Johns, B., Tidal flow and mass transport in a slowly converging estuary, *Geo. J. Roy. Ast. Soc.*, **13** (1967)
[5] McGregor, R. C., The influence of topography and pressure gradients on shoaling in a tidal estuary, *Geo. J. Roy. Ast. Soc.*, **25** (1971)
[6] Heaps, N. S., On the numerical solution of the three-dimensional hydrodynamic equations for tides and storm surges, *Mem. Soc. R. Sci., Liège*, Ser. **6**, 1 (1971)
[7] Lamb, H., *Hydrodynamics*, 6th ed., Cambridge University Press (1932)
[8] Pillsbury, G. Tidal Hydraulics, *Corps of Engineers, US Army Tech. Bull.* Vicksburg, Miss. (1956)
[9] Hunt, J. N., Tidal oscillations in estuaries, *Geo. J. Roy. Ast. Soc.*, **8** (1964)
[10] Ippen, A. T. and Harleman, D. R. F., Investigation of influence of proposed international Passamaquoddy tidal project on tides in the Bay of Fundy, *Rep. to New England Div. Corps of Eng.*, Boston, July (1958)
[11] Harleman, D. R. F., *Estuary and coastline hydrodynamics*, Ch. 10, McGraw-Hill Book Co. Ltd. (1966)
[12] Dronkers, J. J. and Schönfeld, J. C., Tidal computations in shallow water, *Proc. ASCE*, Paper 714, June (1955)
[13] Abbott, M. B., *An introduction to the method of characteristics*, Thames and Hudson (1966)
[14] Liggett, J. A. and Woolhiser, D. A., Difference solutions of the shallow-water equation, *J. of Eng. Mech. Div. Proc. ASCE*, **93**, EM2, April (1967)
[15] Ellis, J., Unsteady flow in channel of variable cross-section, *J. Hyd. Div. ASCE*, **96**, HY10, October (1970)
[16] Burke, C., The distribution of velocity in tidal flows (Mersey Estuary), *M. Eng. Thesis*, Liverpool University, August (1966)
[17] Rossiter, J. R. and Lennon, G. W., Computation of tidal conditions in the Thames Estuary by the initial value method, *Proc. Inst. Civil Eng.*, **31**, May (1963)
[18] Dronkers, J. J., Tidal computations for rivers, coastal areas and seas, *J. Hyd. Div. Proc. ASCE.*, **95**, HY1, January (1969)
[19] Leendertse, J. J., A water quality simulation model for well mixed estuaries and coastal seas, **1**, *Memo RM-6230-RC*, The Rand Corporation, Feb. (1970)

[20] Kamphuis, J. W., Mathematical tidal study of the St. Lawrence River, *J. Hyd. Div. Proc. ASCE*, **96**, no. HY3 (1970)

[21] Henderson, F. M., *Open Channel flow*, Macmillan, New York (1966)

[22] McDowell, D. M. and Prandle, D., Mathematical model of River Hooghly, *J. Water and Harbour and Coastal Eng. Div. ASCE*, **98**, WW2, May (1972)

[23] Ball, D. J. and McDowell, D. M., The dynamic simulation of unsteady frictional flow in tidal estuaries, *Proc. Inst. Civil Eng.*, **50**, October (1971)

[24] Abraham, G., Horizontal jets in stagnant fluid of other density, *J. Hyd. Div. ASCE*, **91**, HY4, July (1965)

[25] Otter, J. R. H. and Day, A. S., Tidal flow computations, *Engineer*, **209**, Jan. (1960)

[26] Farraday, R. V., O'Connor, B. A. and Smith, I. M., A two-dimensional finite-element model for partially-mixed estuaries, *Proc. 16th Cong. IAHR*, Paper C 35, **3**, São Paulo, Brazil (1975)

[27] Sediment transportation mechanics; F.: Hydraulic relations for alluvial streams, *J. Hyd. Div. ASCE*, **97**, HY1, January (1971)

[28] Taylor, C. and Davis, J., Finite element numerical modelling of flow and dispersion in estuaries, *IAHR Int. Sym. on River Mechanics*, Paper C 39, **3**, Bangkok, January (1973)

[29] Grubert, J. P. and Abbott, M. B., Numerical computations of stratified nearly horizontal flows, *J. Hyd. Div. ASCE*, **98**, HY10, October (1972)

7

The study of tidal systems: water quality models

The majority of water quality models are based on analytical, numerical or simulation solutions to diffusion–advection type equations. Other techniques have been suggested involving stochastic processes[24], or hydrodynamic sources and sinks[25], or co-relation techniques[26], but these methods are not discussed here because of their limited range of applicability. Attention is confined to solutions of diffusion–advection equations (see Chapter 3) or simulation techniques involving similar ideas, since such methods are based on the fundamental principle of mass-conservation and are applicable to a wide variety of situations.

Analytical methods usually involve simplified forms of diffusion–advection equations (e.g., equation 3.32) or their equivalent, and have proved useful in analysing sea outfall problems and general pollution problems in homogeneous tidal zones. They have also been used with constant spatial and temporal dispersion coefficients in non-homogeneous estuarine zones. Simulation methods are mainly used in non-homogeneous estuarine zones and are applicable to one- and two-dimensional situations. The great capacity and speed of modern digital computers has, however, encouraged the development of one- and two-dimensional numerical solutions.

In the great majority of models it is assumed that the discharges of pollutant mix rapidly with the surrounding environment so that the combined flow is neutrally buoyant and has little effect on the existing flow field. Mass continuity equations may then be solved independently of the flow field. This may not be accurate in the immediate vicinity of pollution outfalls. However, jet entrainment theory[2] enables concentrations at the start of the neutrally-buoyant zone to be established and these may then be used as initial conditions for subsequent diffusion–advection analysis. If density differences are significant, the mathematical model must provide for a simultaneous solution of the momentum equations of the flow field and the continuity equations of the flow field and pollutant. This approach may be neccessary for problems involving power station cooling water plants where hot water and/or concentrated brine solutions can be discharged to estuary and coastal waters at temperatures of 15 °C and concentrations of 15 ‰ above ambient values. Problems involving such large density changes are difficult to solve numerically[27], especially in coastal areas, since the problem is three-dimensional and a large amount of computer store is usually required to study the area. Hydraulic model studies may, therefore, be preferred for such problems.

Most analytical mathematical models are one-dimensional representations of the physical situation and have proved adequate for examining conditions in the homogeneous fresh water tidal zone. They have also been used in the non-homogeneous estuarine zone when conditions are only slightly stratified, that is, when surface and bed salinities differ by not more than 2–3 ‰. If, however, significant vertical stratification exists such that surface and bed salinities differ by some 10–15 ‰, a two-dimensional representation of the vertical circulation process is required, particularly if dealing with non-conservative substances. This calls for a two-dimensional model which incorporates longitudinal and vertical velocities. A one-dimensional model may, however, be used to indicate cross-sectional mean values when dealing with a conservative substance such as salt (assuming evaporation and grains from bed deposits are insignificant) but it should be remembered that surface and bed values may differ by large amounts from mean values. Wide shallow estuaries and bays are nearly always vertically homogeneous, and as such can be adequately represented by a two-dimensional model which covers lateral and longitudinal dimensions.

Particular water quality models can also be grouped into three categories based on the time scale used in the model.

1. *Steady-state models* in which the longitudinal distribution of any water quality parameter (ϕ') is assumed to be identical at corresponding points and times over successive tidal cycles. Such models are usually one-dimensional and are assumed to indicate longitudinal pollutant distributions either at times of slack water ($\partial\phi'/\partial t = 0$) or as a tidal average ($\partial\phi'_T/\partial t = 0$) (see Chapter 3). Advective motion is thus taken to be caused by the fresh water velocities or an effective velocity based on fresh water flow and tidal motion (see Chapter 3, equation 3.35 or 3.37). Models of this type are generally applied to fresh water flow zones or long estuaries with continuous fresh water inflow.

2. *Tidal-averaged time-varying models* in which water quality parameters are allowed to vary from one tidal cycle to the next ($\partial\phi'_T/\partial t \neq 0$).Conditions within a tidal cycle are not reproduced. Advective motion is again attributed to fresh water flows as in category 1. This class of model is applicable to intermittent injection of pollutants or to continuous injections which vary slowly with a time scale that is larger than a tidal cycle. Models in the present category have been used in the fresh water flow zone of long estuaries as well as in near-shore coastal zones involving inter-connected tidal channels such as occur, for example, in San Francisco Bay (USA).

3. *General time-varying models* in which water quality parameters are allowed to vary from one instant of time to the next ($\partial\phi'/\partial t \neq 0$). Conditions within a tidal cycle are therefore reproduced. Advective motion is represented as the instantaneous result of tidal and fresh water flows. Such models are essential for short estuaries or for those situations in which discharged quantities vary significantly during the course of a tidal cycle. For example, fresh water inflows can vary rapidly from flashy catchments, while the pro-

duction of dissolved oxygen by photosynthesis, or its depletion by algal growth can show strong diurnal variations.

Some examples of these models and their various solution methods are now considered. For further details the reader should consult the references.

7.1 ANALYTICAL METHODS

The majority of analytical models are based on direct solutions to diffusion–advection equations, including source and sink terms. For example, the cross-sectional average equation describing the rate of change of a neutrally-buoyant non-conservative organic pollutant in a river or estuary is given by Dobbins[1]:

$$\frac{\partial L'}{\partial t} = \frac{1}{A}\frac{\partial}{\partial x}\left(EA\frac{\partial L'}{\partial x}\right) - \frac{U\partial L'}{\partial x} - (K_1+K_3)\,L' + L'_a \qquad (7.1)$$

where, for example

L' is the first stage Biochemical Oxygen Demand (BOD) usually expressed in g/m^3;

K_1 is a rate constant for the removal of BOD by biochemical oxidation;

K_3 is a rate constant for the removal of BOD by sedimentation and adsorption of organic material on bed deposits;

L'_a is the rate of addition of BOD by local runoff or from suspended matter derived from bed deposits.

A similar parallel equation exists to describe the longitudinal distribution of dissolved oxygen (DO)[1]:

$$\frac{\partial \phi'}{\partial t} = \frac{1}{A}\frac{\partial}{\partial x}\left(EA\frac{\partial \phi'}{\partial x}\right) - \frac{U\partial \phi'}{\partial x} - K_1 L' + K_2(\phi'_s - \phi') \pm D_B \qquad (7.2)$$

where ϕ' is the concentration of dissolved oxygen, assumed non-zero and is usually expressed in mg/l;

K_2 is a rate constant allowing for addition of oxygen to the system by re-aeration through the water surface;

ϕ'_s is the saturated concentration of oxygen in water at the system temperature;

D_B is a general sink (negative sign) or source (positive sign) term. For example, it can account for oxygen loss by algal growth, or for oxygen addition by photosynthesis, or both.

Similar equations to 7.1 and 7.2 result if the individual components of the BOD level are considered. For example, five equations can be solved sequentially in order to determine the concentration distribution of organic carbon, organic nitrogen, ammoniacal nitrogen, oxidised nitrogen and dissolved oxygen. The magnitude of the coefficients in these latter equations will, of course be different to those appearing in 7.1 and 7.2.

In order to obtain analytical solutions to equations such as 7.1 and 7.2 it is nearly always necessary to assume constant spatial and temporal dispersion coefficients and advective motions, that is, models in categories 1 and 2 result. However, analytical models in category 3 have been used with constant spatial dispersion coefficients but advective motion, which is variable in time.

Analytical solutions are, therefore, best suited to homogeneous tidal zones in narrow estuaries where dispersion coefficients are relatively small (viz. Chapter 3). However, acceptable engineering answers have been produced in slightly stratified non-homogeneous flow zones, due no doubt, to the shallowness of pollutant gradients, which tend to reduce dispersive effects and therefore the importance of the dispersion coefficient. If significant stratification is present, the vertical circulation pattern must be included in any model and this invariably means that a numerical approach must be adopted.

In large bodies of water, such as the oceans, it is advisable to use analytical solutions based on variable coefficients (see equation 3.46). However, the vagaries of wind and wave action make even these solutions unreliable unless dispersion coefficients are obtained from *in situ* dye tests. Many long sea outfalls have been successfully designed on such a basis[2].

Some particular examples of analytical solutions will now be considered which illustrate models in categories 1–3; more detailed information is given in the references cited.

7.2 CATEGORY 1 MODELS

Consider the point discharge of an organic pollutant into a long, narrow, homogeneous, constant area estuary. The steady-state longitudinal distribution of BOD and DO is then obtained from equation 7.1 and 7.2 with $\partial\phi'/\partial t$ and $\partial L'/\partial t$ equal to zero. The result for the case of constant coefficients ($E = E_{xTs}$ and U_f) and in the absence of algae, photosynthesis effects, local runoff and sedimentation effects is given by O'Connor[3] as

$$L'(x) = \frac{W}{Q_f\alpha_1}\exp\left[\frac{U_f x}{2E}\cdot(1\pm\alpha_1)\right] \tag{7.3a}$$

where

$$\alpha_1 = \left[1+\frac{4K_1E}{U_f^2}\right]^{\frac{1}{2}} \tag{7.3b}$$

$$\phi'(x) = \frac{K_1 W}{(K_2-K_1)\,Q_f}\left\{\frac{\exp}{\alpha 1}\left[\frac{U_f x}{2E}\cdot(1\pm\alpha_1)\right] - \frac{\exp}{\alpha_2}\left[\frac{U_f x}{2E}(1\pm\alpha_2)\right]\right\} \tag{7.3c}$$

$$\alpha_2 = \left[1+\frac{4K_2E}{U_f^2}\right]^{\frac{1}{2}} \tag{7.3d}$$

in which W is mass of pollutant released per unit time, Q_f is the fresh water discharge (AU_f) and $\phi'(x)$ is the DO deficit ($\phi_s'-\phi'$) as a function of distance (x).

The negative signs in the exponents are taken for points downstream of the discharge point ($x > 0$), and the positive signs for upstream points ($x < 0$).

This analytical model has been applied by O'Connor[3] to the James River (USA) in order to predict the longitudinal distribution of BOD and DO resulting from two isolated sources: 22·7 tonnes/day at Richmond and 79·4 tonnes/day at Hopewell. The method showed good agreement (figure 7.1) with field observations using the typical parameters

Q_f = 82·1 m³/s, ϕ_s' at 25 °C = 8·2 g/m³, E = 375 m²/s;

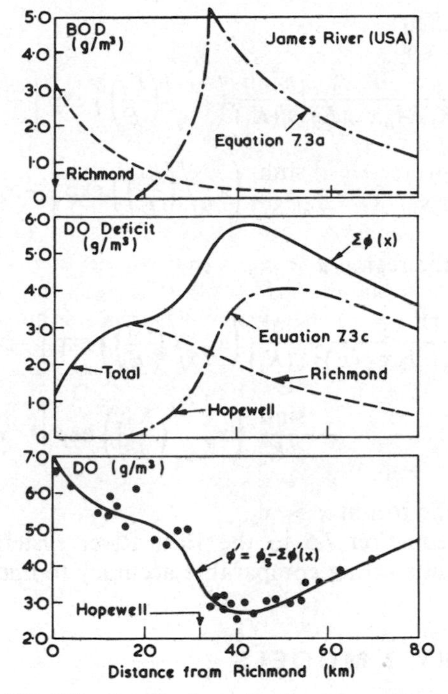

Figure 7.1. BOD–DO distributions in the James Estuary (based on figure 5 from reference[3], courtesy the American Society of Civil Engineers)

and (1) upstream of Hopewell in unidirectional flow conditions:

$$A = 1068 \text{ m}^2;\ K_1 = 0{\cdot}59 \text{ per day};\ K_2 = 0{\cdot}18 \text{ per day},$$

and (2) downstream of Hopewell in homogeneous tidal conditions:

$$A = 2508 \text{ m}^2;\ K_1 = 0{\cdot}18 \text{ per day};\ K_2 = 0{\cdot}16 \text{ per day}.$$

In some estuarine systems, the tidal prism is many orders of magnitude greater than the fresh water discharge and its advective effect can be neglected ($U_f = 0$). In other systems, the cross-sectional area shows a regular change. For example, the cross-section of the Upper East River (USA) changes according to the relationship

$$A = A_0\left(\frac{x}{x_0}\right)^2 \tag{7.4}$$

where A_0 is the area of cross-section at $x_0 (x = 0)$. Equations 7.1 and 7.2 may also be solved analytically for such cases. For example, if U_f, L'_a, K_3 and D_B are all zero, the steady-state solution of equations 7.1 and 7.2 for constant coefficients and a point discharge is given by the equation[3]:

$$L'(x) = \frac{W'x_0}{2A_0 x\sqrt{(EK_1)}}\left\{\exp\left[-\sqrt{\left(\frac{K_1}{E}\right)}\,(x-x_0)\right] - \exp\left[-\sqrt{\left(\frac{K_1}{E}\right)}(x+x_0)\right]\right\} \tag{7.5a}$$

and for the DO deficit:

$$\phi'(x) = \frac{K_1}{(K_2-K_1)} \frac{W'x_0}{A_0 x \sqrt{(E)}} \left\{ \frac{\sinh}{\sqrt{(K_1)}} \left(x \sqrt{\left(\frac{K_1}{E}\right)} \right) \exp\left[-x_0 \sqrt{\left(\frac{K_1}{E}\right)} \right] - \frac{\sinh}{\sqrt{(K_2)}} \left(x \sqrt{\left(\frac{K_2}{E}\right)} \right) \exp\left[-x_0 \sqrt{\left(\frac{K_2}{E}\right)} \right] \right\} \quad (7.5b)$$

which applies for the region $x < x_0$.

$$\phi'(x) = \frac{K_1 W' x_0}{(K_2-K_1) A_0 x \sqrt{(E)}} \left\{ \frac{\sinh}{\sqrt{(K_1)}} \left(x_0 \sqrt{\left(\frac{K_1}{E}\right)} \right) \exp\left[-x \sqrt{\left(\frac{K_1}{E}\right)} \right] - \frac{\sinh}{\sqrt{(K_2)}} \left(x_0 \sqrt{\left(\frac{K_2}{E}\right)} \right) \exp\left[-x \sqrt{\left(\frac{K_2}{E}\right)} \right] \right\} \quad (7.5c)$$

which applies for the region $x > x_0$.

Application of equation 7.5 to the East River system[3] by O'Connor produced good results with a comparable accuracy to figure 7.1.

7.3 CATEGORY 2 MODELS

Consider the instantaneous point release of a general pollutant which decays exponentially with time into a long, narrow, homogeneous tidal zone during a period of constant river flow. The concentration of the pollutant (C) is defined as the mass of the pollutant divided by the mass of the pollutant plus water, and its decay rate constant is K_0. Equation 7.1 with $L'_a = 0$, $(K_1+K_3) = K_0$ and L' replaced by C may now be used to describe the longitudinal distribution of cross-sectional average pollutant concentration at times of slack water. The solution for a constant area estuary with a constant dispersion coefficient is:

$$C(x, t) = \frac{M'}{\rho A_0 \sqrt{(4\pi E t)}} \exp\left[-\left\{ \frac{(x-U_f t)^2}{4Et} + K_0 t \right\} \right] \quad (7.6)$$

where $C(x, t)$ is the concentration at position x after time t
M' is the instantaneous mass of pollutant released at $x = 0$, $t = 0$
ρ is the density of the polluted flow
t is the time elapsed from pollutant release.

Equation 7.6 simply indicates that the pollutant spreads out as an exponentially decaying Gaussian distribution which is advected through the estuary at speed U_f per tidal cycle. Two points should be remembered, however, when using equation 7.6. First, the use of U_f as the advective velocity implies that successive tides are identical, and secondly that cross-sectional mixing is complete. Clearly, some finite time interval is required to achieve lateral, or vertical, or cross-sectional mixing of a cloud of pollutant

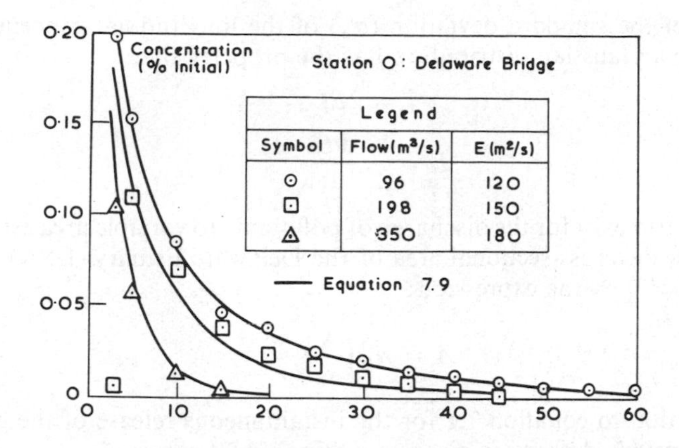

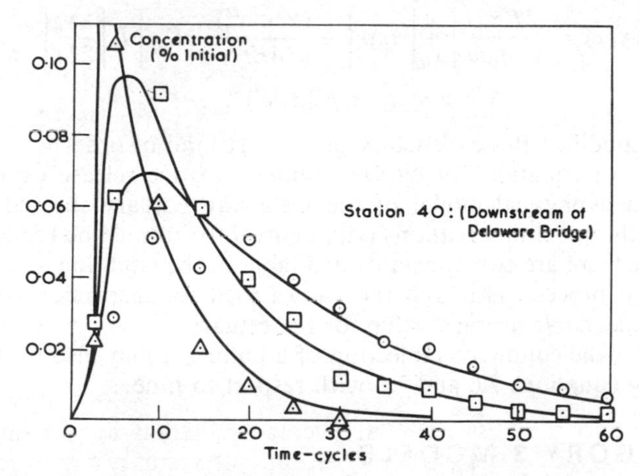

Figure 7.2. Variation of tidal-average concentrations with time for the Delaware estuary model (based on figure 2 from reference[3], courtesy the American Society of Civil Engineers)

released at the bed, or side, of a flow system. One- and two-dimensional models can only be used, therefore, after some initial mixing time. The time required to achieve near vertical or lateral homogeneity is typically the order of $0{\cdot}5T_i$ where T_i is H^2/ϵ_z or B^2/ϵ_y, and all quantities are depth- or laterally-averaged values respectively. For example, equation 7.6 could be expected to apply in a homogeneous tidal zone after 1 or 2 tidal cycles, assuming typical estuary width, depth and U_* values of 100 m, 7 m, 0·05 m/s respectively, and an ϵ_y value increased by a factor of three (see Chapter 3). Narrow tortuous flow conditions as well as shallow water depths will increase ϵ_y, ϵ_z values such that T_i is almost zero. One- and two-dimensional models may then be assumed to apply immediately after pollutant injection.

Equation 7.6 can also be used in the homogeneous tidal zone to estimate the longitudinal dispersion coefficient ($E = E_{xT}$) either directly or by writing it in terms of the standard deviation (σ_x) of the longitudinal concentration curves, since a Gaussian distribution has the property that

$$\sigma_x^2 = 2Et \tag{7.7a}$$

i.e.

$$E = \frac{1}{2} \cdot \frac{d\sigma_x^2}{dt} \tag{7.7b}$$

Solutions also exist for the discharge of pollutants to variable area estuaries. For example, the cross-sectional area of the Delaware Estuary (USA) can be approximated[3] by the expression:

$$A = A_0 \left(\frac{x}{x_0}\right) \tag{7.8}$$

and the solution to equation 7.1 for the instantaneous release of the general pollutant considered above is given by O'Connor[3] as:

$$C(x, t) = \frac{M' x_0}{2EA_0 \rho t} \left[\frac{x}{x_0}\right]^{\nu} \exp\left[-\frac{(x^2 + x_0^2)}{4Et} - K_0 t\right] I_\nu \left[\frac{x x_0}{2Et}\right] \tag{7.9a}$$

$$\nu = Q_f x_0 / (2A_0 E) \tag{7.9b}$$

where I_ν is a modified Bessel function of the first kind of order ν.

Application of equation 7.9 by O'Connor[3] to dye release experiments carried out in a physical model of the Delaware Estuary showed a good agreement with model observations (viz. figure 7.2): it must be remembered, however, that there are two constants available in the equation to help with the calibration process, although the values used for dispersion coefficient represent a reasonable average value for the estuary.

Solutions for the continuous injection of a pollutant may also be obtained by integrating equations 7.6 and 7.9 with respect to time.

7.4 CATEGORY 3 MODELS

Consider the instantaneous release of the same general pollutant into a long narrow homogeneous estuary in which the maximum tidal velocity is spatially constant; such conditions are approximated in the Delaware Estuary (USA). In the absence of local fresh water inflows and for identical successive tides, the instantaneous cross-sectional average velocity can be written as:

$$U = U_f + U_m \sin \sigma(t - \delta_0) \tag{7.10}$$

where U_m is a maximum tidal velocity and is constant spatially; δ_0 is the time phase shift between $t = 0$ and slack water; $\sigma = 2\pi/T$.

The solution of equation 7.1 for the instantaneous point release of a general pollutant which has a decay rate constant K_0 and which is discharged into the homogeneous tidal zone ($E = E_x$ assumed constant over space and time) is given by Holley and Harleman[4] as:

$$C(x, t) = \frac{M'}{\rho A_0 \sqrt{(4\pi Et)}} \exp\left\{\frac{-[x - U_f t + U_m/\sigma(\cos \sigma(t - \delta_0) - \cos(\sigma\delta_0))]^2}{4Et} - K_0 t\right\} \tag{7.11}$$

The solution for a continuous injection follows by integration of equation 7.11 with respect to time and has been found by Holley[5] to give essentially the same answers as a simulation model based on the mixed-segment approach[6] (see section 7.5).

Solutions to the modified form of equation 7.1 also exist for exponentially shaped estuaries and have been found by Zargar[7] to give better agreement with physical model tests than the slack-time theory of equation 7.9.

7.5 NUMERICAL TECHNIQUES

The one- and two-dimensional mass continuity equations can be solved numerically using both explicit and implicit finite difference, or Galerkin finite element techniques. Implicit methods are best for most problems, but explicit methods[8, 14] have been used in some two-dimensional problems, as also has the Galerkin approach[9, 10].

The difficulty with numerical solutions to the diffusion–advection equation is that the numerical scheme itself may cause the pollutant to experience more advection and dispersion than indicated by the physical parameters of a particular problem. This effect is often described by the terms *pseudo-advection* or *pseudo-dispersion*. For example, consider the simple first-order accurate (Δt, Δx) explicit formulation of the simple advection equation, i.e.

$$\frac{\partial C}{\partial t} + U\frac{\partial C}{\partial x} = 0. \tag{7.12}$$

The concentration at point 1,1 in the space–time plane of figure 6.13 is written in terms of $C_{0,0}, C_{1,0}$ using similar equations to 6.25 and 6.26. The result is:

$$C_{1,1} = C_{1,0} - \frac{U\Delta t}{\Delta x}(C_{1,0} - C_{0,0}) + \frac{\Delta t^2}{2}\left(U\frac{\partial U}{\partial x} - \frac{\partial U}{\partial t}\right)\frac{\partial C}{\partial x}$$
$$+ \frac{U\Delta t}{2}(U\Delta t - \Delta x)\frac{\partial^2 C}{\partial x^2} + O(\Delta t^3, \Delta x^2 . \Delta t) \tag{7.13}$$

where U, C and their derivatives are evaluated at point 1,0.

If U is constant in space and time and is equal to $\Delta x/\Delta t$, equation 7.13 reduces to the simple form

$$C_{1,1} = C_{0,0} + O(\Delta t^3, \Delta x^2 . \Delta t) \tag{7.14}$$

which is correct as Δx, $\Delta t \to 0$. However, if U is a function of space and time equation 7.13 shows that the numerical solution causes extra advection ($\partial C/\partial x$) and dispersion ($\partial^2 C/\partial x^2$) of the pollutant. Pseudo-advective and -dispersive terms are also contained in the $O(\Delta t^3, \Delta x^2 . \Delta t)$ term. The pseudo terms will clearly be least for small space and time steps, and for small variations in U over space and time and for small spatial concentration gradients.

If the explicit scheme had included point 0, 2, the pseudo-dispersion error would have been reduced, since the term $U\Delta x\Delta t/2$ would not be present in equation 7.13. It is therefore essential to have differences centred correctly in space and time about the point of application of the differential equation. For example, an optimised implicit finite-difference scheme for the solution of equation 7.1 with K_1, K_3 and L'_a set to zero, L' replaced by C, and with U, E and A constant over space and time is:

$$\frac{1}{6}\left[\frac{C_{0,1}-C_{0,0}}{\Delta t}\right]+\frac{2}{3}\left[\frac{C_{1,1}-C_{1,0}}{\Delta t}\right]+\frac{1}{6}\left[\frac{C_{2,1}-C_{2,0}}{\Delta t}\right]$$

$$+\frac{U}{4\Delta x}[C_{2,1}-C_{0,1}+C_{2,0}-C_{0,0}]$$

$$=\frac{E}{2\Delta x^2}[C_{2,1}-2C_{1,1}+C_{0,1}+C_{2,0}-2C_{1,0}+C_{0,0}]+O(\Delta t^2, \Delta x^2) \quad (7.15)$$

which follows from the use of similar equations to 6.25 and 6.26 and figure 6.13 with the time derivative weighted over adjacent space points. This implicit scheme uses a Crank–Nicolson difference approximation (derivatives weighted equally at present (0) and future time levels (1)) in the space domain, while the time derivatives are spread spatially with weighting coefficients of $\frac{1}{6}$, $\frac{2}{3}$ and $\frac{1}{6}$. The terms in equation 7.15 are thus properly centred at $x = 1$, $t = \frac{1}{2}$ (figure 6.13) and their combination ensures that the low-frequency Fourier components of the solution are advected at the mean flow speed for $U\Delta t/\Delta x \ll 1$[11]. Ideally, all Fourier components of the solution should be advected at the mean flow speed. The high-frequency components are thus not correctly represented. Numerical errors are acceptable, however, if E has a large value as in the non-homogeneous part of an estuary, since the dispersion term in the differential equation has precisely the effect of damping out high-frequency components. Accuracy may also be improved by a cyclic use of similar equations to 7.15 with slightly different numerical coefficients[11].

Finite-difference schemes for advection–diffusion equations contain the same types of error as for the water flow models discussed earlier (see Chapter 6, Section 9). Numerical diffusion and conservation errors can be removed by the use of suitably centred finite-difference schemes and by integrating the diffusion–advection equation over space and time prior to use of numerical techniques[12]. However, all schemes suffer to a greater or lesser degree from dissipation errors due to large concentration gradients. The errors have been shown[12] to be acceptable for high order accuracy finite-difference schemes (Δt^2, Δx^4 see figure 7.3), particularly if dispersion coefficients are large. It is worth remembering that solution techniques based on the Galerkin finite-element approach appear to offer the advantage of such high order accuracy as well as the opportunity of easier model schematisation[9]. This capability is a distinct advantage in two-dimensional problems. It should also be noted that the Galerkin approach gives precisely the same result as equation 7.15 when used with linear shape functions (these are linearly dependent upon the spatial co-ordinates), constant spatial intervals and flow and dispersion parameters and a Crank–Nicolson finite-difference time schematisation.

The numerical approach can also be applied to diffusion–advection equations to produce models in categories 1–3. It is best used, however, in combination with the one- and two-dimensional time-varying water flow models mentioned in Chapter 6. Such a combined model has been found to be

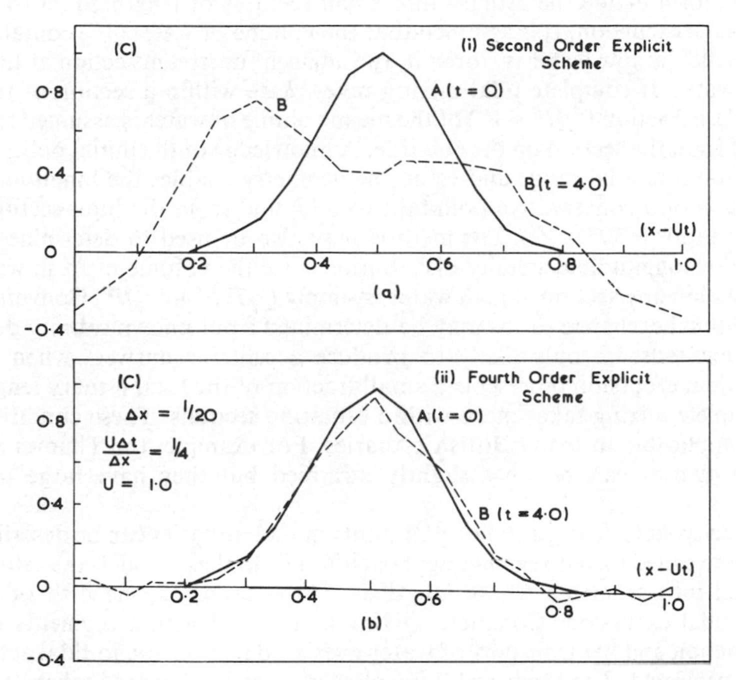

Figure 7.3. The effects of dispersion errors as shown by the advection of a Gaussian profile. (Reprinted with the permission of the Publisher of American Mathematical Society from *Mathematics of Computation*, copyright © 1966, volume xx, no. 94, p. 285)

extremely useful in describing the longitudinal distribution of salinity in the Wear Estuary (UK)[13] as well as the plan distribution of pollutants in Jamaica Bay (USA) due to the combined effects of wind, tide and Coriolis force[14].

Two-dimensional numerical models are also essential in stratified estuaries and have been used with success by Hobbs and Fawcett[8] and Farraday *et al.*[15] where details of the flow field were provided by field observations. However, recent (1976) research[23] enables the flow field to be determined by the solution of the relevant momentum and turbulent kinetic energy equations.

7.6 SIMULATION METHODS

These represent the traditional form of solution for water quality problems, but are slowly being replaced by methods based on the numerical solution to the equations of fluid flow and/or mass continuity.

Early models attempted to predict the steady-state distribution of pollutants in an estuary by replacing diffusive and advective processes by a single mixing operation[16]. The simplest models of this type are one-dimensional and in their best form divide the estuary into small sections of length equal to the average tidal excursion. It is assumed that the volume of water (V') contained in any section at low water is stored in the adjacent upstream section at times of high water. If complete tidal mixing takes place within a section of tidal prism (P'), a fraction $P'/(P'+V')$ of the mean volume of water is assumed to be removed from the section on the ebb tide. A knowledge of the initial pollutant distribution, the tidal range and estuarine geometry enables the longitudinal distribution of a conservative pollutant to be found from the inter-sectional exchange ratios $P'/(P'+V')$. The method may also be used to determine the steady-state longitudinal salinity distribution since the volume of fresh water present within any section at high water is simply $Q_f T(P'+V')/P'$. Conversely inter-sectional exchange ratios may be determined from known salinity data.

Such methods are only likely to produce acceptable answers when the average tidal excursion represents a small fraction of the total estuary length, and complete mixing takes place within estuarine sections. These conditions are not applicable to many British estuaries. For example, the Thames and Severn estuaries can be only slightly stratified but they have large tidal excursions.

The steady-state distribution of pollutants in real estuaries can be described using the so called mixed-segment approach [6, 17]. In these models the estuary is divided into segments whose length is of the order of one fifth of the average tidal excursion. Complete mixing is assumed within segments due to tidal action and net transport of water over a tidal cycle due to tidal action alone is neglected. The landward limit of such models is usually taken to be the tidal limit above which the flow is unidirectional and has zero salinity. The seaward limit is taken at a point where the tidal averaged salinity remains almost constant at full ocean value (34 ‰). The volume downstream of the seaward boundary is regarded as an infinite reservoir of salt water at full ocean salinity. Model water level may be set at any value but is usually taken at mean tide level. Inter-tidal mixing is simulated by exchanging equal volumes of water between segments. Fresh water flows to the estuary are assumed to be steady and are allowed to flow into the appropriate model segment and through downstream segments. The spreading effect of the tidal velocities on a pollutant discharging from a fixed outfall is simulated by proportioning the discharged volume into upstream and downstream segments. Pollutant generation and decay may also be allowed within segments. The tidal model is illustrated schematically in figure 7.4 for two segments immediately downstream of the tidal limit.

The exchange volumes (F_1, F_2, etc.) are determined by fitting the model to a known longitudinal salinity distribution. For example, a salt balance equation can be written for the two segments shown in figure 7.4, i.e.

$$F_1 S_2 = (Q+Q_1+F_1)\, S_1 \tag{7.16a}$$

$$(Q+Q_1+F_1)\, S_1 + F_2 S_3 = (Q+Q_1+Q_2+F_1+F_2)\, S_2 \tag{7.16b}$$

from which F_1 and F_2 can be determined.

The distribution of a pollutant is found by writing similar mass continuity

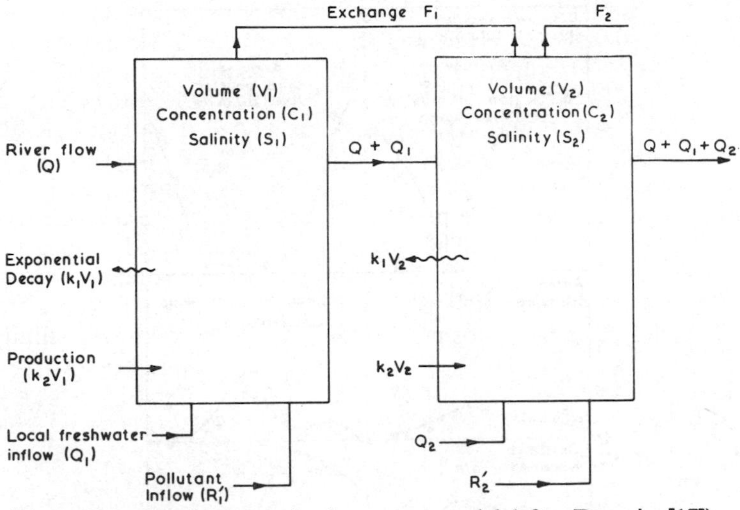

Figure 7.4. Steady-state mixed-segment model (after Downing[17])

equations for the pollutant, i.e.

$$F_1C_2+R_1'+Q_1Co_1+QCo_0+k_2V_1C_1-k_1V_1C_1-(Q+Q_1+F_1)\,C_1 = 0 \quad (7.17a)$$

$$F_2C_3+R_2'+(Q+Q_1+F_1)\,C_1+k_2V_2C_2-k_1V_2C_2+Q_2Co_2 -(Q+Q_1+Q_2+F_2)\,C_2 = 0 \quad (7.17b)$$

where Co, Co_1, Co_2 and k_1, k_2 represent concentrations of the pollutant in the fresh water sources, and reaction rate coefficients, respectively. The final concentrations for any segment are determined by a simultaneous solution of equations similar to equation 7.17 which arise from an application of mass continuity equations to the other segments.

Such mixed-segment models produce acceptable engineering solutions to pollution problems mainly through the fitting process for the exchange coefficients. However, the model has a reasonable physical basis since the method is an approximation (see equation 3.39 and 7.16*a*) to the use of mass continuity equations involving dispersion coefficients. The methods tend to be equivalent as the segment length tends to zero. The model would not, however, be expected to reproduce pollutant distributions involving very steep concentration gradients because of the large step length involved and the inherent smoothing effect of a solution produced from a set of simultaneous equations. Fortunately, the distribution of pollutant constituents in many estuaries and rivers show shallow gradients. The usefulness of the method is clearly demonstrated by figure 7.5 which shows the results of applying the model to the Thames Estuary[17].

Alternative steady-state models based on the simulation of dispersive processes by statistical redistribution functions[18, 19] have also been used (see Gilligan[20] for a good summary). They have been found[17], however, to give a very similar numerical result to the mixed-segment approach, which is a much less involved method.

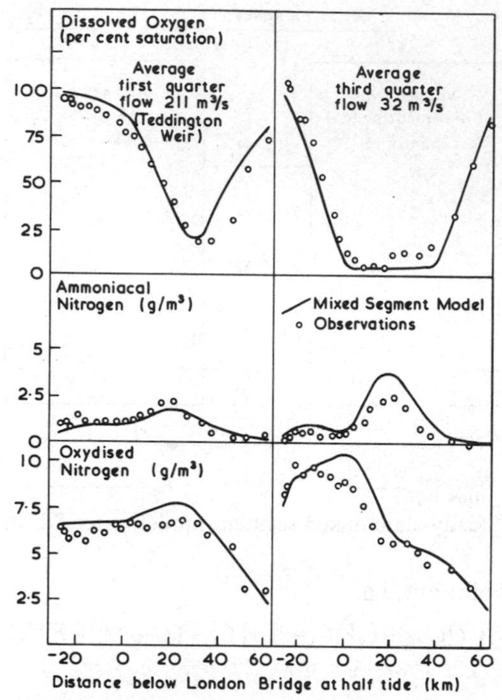

Figure 7.5. Comparison of field and mixed-segment model results for the Thames Estuary (based on figure 2, ref. 17, courtesy of the Controller of Her Majesty's Stationery Office)

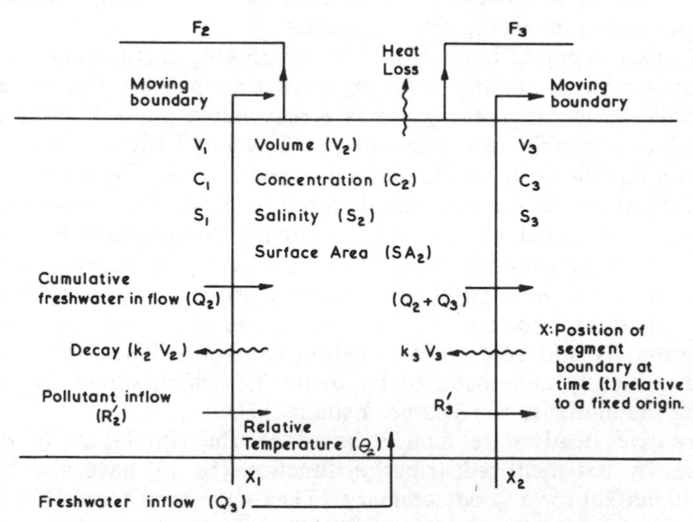

Figure 7.6. Time-varying mixed segment model (after Barrett and Mollowney[21])

Recently, a modification to the mixed-segment approach has been suggested by Barrett and Mollowney[21] which enables the tidal variation of a pollutant to be described (Category 3 model). The estuary is again divided into a number of uniformly mixed segments, the boundaries of which are assumed to move along the estuary with the tidal currents. A seaward flow is allowed through the segment boundaries and is made equal to the upstream cumulative fresh water flow. The volume of water upstream of any boundary is therefore assumed to be constant. Longitudinal dispersion is again simulated by allowing equal and opposite exchange flows between segments (see figure 7.6).

The equation for the time rate of change in mass of a pollutant with an exponential rate of decay and no production is:

$$V_2 \frac{dC_2}{dt} = Q_2 C_1 + Q_3 Co_3 + F_3(C_3 - C_2) + F_2(C_1 - C_2) + R_2' - k_2 V_2 C_2 - (Q_2 + Q_3)\, C_2 \quad (7.18)$$

The exchange coefficients are found by assuming a balance between the salt transported between segments by the fresh water velocities and the exchange flow. For example

$$F_3 = \frac{(Q_1 + Q_2)\, S_2}{(S_3 - S_2)} \quad (7.19)$$

Equations similar to equation 7.18 and 7.19 can be written for each segment and solved simultaneously to give the instantaneous longitudinal distribution of the pollutant. A convenient method used by Barrett and Mollowney[21] is to express the equations in finite-difference form using the Crank–Nicolson approximation. For example the concentration in segment two ($C_{2,1}$) after one time step ($t = \Delta t$) is related to the initial concentration ($C_{2,0}$) at time ($t = 0$) by the equation (see equation 7.18).

$$\begin{aligned} V_2[C_{2,1} - C_{2,0}] = {} & \tfrac{1}{2}\Delta t\{Q_2 C_{1,1} + Q_3 Co_{3,1} + F_3(C_{3,1} - C_{2,1}) + F_2(C_{1,1} - C_{2,1}) \\ & - k_2 V_2 C_{2,1} - (Q_2 + Q_3)\, C_{2,1}\} \qquad (7.20a) \\ & + \tfrac{1}{2}\Delta t\{Q_2 C_{1,0} + Q_3 Co_{3,0} + F_3(C_{3,0} - C_{2,0}) + F_2(C_{1,0} - C_{2,0} \\ & - k_2 V_2 C_{2,0} - (Q_2 + Q_3)\, C_{2,0}\} \\ & + R_2' \qquad (7.20b) \end{aligned}$$

where $C_{1,1}$ represents the pollutant concentration at time $t = \Delta t$ in segment 1
$Co_{3,1}$ represents the pollutant concentration at time $t = \Delta t$ in inflow Q_3
Q_2, F_2 represent average values over the time interval Δt
R_2' represents the total influx of pollutant into segment 2 in time Δt

Rearrangement of equation 7.20 and those arising from the other segments produced a set of linear simultaneous equations which can be solved by standard techniques to produce concentration values at successive time intervals provided initial concentration values are known. If the fresh water inflows, tidal range, etc. are constant, then a quasi-steady state will be reached after several tidal cycles of model operation and quite arbitrary initial values may be used.

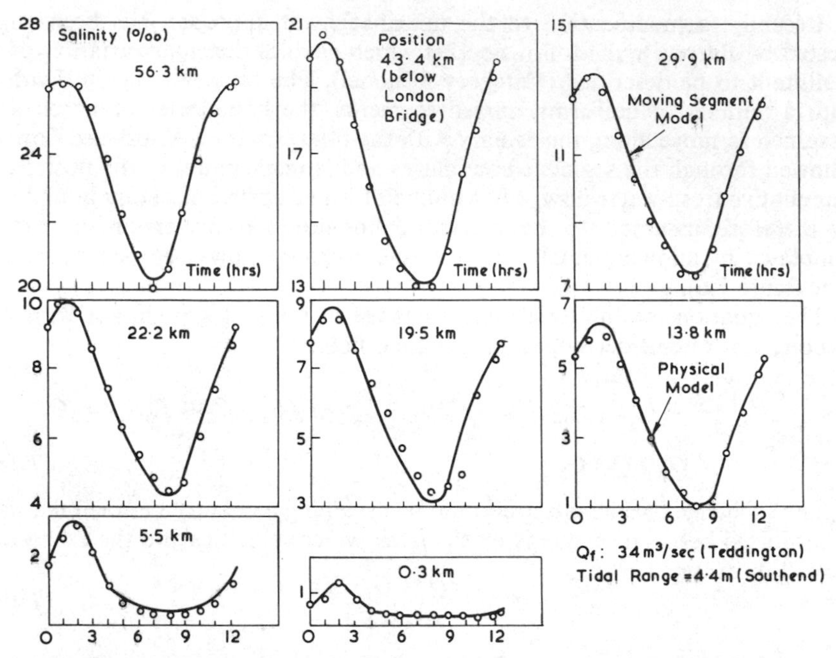

Figure 7.7. Comparison of moving-segment model and a physical model of the Thames Estuary (based on figure 5 from reference[21], courtesy the Royal Society and the Controller of Her Majesty's Stationery Office)

This model would be expected to give acceptable engineering results since the model is almost equivalent to that based on the advection–dispersion equation. Tidal advection is directly reproduced by moving the segment boundaries while the exchange coefficient method is equivalent to the use of dispersion coefficients. The model does, however, assume that the salinity of the volume of water upstream of segment boundaries is constant, i.e. a balance exists at all times between upstream dispersive flux and downstream net fresh water advective flux. This is not so in a dynamic model since dispersion exceeds advection at low water slack, and vice versa near high water slack. However, the tidal average salinity of an upstream volume is constant. Field observations in estuaries such as the Thames in the UK suggest that the tidal variation in salt content of moving upstream volumes is a small percentage of the tidal average value. Errors introduced from this source as well as those resulting from the assumption of complete inter-segment mixing are minimised by the fitting process used to determine the exchange coefficients. Application of the model to other estuaries may involve larger errors in terms of inter-segment salt transfer, but again the fitting process should ensure acceptable engineering answers. It must be remembered that answers from any model should only be accepted if the model has been shown to be capable of reproducing existing prototype observations.

A measure of the closeness of fit of this mathematical model to physical model results is provided by figure 7.7, which shows the spatial and temporal distribution of salinity for the Thames Estuary[21]. The exchange coefficients obtained from this fitting process were then used to determine the tidal variation of dissolved oxygen, ammoniacal nitrogen, oxidised nitrogen and temperature rise. If temperature changes are to be studied, it is necessary to account for the loss of heat through the water surface. This requires the addition of the term $k_3SA_2\theta_2$ to the right hand side of equation 7.18 where k_3 is a heat-exchange coefficient and has a probable value of 0·037 m/h in the case of the Thames Estuary (UK), and θ_2 is the water temperature relative to atmospheric values.

Previous simulation models have been applied to well mixed or slightly stratified estuaries. They could also be used for fully stratified situations provided the model geometry etc. was modified to that of the appropriate upper or lower flow layer. In partially-mixed estuaries it is essential that any model includes the major spreading effect of the internal density circulation as well as the vertical variation of the pollutant. This can be done at present in steady-state simulation models[20, 22], by dividing the estuary into longitudinal and vertical (usually two in number) segments. Exchange and residual flows may then be allowed in both horizontal and vertical directions.

7.7 SUMMARY

Water quality models have proved invaluable for predicting the effects of polluting discharges on coastal, estuarial and river environments and usually fall into three categories: 1, steady-state models; 2, single or multiple tidal-average time-varying models; 3, general time-varying models. Such models are based on analytical, numerical or simulation type solutions to appropriate mass continuity equations. However, problems involving pollutants which modify environmental flows must either be solved by a simultaneous solution of momentum and mass continuity equations or by the use of physical modelling techniques.

Traditional mathematical models use analytical and simulation techniques for models in categories 1 and 2 but the provision of improved computing facilities has led to the development of numerical models in category 3 based on finite difference or finite element techniques. Care is required, however, when using numerical techniques to avoid pseudo-dispersion or -advection errors.

Models are usually limited, due to cost and computer storage restrictions, to one or two space dimensions, but such approximations are found to be adequate for many problems since tidal rivers and estuaries are often cross-sectionally well mixed, or narrow and laterally well mixed but with some degree of vertical stratification, or vertically well mixed. Model accuracy is usually better than in a physical model since problems of scaling pollutant generation and decay rates and turbulence characteristics are avoided.

NOTATION

k decay rate constant for a pollutant
t time co-ordinate
x Longitudinal distance co-ordinate

x_0 Origin of x, i.e. $x = 0$
A Flow cross-sectional area
A_0 Flow cross-sectional area at $x = x_0$
C Cross-sectional average pollutant concentration
Co Pollutant concentration in fresh water inflows
$C(x, t)$ Cross-sectional average pollutant concentration at position (x) and time (t)
D_B General generation or decay rate constant for dissolved oxygen
E One-dimensional dispersion coefficient
F Exchange flow for a box model
I_ν Modified Bessel function of first kind and order ν
K_0 General decay constant for a pollutant
K_1 Decay rate constant to allow for removal of BOD by biochemical oxidation
K_2 Time rate constant to allow for the addition of oxygen by aeration through the water surface
K_3 Decay rate constant to allow for BOD loss by sedimentation and adsorption by bed deposits
L' Biochemical oxygen demand (BOD)
L'_a Rate of addition of BOD by surface run-off or from re-suspended bed deposits
$L'(x)$ Value of BOD at position x
M' Mass of pollutant added to a flow
$O(\ )$ Error terms contained in a finite-difference equation
P' Tidal prism
Q_f Cross-sectional fresh water discharge rate
Q Local discharge rate of fresh water source
R' Local pollutant discharge rate
S Cross-sectional salinity value
SA Surface area of a box model segment
T Tidal period
T_i Time scale for vertical or cross-sectional mixing
U Cross-sectional flow velocity
U_f Cross-sectional average fresh water velocity
U_m Maximum tidal cross-sectional velocity
V' Volume of a section of an estuary
W' Weight of pollutant added to a flow
$\alpha_1; \alpha_2$ Numerical constants defined by equations 7.3(b), 7.3(d)
δ_0 Time phase shift
θ Water temperature relative to atmospheric temperature
ν Numerical constant defined by equation 7.9(b)
ρ Fluid density
σ Tidal wave frequency ($= 2\pi/T$)
ϕ' Any general water quality parameter
ϕ'_s Saturation value of dissolved oxygen in water
$\phi'(x)$ Cross-sectional dissolved oxygen deficit ($\phi'_s - \phi'$) as a function of x
Δx, Δt Space and time intervals

REFERENCES

[1] Dobbins, W. E. Diffusion and Mixing, *J. Boston Soc. Civil Eng.*, **52**, Nov. (1965)

[2] Baumartner, D. J. and Trent, D. S., Ocean Outfall Design, (Pt. 1), *FWQA Report*, Corvallis, Oregon, April (1973)

[3] O'Connor, D. J., Estuarine Distribution of Non-Conservative Substances, *J. San. Eng. Div. ASCE*, **91**, SA1, Feb. (1965)

[4] Holley, E. R. and Harleman, D. R. F., Dispersion of Pollutants in Estuary-Type Flows, *Hyd. Lab. Report No. 74*, MIT, Cambridge, Mass., Jan. (1965)

[5] Holley, E. R., Discussion of Ref. 6, *J. San. Eng. Div. ASCE*, **95**, SA5, Oct. (1969)

[6] Bella, D. A. and Dobbins, W. E., Difference Modelling of Stream Pollution, *J. San. Eng. Div. ASCE*, **94**, SA5, Oct. (1968)

[7] Zargar, D., Dispersion of Pollutants in a Variable Area Estuary, Master of Science Thesis, MIT Cambridge, Mass., (1966)

[8] Hobbs, C. and Fawcett, A., *Symposium on Mathematical and Hydraulic Modelling of Estuarine Pollution*, WPRL (UK), Stevenage, Herts. April (1972)

[9] Smith, I. M., Farraday, R. V. and O'Connor, B. A., 'Rayleigh-Ritz and Galerkin Finite Elements for Diffusion–Convection Problems', *Water Research*, **9**, 3, June (1973)

[10] Taylor, C. and Davis, J., Finite Element Numerical Modelling of Flow and Dispersion in Estuaries, *IAHR Int. Sym. on River Mechanics*, Paper C 39, Jan. (1973)

[11] Stone, H. P. and Brian, P. L. T., Numerical Solution of Convective Transport Problems, *A.I.Ch.E. Journal*, **9**, 94 (1963)

[12] Roberts, K. V. and Weiss, N. O., Convective difference Schemes, Mathematics of Computation, **XX**, 94, April (1966)

[13] O'Connor, B. A. and Thompson, G., A mathematical Model of Chloride Levels in the Wear Estuary, *Int. Sym. Unsteady flow in Open Channels*, Univ. Newcastle, April (1976)

[14] Leendertse, J. J., A Water Quality Simulation Model for well-mixed estuaries and Coastal Seas, **1**, *Memo RM-6230-RC*, The Rand Corporation, Feb. (1970)

[15] Farraday, R. V., O'Connor, B. A. and Smith, I. M., A two-dimensional finite-element model for Partially-Mixed Estuaries, *Proc. 16th Cong. IAHR*, Paper C 35, **3**, São Paulo, Brazil (1975)

[16] Ketchum, B. H., The Exchanges of Fresh and Salt Water in Tidal Estuaries, *Proc. Col. on Flushing of Estuaries*, Woods Hole Ocean. Inst., Mass. (1950)

[17] Downing, A. L., Forecasting the effects of Polluting Discharges on Natural Waters, II, *Int. J. Environmental Studies*, **2**, Part II (1971)

[18] Preddy, W. S., The Mixing and Movement of Water in the Estuary of the Thames, *J. Mar. Biol. Ass. U.K.*, **33** (1954)

[19] Di Toro, D. M., Maximum entropy mixing in estuaries, *J. Hyd. Div. ASCE*, **95**, HY4, July (1969)

[20] Gilligan, R. M., Forecasting and effects of polluting discharge on estuaries, Pt. II, *Chem. and Ind.*, Dec. (1972)
[21] Barrett, M. J. and Mollowney, B. M., Pollution problems in relation to the Thames Barrier, *Phil. Trans. Roy. Soc., London.* A, **272** (1972)
[22] Pritchard, D. W., Dispersion and flushing of pollutants in estuaries, *J. Hyd. Div. ASCE*, **95**, HY1 (1969)
[23] Smith, T. J., A mathematical model for partially-mixed estuaries, Ph.D. Thesis, Manchester University, October (1976)
[24] Custer, S. K. and Kruthkoff, R. G., Stochastic model for BOD and DO in estuaries, *J. San. Eng. Div., ASCE*, **95**, SA5, Oct. (1969)
[25] Price, R. K., Dalrymple, R. A. and Dean, R. G., Recirculation in shallow bays and rivers, *Proc. 12th Coastal Eng. Conf.*, **3**, pt. 4, ch. 110, Washington, Sept. (1970)
[26] Mackay, D. W. and Fleming, G., Correlation of dissolved oxygen levels, freshwater flows and temperatures in a polluted estuary. *Water Research*, **3** (1969)
[27] Wada, A., Numerical analysis of the distribution of flow and thermal diffusion caused by outfall of cooling water, *Proc. 13th Congress IAHR*, **3**, Paper C 35, Kyoto, Sept. (1969)

8

The study of tidal systems: hydraulic models

The mathematical and electrical models that were described in the last two chapters can be set up and operated quickly and cheaply, but they represent an estuary as a connected system of elements, each of which has physical characteristics that are either averaged over its length or are lumped together at its interfaces with adjacent elements. Major changes of cross-section, reclamation or tidal capacity can be reproduced by changing the value of the parameters describing each element; but changes in flow resistance, bed form and sediment transport in a three-dimensional situation cannot be forecast adequately. Moreover, the number of finite sized elements is limited by the size of available computer. As a result, the possible accuracy of prediction is quite low. Although it is theoretically possible to reproduce behaviour of flow in three dimensions, the practical difficulties of doing so with sufficient accuracy are considerable.

Physical models have the advantage that they have continuous boundaries, the shape of which can be varied at will. Moreover, the tidal input to the model and river flows can be made to vary continuously with time in any desired manner. Flow is three-dimensional and can be examined in detail.

This was a particular virtue of the earliest models, which represented a complete estuary to a very small scale. The whole estuary was reproduced on a table top and water was made to rise and fall at the seaward end by means of a plunger operated by cams. The water flowed into and out of the model due to the force of gravity just as it did in the prototype, and the changing flow patterns could be observed at will.

Later models become more sophisticated. Osborne Reynolds[1, 3] had shown that it was necessary to use a vertical scale big enough to ensure turbulent flow during the greater part of the tidal cycle. In the first models, the horizontal scale (x_s), and vertical scale (z_s), were very different; in a model of the Mersey[2, 3] they were 1:10560 and 1:396 respectively. The high distortion of scale made it impossible to represent the movement of sand except in the crudest manner because side-slopes in the model were much steeper than in the prototype. Clearly any distortion of scale falsifies side-slopes, but in models in which sediment movement was to be studied, it become accepted that a relatively small distortion, say than $x_s/z_s = 1/10$, could give useful results. This inevitably meant larger horizontal scales and larger and more sophisticated models. Improvements in measuring instruments and photographic recording of current patterns made visible by means of floats have offset, to some degree, the unwieldy size of the larger models.

The enthusiasm of early model engineers led them to attempt to reproduce all aspects of estuarine behaviour. At various times, models have been built to include the effects of wind, waves, Coriolis force, thermal effects, floculation of fine sediment, bank erosion, and transport of sand and silt[3, 4]. Models nearly always gave unexpected insights into the behaviour of estuaries. Unfortunately the laws of model scaling were not clearly understood and some of the effects that were represented gave false results.

A comparatively recent development has been the reproduction of density gradients due to salinity. It has been shown that models cannot be made to reproduce the three-dimensional detail of currents in parts of estuaries affected by saline gradients unless density variations are incorporated in the model, as shown in figure 1.14. In practice, this necessitates maintaining the density of the sea at the seaward end by addition of dissolved salt and using fresh water at the landward end. The complete estuary must therefore be reproduced, usually to a large horizontal scale ($> 1:1000$). The cost of building and operating such a sophisticated model is very high. The consumption of salt alone may be several tonnes a week.

It is often necessary to study a short length of an estuary in some detail, for example when there are local problems of navigation, or flow disturbance by construction of jetties. This can be done by using a tidal generator at each end of the reach in a manner that reproduces both the correct rise and fall of level, and the ebb and flow of current. It is usual to monitor level and flow at a point within the model in order to compare model performance with corresponding field measurements. Conditions in the model are then corrected by an automatic servo mechanism; but the time delay between detection of an error and its correction presents problems of stability[5]. It is not possible to reproduce variations of salinity under these conditions and care must be taken in interpreting results if density effects are appreciable in the estuary.

The laws of scaling of hydraulic models can be determined either by dimensional analysis as described in many textbooks on hydromechanics, or from the equations of motion (Yalin[6], Birkhoff[7] and Langhaar[8]).

In this chapter, the equations of motion that were developed in Chapter 2 will be used. They have the advantage over dimensional analysis that they give a physical insight into the processes that must be reproduced. They can also be used to determine the order of magnitude of the variables in such a way that their significance can be established and the consequences of their omission assessed.

8.1 SIMILARITY OF BEHAVIOUR IN ESTUARY MODELS: ONE-DIMENSIONAL SYSTEM

Physical models have channels that conform in some way to the geometry of the real channels that they represent. They are not geometrically similar; a different scale is generally used for the vertical and horizontal dimensions,

Figure 8.1. Part of model of Rhine Delta, looking upstream. Copperbars are used to increase the hydraulic roughness (courtesy Delft Hydraulics Laboratory)

and details of surface roughness and bed forms are not reproduced to scale; in fact, friction would not be reproduced correctly if they were, because roughness Reynolds numbers would be too small in the model. Dissimilar material is used on the bed and friction may have to be further corrected in the model by provision of roughness elements (figure 8.1). One of the main objectives in scaling a model is to establish a satisfactory relationship between the geometry of the real estuary and the model.

Similarity of flow can be achieved if variations of local velocity with time in the model are simply related to velocities at corresponding points and times in the prototype. If flow conditions can be reproduced correctly at a given instant, similarity of flow will occur at all subsequent times if accelerations are reproduced correctly. This will only happen if accelerations at all corresponding points and times bear the same model-to-prototype ratio.

Accelerations are the consequence of force acting on mass and are described by the dynamic equations. When applied at a point in an estuary, these equations relate forces and the resultant accelerations at any instant, including local fluctuations due to turbulence. For strict similarity of behaviour, turbulence should be reproduced in all its detail. This is impossible because of the random nature of turbulent fluctuations. For the sake of simplicity in deriving scale ratios it is assumed, at first, that the temporally smoothed and spatially averaged equations can be used to describe the flow at any cross-section and that friction can be represented as concentrated at the bed. It is also assumed that the liquid is homogeneous, with no significant variations in density; and that the effects of vertical accelerations and rotation of the earth can be ignored. Equation 2.20 will then apply, with the density term omitted.

Assume that the x-co-ordinate lies along the axis of the curved channel and that

$$\frac{\partial U}{\partial t} + U\frac{\partial U}{\partial x} + g\frac{\partial z_1}{\partial x} - \frac{\tau_{z_0 x}}{\rho H} = 0 \tag{8.1}$$

This equation can be applied at any cross-section, or time, within an estuary, and at corresponding points in a model.

Similarity of behaviour will occur if the model:prototype ratio of each term in equation 8.1 is the same at all corresponding cross-sections and times. This will be the case if

$$\frac{\left(\frac{\partial U}{\partial t}\right)_{\rm m}}{\left(\frac{\partial U}{\partial t}\right)_{\rm p}} = \frac{\left(U\frac{\partial U}{\partial x}\right)_{\rm m}}{\left(U\frac{\partial U}{\partial x}\right)_{\rm p}} = \frac{\left(\frac{\partial z_1}{\partial x}\right)_{\rm m}}{\left(\frac{\partial z_1}{\partial x}\right)_{\rm p}} = \frac{\left(\frac{\tau_{z_0 x}}{\rho H}\right)_{\rm m}}{\left(\frac{\tau_{z_0 x}}{\rho H}\right)_{\rm p}} \tag{8.2}$$

Suffix 'm' refers to conditions in the model and suffix 'p' refers to the prototype.

Consider two graphs, each representing the same velocity but plotted to different scales of length or time as in figure 8.2. The components of acceleration $\frac{\partial U}{\partial t}$ and $U\frac{\partial U}{\partial x}$ bear the same ratio to each other at corresponding points and times if $\frac{U_{\rm m}}{t_{\rm m}} \Big/ \frac{U_{\rm p}}{t_{\rm p}}$ and $\frac{U_{\rm m}^2}{x_{\rm m}} \Big/ \frac{U_{\rm p}^2}{x_{\rm p}}$ are constant at all corresponding points. Introducing suffix 's' to indicate the model:prototype scale ratio,

$$\left(\frac{\partial U}{\partial t}\right)_{\rm m} \Big/ \left(\frac{\partial U}{\partial t}\right)_{\rm p} = \left(\frac{U}{t}\right)_{\rm m} \Big/ \left(\frac{U}{t}\right)_{\rm p} = \frac{U_s}{t_s}$$

for similarity of flow patterns at all corresponding points.

Applying the same method to the convective acceleration term, the requirement for similarity is that

$$\left(U\frac{\partial U}{\partial x}\right)_{\rm m} \Big/ \left(U\frac{\partial U}{\partial x}\right)_{\rm p} = \frac{U_{\rm m}^2}{U_{\rm p}^2} \cdot \frac{x_{\rm p}}{x_{\rm m}} = \frac{U_s^2}{x_s}$$

at all corresponding points. Extending this argument to the other terms of equation 8.2, similarity will occur if, at all corresponding points and times,

$$\underset{(1)}{\frac{U_s}{t_s}} = \underset{(2)}{\frac{U_s^2}{x_s}} = \underset{(3)}{\frac{h_s}{x_s}} = \underset{(4)}{\frac{(\tau_{z_0 x})_s}{\rho_s z_s}} \tag{8.3}$$

$$\text{Term (2)} = \text{term (3) if } U_s^2 = z_s \tag{8.4}$$

$$\text{Term (2)} = \text{term (4) if } (\tau_{z_0 x})_s/\rho_s = z_s^2/x_s \tag{8.5}$$

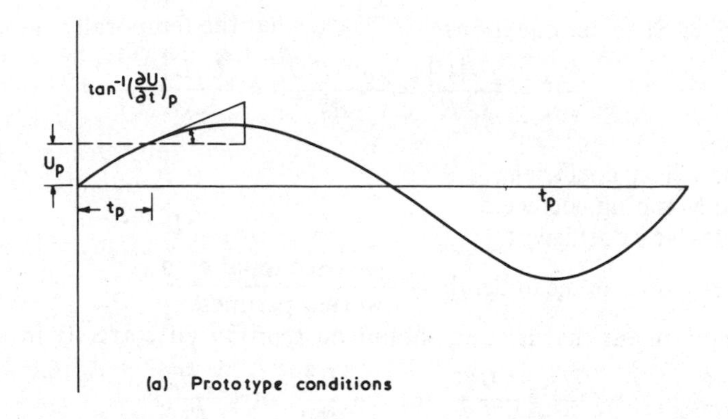

(a) Prototype conditions

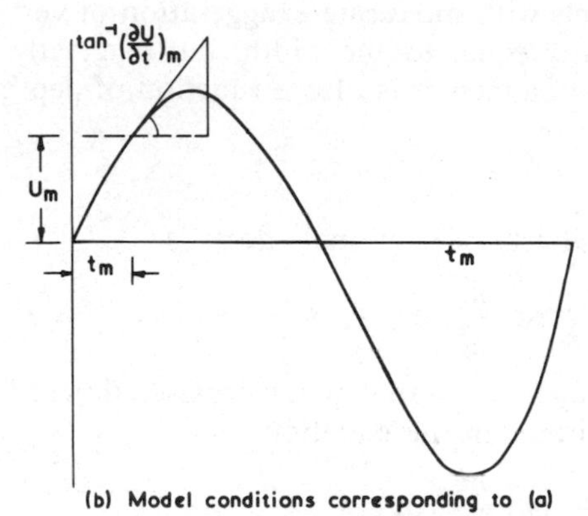

(b) Model conditions corresponding to (a)

Figure 8.2. Similarity of differential coefficients in model and prototype

(8.4) and (8.5) are the requirements for similarity of steady flows, as in models of rivers or canals.

$$\text{Term (2)} = \text{term (1) if } x_s = U_s t_s \tag{8.6}$$

This is the additional requirement that takes into account the unsteady nature of the flow. If equations 8.5 and 8.6 are combined, the requirement for similarity of reproduction of shear stress in unsteady flow is that

$$(\tau_{z_0x})_s/\rho_s = z_s^2/U_s t_s = \frac{z_s^2}{x_s}\frac{x_s}{U_s t_s} = \frac{z_s^2}{x_s S_s},$$

where S is a Strouhal number, $(Ut)/x$. According to equation 8.5, similarity will only occur if $S_s = 1$.

In equation 8.1, $\partial z_1/\partial x$ is the mean surface slope in the direction of mean flow. $(\tau_{z_0x}/\rho H)$ is the mean value of force per unit mass due to friction from all causes and does not, in this case, represent only the effect of local shear stress at the bed. To make this clear, the friction term can be replaced by one of the standard equations for friction due to steady flow, such as the Chezy,

Manning, or Strickler equations:

$$\frac{\tau_{z_0x}}{\rho H} = \frac{gU|U|}{C^2R} = \frac{gn^2U|U|}{R^{\frac{4}{3}}} = \frac{fU|U|}{8R}$$

where

C is the Chezy coefficient
n is the Manning coefficient
f is Strickler's coefficient

$$R \text{ is the hydraulic mean depth} = \frac{\text{cross-sectional area}}{\text{wetted perimeter}}$$

The requirement that friction should be reproduced correctly in a model is that

$$\frac{U_s^2}{x_s} = \frac{U_s^2}{C_s^2R_s} \quad \text{or} \quad \frac{n_s^2U_s^2}{R_s^{\frac{4}{3}}} \quad \text{or} \quad \frac{f_sU_s^2}{R_s}.$$

In models of wide channels with moderate exaggeration of vertical scale, the wetted perimeter is nearly equal to the width, but in greatly distorted models of relatively narrow channels it is also a function of depth. For the former case,

$$R_s \approx \frac{x_sz_s}{x_s} \approx z_s$$

and the requirement for similarity of frictional effects is

$$\frac{1}{C_s^2} \quad \text{or} \quad \frac{n_s^2}{z_s^{\frac{1}{3}}} \quad \text{or} \quad f_s = \frac{z_s}{x_s}.$$

In many practical cases R may be estimated at any cross-section in the model, and prototype and R_s substituted in the equation

$$\frac{1}{C_s^2} \quad \text{or} \quad \frac{n_s^2}{R_s^{\frac{1}{3}}} \quad \text{or} \quad f_s = \frac{R_s}{x_s}$$

This derivation of requirements for similarity from the simplified one-dimensional equation of motion has led to all the basic scale relationships that can be derived by dimensional analysis plus the relationship between velocity, shear or total rate of energy loss and the horizontal and vertical scales. The equation used in this analysis, however, was one-dimensional and described values averaged over space and time. The same method can be applied to equations of any desired complexity, leading to more comprehensive results and giving more information about scale ratios.

8.2 SIMILARITY OF THREE-DIMENSIONAL TURBULENT FLOWS

Consider the equations of motion of liquid at a point expressed in Cartesian coordinates and include in the equations spatial variations of density and the effect of rotation of the Earth:
In the x-direction,

$$\frac{\partial u}{\partial t} + u\frac{\partial u}{\partial x} + v\frac{\partial u}{\partial y} + w\frac{\partial u}{\partial z} + \frac{1}{\rho}\frac{\partial p}{\partial x} - 2\omega v \sin\phi - \frac{1}{\rho}\left(\frac{\partial \tau_{zx}}{\partial z} + \frac{\partial \tau_{yx}}{\partial y}\right) = 0 \quad (8.7)$$

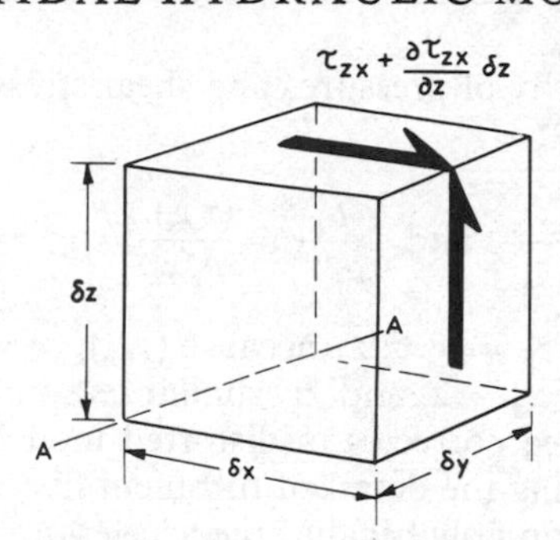

Figure 8.3. Moments caused by shear stresses

In the y-direction,

$$\frac{\partial v}{\partial t}+u\frac{\partial v}{\partial x}+v\frac{\partial v}{\partial y}+w\frac{\partial v}{\partial z}+\frac{1}{\rho}\frac{\partial p}{\partial y}+2\omega u \sin\phi-\frac{1}{\rho}\left(\frac{\partial \tau_{zy}}{\partial z}+\frac{\partial \tau_{xy}}{\partial x}\right)=0 \quad (8.8)$$

In the z-direction

$$\frac{\partial w}{\partial t}+u\frac{\partial w}{\partial x}+v\frac{\partial w}{\partial y}+w\frac{\partial w}{\partial z}+\frac{1}{\rho}\frac{\partial p}{\partial z}-2\omega u_e \cos\phi+g-\frac{1}{\rho}\left(\frac{\partial \tau_{xz}}{\partial x}+\frac{\partial \tau_{yz}}{\partial y}\right)=0 \quad (8.9)$$

These equations can be simplified considerably when applied to real conditions in an estuary. If applied rigorously they relate to instantaneous conditions, so that for example $u = \bar{u}+u'$ where $\bar{u}$ represents a temporal mean value over an interval long compared with turbulent fluctuations, but short compared with a tidal period, e.g. 5–10 min.; and u' represents an instantaneous departure from this mean. Reynolds has shown that the shear stress terms would then have components due to viscous shear, such as $(\mu/\rho)\, \partial^2 u/\partial z^2$ and due to turbulence such as $\partial(\overline{u'v'})/\partial z$, as discussed in Chapter 3.

Turbulence cannot be reproduced correctly in small-scale models, according to the requirements for similarity that would satisfy equations 8.7 to 8.9.

Consider moments due to shearing stresses. It is simplest to consider the particular case of steady flow in which moments due to local mean shear stress are exactly in balance (figure 8.3):

$$\left(\tau_{zx}+\frac{\partial \tau_{zx}}{\partial z}\delta z\right)\delta x \delta y . \delta z \approx \left(\tau_{xz}+\frac{\partial \tau_{xz}}{\partial x}\delta x\right)\delta y \delta z . \delta x$$

$$\therefore \tau_{zx}+\frac{\partial \tau_{zx}}{\partial z}\delta z \approx \tau_{xz}+\frac{\partial \tau_{xz}}{\partial x}\delta x$$

It can be seen that $\tau_{zx} \approx \tau_{xz}$ and their scale ratios must also be equal, regardless of the linear scales that are chosen.

Similarly,

$$\tau_{xy} \approx \tau_{yx}$$

$$\tau_{yz} \approx \tau_{zy}$$

Requirements for similarity of pressures and shear stresses can be obtained from equations 8.7 and 8.9:

$$\frac{p_s}{\rho_s x_s} = \frac{(\tau_{zx})_s}{\rho_s z_s} \quad \text{and} \quad \frac{p_s}{\rho_s z_s} = \frac{(\tau_{xz})_s}{\rho_s x_s} = g_s = 1$$

Thus $p_s = \rho_s z_s = (\tau_{zx})_s x_s / z_s = \rho_s x_s^2 / z_s$ because $(\tau_{xz})_s \approx (\tau_{zx})_s$. These conditions can only be satisfied if $x_s = z_s$ and, by similar reasoning, if $x_s = y_s = z_s$. Turbulence cannot be scaled correctly in distorted models. Although this is strictly true when considering the detail of turbulent fluctuations, it has little significant effect on the mean flow because the acceleration terms in equation 8.9 are very small compared with gravitational force per unit mass.

Consider next the requirement for similarity of the shear stress over a range of flow conditions that may include laminar and turbulent flow. In equation 8.7

$$\frac{1}{\rho}\frac{\partial \tau_{zx}}{\partial z} = \frac{\partial(\overline{u'w'})}{\partial z} \text{ in turbulent flow or } \nu \frac{\partial^2 w}{\partial z^2} \text{ in laminar flow}$$

$$\frac{1}{\rho}\frac{\partial \tau_{yx}}{\partial y} = \frac{\partial(\overline{u'v'})}{\partial y} \text{ in turbulent flow or } \nu \frac{\partial^2 v}{\partial y^2} \text{ in laminar flow.}$$

The requirements for similarity of behaviour, according to these relationships, are

$$\frac{x_s}{z_s} = \frac{u_s w_s}{z_s} = \nu_s \frac{w_s}{z_s^2} \quad \text{and} \quad \frac{z_s}{x_s} = \frac{u_s v_s}{y_s} = \nu_s \frac{v_s}{y_s^2}$$

If the same liquid is used in model and prototype such that $\nu_s = 1$, this can only be satisfied if $x_s = y_s = z_s = 1$. Only full-sized, undistorted models can reproduce turbulent processes, including dispersion of sediment in suspension, salt and pollutants.

Flows in estuaries and estuary models pass through the whole range of possible flow conditions from slack water to fully developed turbulent flow in each tidal cycle. It is not possible to obtain true similarity of behaviour under all flow conditions in reduced scale models. Every model is therefore a compromise between the desirable and the attainable. The real issue is not how to design a perfect model, but how to interpret its results in the knowledge of its limitations.

8.3 SCALE EFFECTS AND DEPARTURES FROM SIMILARITY IN MODELS OF ESTUARIES

Although it is not possible to reproduce all the features of turbulent and laminar flows in reduced scale models, acceptable results can be achieved in practice. The reason is that some physical effects are relatively unimportant. The result of ignoring them can be examined through the equations of motion. It is assumed that the equations can be written in terms of the local mean values; that is, with values at a point averaged temporally to eliminate fluctuations due to turbulence. Any errors caused by this averaging process are assumed to be assimilated in the terms describing shear stress, as in equation 3.9.

The vertical components of velocity and acceleration are small in estuaries because they are shallow compared with their length and breadth. They are negligible when compared with the gravitational force per unit mass in real estuaries. The only exceptions occur when vertical accelerations become large, as in a tidal bore: but such effects are localised in space and time and usually have little effect on overall behaviour.

Equations 8.7 to 8.9 can therefore be re-written:

$$\frac{\partial \bar{u}}{\partial t}+\frac{\bar{u}\,\partial \bar{u}}{\partial x}+\bar{v}\frac{\partial \bar{u}}{\partial y}+\frac{1}{\rho}\frac{\partial p}{\partial x}-2\omega\bar{v}\sin\phi-\frac{1}{\rho}\frac{\partial(\tau_{zx})}{\partial z}-\frac{1}{\rho}\frac{\partial(\tau_{yx})}{\partial y}=0 \qquad (8.10)$$

$$\frac{\partial \bar{v}}{\partial t}+\frac{\bar{u}\,\partial \bar{v}}{\partial x}+\frac{\bar{v}\,\partial \bar{v}}{\partial y}+\frac{1}{\rho}\frac{\partial p}{\partial y}+2\omega\bar{u}\sin\phi-\frac{1}{\rho}\frac{\partial(\tau_{zy})}{\partial z}-\frac{1}{\rho}\frac{\partial(\tau_{xy})}{\partial x}=0 \qquad (8.11)$$

$$-\frac{1}{\rho}\frac{\partial p}{\partial z}=g \qquad (8.12)$$

From equation 8.12 $\qquad p_1-p=\int_z^{z_1}\rho g\,\mathrm{d}z$

and, if the surface pressure p_1 is atmospheric,

$$\frac{1}{\rho}\frac{\partial p}{\partial x}=-\frac{1}{\rho}\,\frac{\partial\left(\int_z^{z_1}\rho g\,\mathrm{d}z\right)}{\partial x}$$

One requirement for similarity of behaviour according to equation 8.10 is that

$$\frac{\bar{u}_s^2}{x_s}=\left(\frac{1}{\rho}\frac{\partial p}{\partial x}\right)_s=\frac{\left(\int\rho g\,\mathrm{d}z\right)_s}{\rho_s x_s}=\frac{z_s}{x_s}\quad\text{or}\quad \bar{u}_s^2=z_s\ \text{ if }\ \int\rho g\,\mathrm{d}z$$

is similar at corresponding points. This will be achieved if $\rho_s = 1$ at all points.

Similarly, from equation 8.11, $\bar{v}_s^2 = z_s$.

It follows from equation 8.10 and 8.11 that $x_s = y_s = \bar{u}_s t_s$.

Dealing with the terms involving shear stress in a similar manner,

$$\frac{\bar{u}_s^2}{x_s}=\frac{(\tau_{zx})_s}{\rho_s z_s}=\frac{(\tau_{yx})_s}{\rho_s y_s}=\frac{\bar{v}_s^2}{y_s}=\frac{(\tau_{zy})_s}{\rho_s z_s}=\frac{(\tau_{xy})_s}{\rho_s x_s}$$

Putting $x_s = y_s$, $\bar{u}_s^2 = z_s$ and $\rho_s = 1$,

$$(\tau_{yx})_s \approx (\tau_{xy})_s = z_s \quad\text{and}\quad (\tau_{zx})_s \approx (\tau_{xz})_s = z_s^2/x_s.$$

It is evidently permissible, when using local mean values of parameters rather than instantaneous values, to adopt different horizontal and vertical scales for estuary models. A consequence of doing so is that the scales of shear stresses acting on vertical and horizontal planes will differ. It is worth considering the order of magnitude of these differences.

Model scales differ widely, but representative values are $z_s = 10^{-2}$ and $x_s = 10^{-3}$. For these scale ratios

$$\frac{(\tau_{yx})_s}{\rho_s} = z_s = 10^{-2}$$

$$\frac{(\tau_{zx})_s}{\rho_s} = \frac{z_s^2}{x_s} = 10^{-1}$$

The scale ratios of shear stresses acting on horizontal planes, that is parallel to the bed, are usually an order of magnitude greater than those acting on vertical planes. As the former stresses are also much larger, it is particularly important to reproduce them correctly. It is usually possible to obtain good similarity of flow patterns as viewed in plan by adjustment of the bed roughness of a model. This greatly simplifies the adjustment and proving of distorted scale models.

So far, the Coriolis terms $2\omega\bar{v}\sin\phi$ and $2\omega\bar{u}\sin\phi$ have not been considered. The requirement for similarity of behaviour is satisfied if

$$\frac{\bar{u}_s^2}{x_s} = \bar{u}_s(\omega\sin\phi)_s$$

i.e., if

$$(\omega\sin\phi)_s = \frac{u_s}{x_s} = \frac{z_s^{\frac{1}{2}}}{x_s}$$

$\omega\sin\phi$ is the component of the Earth's angular velocity about an axis normal to the surface; it is zero at the equator and equal to ω at the poles. Evidently the scale ratio gives the angular velocity of the model needed to reproduce the effects of the Earth's rotation. Using as typical model scales $x_s = 10^{-3}$ and $z_s = 10^{-2}$,

$$(\omega\sin\phi)_s = 10^2$$

The angular velocity required of the model is so much larger than the effective local value for the Earth that the direct effect of the latter on the model can be ignored.

Let ω_m be the angular velocity required of the model and $\omega\sin\phi$ be the corresponding value of the Earth.

$$\frac{\omega_{\mathrm{m}}}{\omega\sin\phi} = \frac{z_s^{\frac{1}{2}}}{x_s}$$

$$\omega_{\mathrm{m}} = \frac{2\pi\sin\phi}{86400} z_s^{\frac{1}{2}}/x_s$$

$$= 7.3\times10^{-5}\sin\phi z_s^{\frac{1}{2}}/x_s$$

For $x_s = 10^{-3} z_s = 10^{-2}$ and $\phi = 55°$

$$\omega_{\mathrm{m}} = 6\times10^{-3}\ \mathrm{rad/s}.$$

It is seldom practical to rotate a complete hydraulic model, though this has been done with some very small-scale models [9]. One compromise solution has been to place tops rotating about a vertical axis at intervals throughout a model, the intention being to superimpose a nearly constant angular velocity on the water by generation of vorticity at many points. Figure 8.4 shows this

Figure 8.4. Tidal model of the Brouwershavense Gat. To reproduce geostrophic acceleration, the model is fitted with 156 revolving Coriolis tops (courtesy Delft Hydraulics Laboratory)

technique in use in a model of Brouwershavense Gat at the Delft Hydraulics Laboratory. The tops also cause vertical mixing, however, and cannot be used if density variations are reproduced.

It is sometimes argued that no significant errors will result if Coriolis effects are omitted from estuary models. This may be so if the models have a fixed bed. The channels pre-formed in the bed will then have a bigger effect in guiding the flow, through pressure gradients, than the Coriolis forces. If, however, a large part of the bed is moulded initially in erodible material, the minor modification of flow by Coriolis force has a continuous effect on local sediment transport and it may be difficult, or impossible, to reproduce the correct pattern of channels for more than a few tidal cycles.

8.4 SIMILARITY OF SPATIALLY AVERAGED FLOWS. BOUNDARY SHEAR STRESS

The scale relationships that follow from equations 8.10 to 8.12 apply at corresponding points and times throughout the water in an estuary, or shallow sea, and its model. It is usually necessary to consider boundary conditions such as water surface levels and shear stresses at the bed of the prototype. For this purpose it is necessary to average the terms of the equations over the depth of the stream, or over a cross-section of a channel. The equations averaged over a vertical become:

$$\frac{\partial U}{\partial t}+U\frac{\partial U}{\partial x}+V\frac{\partial U}{\partial y}-gi_x+\frac{gH}{2\rho}\frac{\partial \rho}{\partial x}-2\omega V\sin\phi+\frac{\tau_{z_0x}}{\rho gH}+\frac{1}{\rho gH}\int_{z_2}^{z_1}\frac{\partial(\tau_{yz})}{\partial y}\,dz=0 \tag{8.13}$$

$$\frac{\partial V}{\partial t}+U\frac{\partial V}{\partial x}+V\frac{\partial V}{\partial y}-gi_y+\frac{gH}{2\rho}\frac{\partial \rho}{\partial y}+2\omega U\sin\phi+\frac{\tau_{z_0y}}{\rho gH}+\frac{1}{\rho gH}\int_{z_2}^{z_1}\frac{\partial(\tau_{xy})}{\partial x}\,dz=0 \tag{8.14}$$

where τ_{z_0x} and τ_{z_0y} are components of shear stress at the stream bed in the x and y directions respectively.

The requirements for similarity are exactly the same as those obtained for a point in the flow.

The order of magnitude of each term in the equations can be determined with the aid of field observations, as described in Chapter 2, Section 6. This gives useful guidance regarding the inclusion or exclusion of individual terms from the analysis. Two of the terms in each of equations 8.13 and 8.14 call for special comment. The surface slope terms i_x and i_y refer to the local components of surface slope at a point (x, y) in plan. They will, in general, vary over the width of an estuary. They are difficult to measure in practice, and are never measured as a matter of routine. However, one of the main purposes of a physical model is to reproduce three-dimensional effects; in fact, there is no other way in which this can be done in any detail. The surface slopes will be reproduced in a properly calibrated model.

The other terms requiring special comment are:

$$\int\frac{\partial(\tau_{yx})}{\partial_y}\,dz \quad \text{and} \quad \int\frac{\partial(\tau_{xy})}{\partial x}\,dz$$

These terms are also difficult to evaluate; they would require knowledge of velocity gradients in the horizontal plane, throughout the depth of the stream.

Scaling of shear stresses on vertical planes presents a difficulty. They must be reproduced correctly in models yet they depend on turbulent mixing processes that can only be reproduced in undistorted, full-sized models. They cannot be reproduced correctly in distorted, small scale models so a correction must be applied in the form of control of flow resistance. Because it is difficult to use field observations to evaluate local shear, it is not possible to calculate the exact amount of correction needed.

In practice, adjustment is done by trial, based on the correct reproduction of flow patterns observed in the prototype. An important feature of any scale model study must therefore be provision of adequate field data. It is seldom possible to forsee the pattern of velocity distribution in plan in estuaries whereas it is readily seen, with the help of flow visualisation techniques, in models. If model calibration and field observation proceed together, each can lead to very significant economies in the other.

Although shear stresses on vertical planes cannot be evaluated in practice their effect on the flow field is often small compared with the shear stresses on horizontal planes. (They are nevertheless important in determining mixing processes). Depths in estuaries are much less than widths – commonly between 1 % and 5 % – so velocity gradients over the vertical are correspondingly greater than those over the width. It is much more important to reproduce shear stresses at the stream bed correctly than it is to reproduce horizontal velocity gradients correctly; in fact, the latter depend strongly on the former. In tortuous river channels, the shear stresses on vertical planes may be relatively much more important than they are in estuaries. In most wide estuaries, however, energy losses due to these shear stresses are comparatively unimportant.

The situation may be different in small-scale models in which there is a large exaggeration of vertical scale, particularly in relatively small channels such as occur in the landward reaches of tidal rivers. Depths in such models may then be equal to, or greater than widths, and friction on the sides is relatively large. However, details of flow in very small distorted channels cannot be reproduced and the requirements of representation of mean tidal flow can be met by modifying general resistance to flow, regardless of its cause.

8.5 SCALING OF ENERGY LOSSES AND FLOW RESISTANCE IN TIDAL STREAMS

Energy losses occur due to changes in cross-section and curvature of channels; due to drag caused by bed forms such as dunes and ripples; and due to shear flow over the plane bed, the roughness of which depends on the presence of granular material on its surface. The rate of energy loss may be affected by material in motion in the stream. Fine material moving in suspension can cause a marked reduction in shear stress at the bed (Chapter 4).

Changes in cross-section and curvature of channels cause local accelerations and decelerations of flow speed and the generation of large-scale eddies.

The resultant energy losses depend, in general, on the square of the flow speed, and the coefficient of proportionality varies but little between a model and its prototype. The loss of energy per unit mass due to a change in flow speed over distance δx can be expressed as

$$\frac{k}{2g}\left[\left(U+\frac{\partial U}{\partial x}\delta x\right)^2 - U^2\right] \doteqdot \frac{k}{g}U\frac{\partial U}{\partial x}\delta x$$

where k is the coefficient of energy loss. The gradient due to this loss is $(k/g)\,U(\partial U/\partial x)$. If k is the same in model and prototype, this energy loss term is identical in form with the convective acceleration term, and the requirements for similarity of behaviour of these terms are the same. Provided that the depth scale of the model is equated to the square of the velocity scale, energy lossess due to these causes should be reproduced correctly in a model. However, these effects of channel shape are usually much smaller than the effects of boundary drag because depth/width ratios in estuary channels are low; less than 1/20 in minor branches, and less than 1/1000 in major estuary channels. Total frictional losses calculated from field measurements of flow

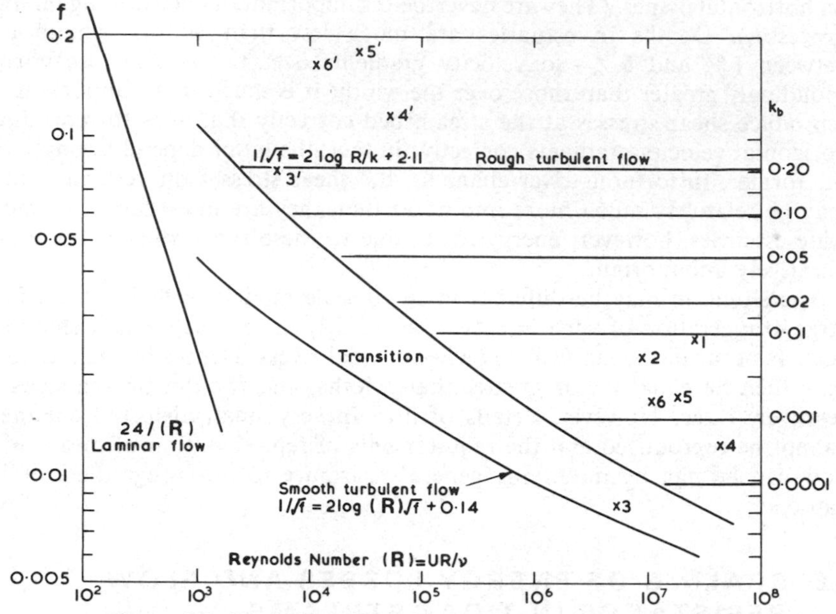

Figure 8.5. Variation of friction coefficient f with Reynolds number **R** showing typical prototype and model conditions

speeds, surface slopes and density gradients therefore depend mainly on bed resistance. Knowledge of the relationship between bed friction coefficients and flow parameters can enable the requirements of small-scale models to be studied.

Friction over plane surfaces due to steady, uniform flows can be quantified with the aid of a modified Moody diagram (figure 8.5) in which hydraulic radius is used as the measure of length. In natural channels, hydraulic radius is nearly equal to the mean depth of flow; but in small-scale models with relatively large vertical scales it will be appreciably smaller than the mean depth.

Essential requirements for correct reproduction of flow in models are that friction should be reproduced correctly at all corresponding points and times. For this to happen, the variation of friction with Reynolds number must be the same at all times. Local Reynolds number $\mathbf{R} = (UH/\nu)$ where U is depth-averaged velocity and H is total depth. The scale of Reynolds number in a model is

$$\mathbf{R}_s = \frac{U_s H_s}{\nu_s} = z_s^{\frac{3}{2}} \quad [\text{if } \nu_s = 1].$$

The scale of shear stress at the bed is $(\tau_{z_0})_s = (R_s z_s/x_s)$ and the scale of friction coefficient, assuming that $(\tau_{z_0}/\rho g R) \approx (fU^2/8gR)$, is $f_s = R_s/x_s$.

Some typical values of total friction coefficient in estuaries are given in table 8.1. These values are all based on knowledge of typical flow conditions and surface slopes at times of maximum velocity ($\partial U/\partial t = 0$). Each of these values

Table 8.1. *Values of total friction coefficient in estuaries*

Estuary	f	$\mathbf{R}\times 10^7$	Remarks
1 Gironde	0·0260	1·40	Typical of flood and ebb
2 Hooghly	0·0225	0·99	Dry season spring tide: flood
3 Hooghly	0·0082	0·55	Dry season spring tide: ebb
4 Mersey	0·0126	3·8	Spring tide: flood
5 Severn	0·0177	1·9	Spring tide: flood and ebb
6 Thames	0·017	1·05	Spring tide: flood and ebb

is plotted in figure 8.5, from which it can be seen that they lie in the range of fully turbulent, rough-boundary flows.

The scale reduction of Reynolds number **R** depends solely on the vertical scale, usually between 1/50 and 1/100. It is clear that reduction of **R** to between 1/350 and 1/1000 of its prototype value without change of friction coefficient f would be impossible; the numbered points in figure 8.5 would fall in the transition zone, or below the minimum possible values corresponding to smooth boundaries. Adoption of horizontal scale smaller than vertical scale could help, because $f_s = R_s/x_s$.

Consider horizontal scale $x_s = 1/1000$

vertical scale $z_s = 1/100 \approx R_s$

then scale of Reynolds number $\mathbf{R}_s = 1/1000$

scale of friction coefficient $f_s = 10$.

Points 3′, 4′, 5′ and 6′ in figure 8.5 correspond to points 3, 4, 5 and 6 according to these scale ratios. They lie within the rough turbulent zone of the diagram and a model with friction coefficients corresponding to these would represent the frictional conditions of the prototype.

The diagram shows that the friction coefficients in the prototype would arise from small roughness elements – roughness height k_b/(hydraulic radius R) between 0·0001 and 0·01; they were probably caused by plane sand beds or small dunes less than 0·1 m high. They would be represented in the model by dunes or ripples exceeding 1/5 of the water depth. In fixed-bed models, such high friction would be produced by roughness elements as illustrated in figure 8.1 which shows part of a model of the Rhine delta built to a horizontal scale of 1/2500 and a vertical scale of 1/64.

Models built to a smaller scale distortion require less roughening of the bed. Figure 8.4 shows roughness elements on the bed of a tidal model of Brouwershavense Gat, horizontal scale 1/300 and vertical scale 1/100. Note that the size and spacing of roughness elements varies in different parts of the model and that regions of similar depth have similar degrees of roughening.

The spacing of roughness elements is adjusted until tides and flow patterns in the model can reproduce known conditions in the prototype with acceptable accuracy.

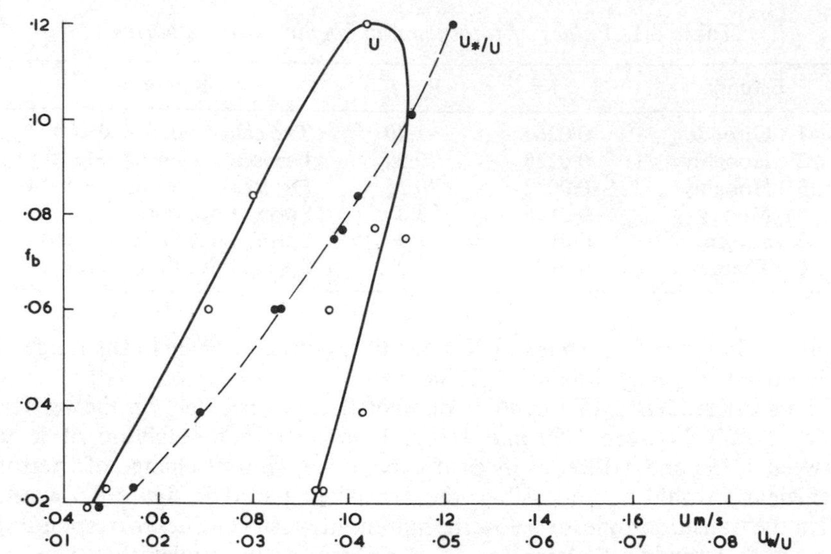

Figure 8.6. Variation of bed friction coefficient f_b with mean velocity and ratio of shear velocity U_* to U_1 according to A. Raudkivi (based on p. 18 of *Proc. A.S.C.E.*, **89**, no. HY6, pp. 15–33, Nov., 1963, courtesy the American Society of Civil Engineers)

Resistance to flow due to varying bed forms of alluvial beds is described in Chapter 4, Section 5. Ripples are the bed form most likely to occur in hydraulic models, whereas dunes are likely to be dominant in prototype situations. Resistance due to ripples reduces with depth but is independent of grain roughness. Resistance due to dunes can increase with depth, particularly if bed forms have enough time to develop fully during a tide. However, full development does not often occur in estuaries. Evidently bed resistance due to ripples in a model will not obey the same laws as resistance due to dunes.[10] However, experiments have shown that friction coefficients due to bed forms in open channels increase from about 0·02 for plane beds to 0·12 or more due to dunes. Raudkivi[11] has shown that there can be more than one friction coefficient at certain velocities, as shown in figure 8.6; while Daranandana, working with Simons and Richardson[12], has shown that material with a small spread of grain sizes can give rise to much higher friction coefficients than material with the same mean grain size but a larger spread of sizes. This accords with experience in models with mobile beds which are often operated with carefully selected material of a size that is chosen on theoretical grounds.

Consideration of figure 8·5 shows that friction factors as measured in estuaries are little higher than occur over plane beds. Even in the upper reaches of estuaries, it rarely exceeds 0·04. In models with exaggerated vertical scales, however, friction factors of order 0·1 to 0·2 are often needed. It appears that such friction can be obtained with free-moving, closely-graded material.

Evidently there is still much to be learned about the behaviour of bed resistance in tidal flows and in models. Careful calibration of a model is essential, and this calibration should be repeated when bed forms have changed during an experiment. Accurate reproduction of frictional resistance in all parts of a mobile bed model is not possible. However, if the friction term in equation 8.1 were to be always smaller than the other terms, some error in the reproduction of friction would be permissible.

In models of estuaries, the large variation of friction factor has some interesting consequences. The slightly higher resistance coefficient found with ripples, as compared with dunes, is useful in helping to increase the friction coefficient in models with exaggerated vertical scales. The biggest help, however, occurs when ripples form in models corresponding to the places where there are only very small dunes in nature. The frictional increase is further helped by the use of uniform sands in models to represent graded sands in nature, but this usually turns out to be a mixed blessing; the resultant free formation of ripples may be so widespread that the small changes of bed profile which may be sought in model tests become completely obscured by sand waves.

There is a further characteristic of the movement of bed material which may have been instrumental in the success of many models. It has been noted that the friction factor can increase by six times due to the formation of sand ripples and dunes. If the maximum flow in an estuary model is near that which would give the highest friction factor, the bed resistance of the model will be, to some extent, self-regulating. Insufficient resistance would give rise to increased flow speeds, which would cause increased resistance. It should be noted that, if the flow required to give maximum friction factor is exceeded, the self-regulating character of the flow will be reversed. If this should happen, the model cannot behave like the prototype; if this happens over extensive areas it will be impossible to calibrate the model, whilc if it happens only locally, the predictions given by the model may be quite wrong locally.

8.6 MODEL SCALING OF SEDIMENT TRANSPORT

An hydraulic model with a mobile bed may be intended to facilitate study of a local problem or to investigate changes in behaviour of a large part of a tidal system There is a fundamental difference in approach between these cases. Local problems may have a negligible effect on the tidal system as a whole and, consequently, it may be unnecessary to make a large area of the bed mobile. Control of bed resistance is much easier if only a limited area is allowed to change its form.

Problems that may affect a large area of the bed are much more difficult to study because changes in rates of sediment transport cause changes in mean bed level and configuration of channels. This causes tidal flow and sediment transport to interact. If changes in tides near the model boundaries are likely to occur, the problem of control becomes much more difficult. All models, whether mathematical, electrical analogues or physical, depend ultimately on the boundary values of water level and flow that are imposed. If these imposed values are falsified by changes within the model, the model indications will be incorrect.

The rate of transport of sediment and the consequent rate of build-up or scour of the bed are very sensitive to changes of velocity. In a perfect tidal model with a large area of mobile bed it would be necessary to reproduce a full sequence of tidal ranges and river hydrographs. In such a model, if perfection of behaviour were to be maintained for periods corresponding to several months or years in the prototype, the volumetric rate of sediment transport would have to follow the same law of scaling as the volumetric rate of water flow. If this was not achieved, the changes of bed profile and consequent changes in tidal propagation would not be reproduced exactly.

In real tidal models, there are several reasons why perfection cannot be reached. Models cannot be designed to include both variations of salinity and Coriolis effects. Models are seldom operated with a true sequence of tides and hydrographs. The major source of imperfection, however, is in the reproduction of sediment transport. The desired rate of sediment transport cannot be achieved at all corresponding points and times. Moreover a mobile bed leads to changes in frictional drag as ripples or dunes develop. Exaggeration of vertical scale increases bed slopes, whereas the natural angle of repose of mobile sands changes only slightly. Flow resistance cannot be reproduced perfectly in fixed bed models; the additional complexity of changes in friction associated with sediment transport makes its reproduction even less accurate in mobile bed models.

8.7 REPRODUCTION OF BEHAVIOUR OF CLAY-SIZED SOLIDS

In real estuaries there is usually a zone in which the bed consists largely of fine grained particles in the silt and clay range. If these form an appreciable part of the total sediment load, their behaviour should be reproduced in the model. The properties of such materials are described in Chapters 1 and 4. Main features are:

1. Clay-sized particles require a higher shear stress to bring them into motion than the maximum shear stress at which they can be maintained in suspension.
2. When the shear stress falls below the latter values, there is a time lag before all particles in suspension can settle onto the bed.
3. Once settled on the bed, fine particles consolidate slowly, their strength to resist shear increasing gradually with time.
4. The behaviour of such fine particles depends more on the degree of flocculation than on the behaviour of individual particles.
5. There is not a unique concentration of solids in suspension for a given local shear stress and velocity distribution.

For strict similarity of behaviour of fine, flocculated particles, movement must occur at corresponding shear stresses and settlement must occur at a rate corresponding to the scale of vertical velocity. Consideration of some typical model scales shown in Table 8.2 shows the difficulty of meeting these requirements.

Actual material from the prototype cannot be used. Its fall velocity would

Table 8.2. *Typical model scales*

Horizontal scale x_s	1/5000	1/2000	1/1000	1/500	1/1000
Vertical scale z_s	1/50	1/100	1/100	1/100	1/200
Vertical velocity scale $w_s = z_s^{1\cdot5}x_s^{-1}$	1/2·8	1/0·5	1/1	1/2	1/0·35
Scale of shear stress $(\tau_{z_0})_s = z_s^2 x_s^{-1}$	1/2·5	1/5	1/10	1/20	1/40

only be correct if $w_s = 1$, when $z_s^3 = x_s^2$, but if this criterion is satisfied, shear stresses in the model would be too low for motion to be initiated at the right time.

Material settling more slowly than the prototype material might be used if the vertical scale were distorted more than that given by $z_s^3 = x_s^2$. With clay-type particles, however, the shear stress needed to initiate movement becomes greater with finer particles, as shown in figure 4.3.

If an attempt is made to use particles that settle faster than those in the prototype, figure 4.3 also shows that a slightly smaller shear stress is needed to initiate motion. The scale relationships show that the available shear stress falls rapidly as scale distortion is reduced.

Two basic requirements for similarity of motion in suspension of clay-sized particles cannot be satisfied. Other requirements such as the need to reproduce rates of flocculation and consolidation are still more difficult to satisfy.

There is also an important practical difficulty that militates against attempts to reproduce silt and clay sized particles in models that also have regions where the clay content is negligible. The presence of a wide range of particle sizes tends to reduce the voids ratio in a sediment mixture. A higher initial shear stress is then required to initiate sediment movement than if sand having only the median grain size were present. When movement does occur, the rate of transport is lower for a given shear stress than for the uniform-sized sand. In small scale models, the low shear stresses are usually inadequate to produce sufficient movement of mixed-sized materials. As a result, the bed material becomes consolidated and often cemented by growth of algae.

The quantitative reproduction of fine flocculated solids in suspension will not be considered further in this book.

8.8 REPRODUCTION OF MOVEMENT OF NON-FLOCCULATED SOLIDS

The requirement for perfect scaling of sediment transport of solids has been shown to be reproduction of the volume rate of transport to the same scale as the volume rate of water flow, at all corresponding points and time. This applies to all solids in motion, regardless of the mode of transport. If an equation can be drawn up between the total volume rate of transport and the parameters that affect it, of such a form that a scale ratio consistent with that for the flow of water can be found, then perfect model scaling could be devised. Such an equation would have to be expressible in the form

$$\frac{q_b}{q} = f\,(\text{linear scales, sediment parameters})$$

where q is the rate of volumetric flow of water per unit width

q_b is the rate of volumetric flow of solids per unit width.

This would be applicable at all times during the flow cycle and to all parts of the area being studied.

Equations of this form have been devised for bed-load transport of uniform-sized material due to steady flows, but none has proved to be applicable to rates of transport found in both prototypes and models. Perhaps the best-known is due to Einstein[13]

$$0{\cdot}465\phi = e^{-0{\cdot}391\psi} \tag{8.15}$$

where $\phi = q_b w_f^{-1} d^{-1}$

where w_f = fall velocity of particles of diameter d

and $\psi = \tau_{z_0}\gamma^{-1}(S_b-1)^{-1}d^{-1}$

$U_* =$ shear velocity $(\tau_{z_0}/\rho)^{0{\cdot}5}$

$\gamma =$ specific weight of water $(=\rho g)$

$S_b =$ relative density of the solids, ρ_b/ρ.

Another equation due to Kalinske[14]

$$\frac{q_b}{U_* d} = f\left[\frac{\tau_{z_0}}{\gamma(S_b-1)d}\right] \tag{8.16}$$

where f is a function dependent on the intensity of turbulence.

Both Einstein's and Kalinske's equations have been shown to be effective at low rates of sediment transport of uniform-sized grains, such as may occur in models. At high rates of transport, both diverge from laboratory and field measurements. It has been suggested that this divergence occurs when many particles begin to be transported in suspension

Modified versions of Einstein's and Kalinske's equations have been proposed for higher rates of flow:

$$q_b = 40\tau_{z_0}^3 wfd^{-2}\gamma^{-3}(S_b-1)^{-3} \quad \text{(Einstein–Brown)} \tag{8.17}$$

and

$$q_b = 10\tau_{z_0}^{2{\cdot}5} g^{-0{\cdot}5}(S_b-1)^{-2}d^{-1}\gamma^{-1{\cdot}5} \quad \text{(Kalinske)} \tag{8.18}$$

Either of these equations can be used in the required manner to derive scales of sediment size and specific weight. Unfortnately neither will apply satisfactorily to model conditions.

Alternative equations based on stream power can, theoretically, include sediment transport in suspension as well as bed load. In practice, however, they depend on empirically determined coefficients and give results which deviate as much from actual measurements as the other equations. Bruk[15] has shown that equations due to Bagnold[16] and Engelund and Hansen[17] can be combined to give

$$q_b = 0{\cdot}1 \frac{U}{U_*} g\theta\tau_{z_0} U(\rho_b - \rho) \tag{8.19}$$

where U is velocity of flow, $\tau_{z_0} U$ is the stream power, or the product of mean shear stress and mean velocity in steady flow, and θ' is $\tau_{z_0}/(S_b - 1)\gamma d$.

It is clear that it is not possible to devise a model scale that leads to correct behaviour over a large area; there is no point in trying to perfect such a model. However, many models have given very useful information that could not be obtained in any other way. A reasonable compromise is evidently possible and is acceptable provided that its limitations are recognised. The problem is to make such a compromise without producing grossly misleading results. There are several steps that can be taken to reduce the consequences of scaling errors.

First, good field data is essential; without this, there is no possible criterion for judging model performance. It should be in such a form that the rate of development of bed changes can be estimated, under known conditions of tide, river flow, temperature and salinity of water. Such data are difficult to obtain and interpret. Given good basic data, the time scale of changes in the model can be estimated.

Second, the mobile part of the model should be limited to the smallest acceptable area, consistent with maintenance of an adequate supply of material in motion throughout the whole area under study.

Third, model scales must be chosen such that flow will be turbulent throughout the greater part of the tidal period, using the principles outlined earlier in this chapter. Some compromise is possible in order to obtain the best possible reproduction of sediment transport.

Fourth, material must be chosen for the bed of the model. It may be possible to design for time scale of sediment transport equal to that of flow of water, but this will depend on the specific weight of material available.

Fifth, the model must be calibrated by making it reproduce known changes before any new schemes are tested. This calibration will reveal the true time scale of transport and will also show the limitations of reproduction that can be expected.

Once the model has been calibrated, testing of changed conditions can be done. The results must be interpreted with care by an experienced engineer who is familiar with the general behaviour of tidal waters and who has a thorough knowledge of the locality being modelled.

8.9 SELECTION OF BED MATERIAL

Preliminary selection of a suitable material may be made by using one of the simpler equations of sediment transport. The use of Shields entrainment function is sometimes advocated, on the ground that the main requirement is that transport in the model should begin and end at the corresponding times in model and prototype. This, however, will not ensure that total volume rates of transport are correct. Moreover, this method has often led to great mobility of bed material, such that the time scale for sediment transport is much smaller than that for tidal behaviour. Bed changes representing a year in nature may occur in only a few tides in the model. This makes nonsense of any attempt to reproduce a sequence of spring and neap tides in the model and reduces the model to a very crude indicator of bed changes. Careful comparison of model and prototype is then the only safeguard against misleading results. Another view is that bed resistance should be correct, to ensure that flow patterns are not changed wrongly as a result of bed changes. This is a fundamental point and may be of overriding importance if a large part of an estuary has to be made mobile[18].

A third view is that, unless rates of transport are nearly correct at all places, but particularly round about the times of maximum velocity when most movement takes place, volume changes can never be correct and channel distortion will follow. Using equations 8.17, 8.18 or 8.19, the scale ratios of discharge of sediment per unit width, $(q_b)_s$, can be made equal to nq_s where n is the chosen factor by which the sediment transport rate is increased over the flow rate[19]. Then

$$nq_s = w_s(\tau_{z_0}^3)_s(S_b-1)_s^{-3}d_s^{-2} \quad \text{(Einstein–Brown)}$$

or

$$nq_s = (\tau_{z_0}^{2\cdot5})_s(S_b-1)_s^{-2}d_s^{-1} \quad \text{(Kalinske)}$$

or

$$nq_s = (\tau_{z_0}^{1\cdot5})_s(S_b-1)_s^{-2}\,U_s^2 d_s^{-1} \quad \text{(Bagnold–Bruk)}$$

and, putting $q_s = z_s^{1\cdot5}$

and $(\tau_{z_0})_s = \beta z_s^2 x_s^{-1}$, where $\beta = R_s/z_s$ and R = hydraulic mean depth,

$$z_s^{4\cdot5}x_s^{-3} = n\beta^{-3}(S_b-1)^3 d_s^2(w_f^{-1})_s \quad \text{(Einstein–Brown)} \qquad (8.20)$$

$$z_s^{3\cdot5}x_s^{-2\cdot5} = n\beta^{-2\cdot5}(S_b-1)_s^2 d_s \quad \text{(Kalinske)} \qquad (8.21)$$

$$z_s^{2\cdot5}x_s^{-1\cdot5} = n\beta^{-1\cdot5}(S_b-1)_s^2 d_s \quad \text{(Bagnold–Bruk)} \qquad (8.22)$$

Knight[20] has pointed out the advantages of using a time scale for changes of bed level instead of a scale of rate of total sediment discharge. This has the practical advantage that direct calibration against field surveys is possible.

The equation of continuity for bed-level changes is similar to that for water:

$$\frac{\partial q_b}{\partial x} + \frac{\partial z_0}{\partial t_b} = 0$$

where z_0 is the level of the bed above datum.

The scale of vertical velocity of the bed is therefore:

$$\frac{(z_0)_s}{(t_b)_s} = \frac{(q_b)_s}{x_s} = n\frac{z_s}{t_s} \quad \text{where } n \text{ is a multiplier} \qquad (8.23)$$

The time scale for bed changes is therefore

$$(t_b)_s = \frac{t_s}{n} = \frac{x_s z_s}{(q_b)_s} \qquad (8.24)$$

The Einstein–Brown equation gives

$$(t_b)_s = \beta^{-3}x_s^4 z_s^{-5}(s_b-1)^3\, d_s^2(w_f^{-1})_s \qquad (8.25)$$

the modified Kalinske equation gives

$$(t_b)_s = \beta^{-2\cdot5}x_s^{3\cdot5}z_s^{-4}(S_b-1)_s^2 d_s \tag{8.26}$$

and the Bagnold–Bruk equation gives

$$(t_b)_s = \beta^{-1\cdot5}x_s^{2\cdot5}x_s^{-3}(S_b-1)_s^2 d_s \tag{8.27}$$

Knight[21] has made an extensive study of several methods of scaling of tidal models. His research took into account both the time scaling of sediment transport and the provision of the correct frictional resistance. He did tests on models of the Tay Estuary using five different bed materials and these showed that several methods gave quite reasonable results though, as might be expected, none was perfect. A method proposed by Zwamborn[22] gave noticeably better results than others, for these particular experiments.

A method developed at the Delft Hydraulics Laboratory is interesting because it includes both the sediment transport rate and the bed resistance. It makes use of Frijlink's equation:

$$q_b = F'(d, g, (S_b-1)).F''((S_b-1), d, \mu, U_*, g) \quad \text{where } F' \text{ and } F'' \text{ are functions}$$

$$\mu = \left(\frac{C_{\text{bottom}}}{C_{\text{grain}}}\right)^{1\cdot5} \quad (C = \text{Chezy coefficient})$$

and $C = 18 \log (12hk^{-1})$

k_{bottom} = roughness height of bed forms (ripples, etc.)

k_{grain} = grain size at which 90 % of bed sand is finer (d_{90})

Bijker[23] suggested that $F' = d^{1\cdot5}g^{0\cdot5}(S_b-1)^{0\cdot5}$

and, if the Kalinske equation modified by Frijlink is used,

$$F'' = 5\mu U_*^2 g^{-1}d^{-1}(S_b-1)^{-1} \exp(-0\cdot27gd(S_b-1)\mu^{-1}U_*^{-2})$$

The most comprehensive method devised so far is one due to Einstein and Chien[24], but it is difficult to apply and has not been widely used. In the experiments on models of the Tay Estuary, Knight found no particular advantage in any of these methods compared with the simpler one due to Zwamborn. In fact, it is not possible to justify very sophisticated methods of scaling, for not only do scale effects cause inaccuracy; field surveys lack precision, and variability of natural forces limit the extent to which useful comparisons can be made.

There are several instances of hydraulic models that have been carefully calibrated by comparison with known changes. A selection of these was used to compare the predictions of the bed-movement scale multiplier n with model performance using equations 8.25–8.27. The results are shown in table 8.3. The well-tried Einstein–Brown and Kalinske bed-load equations failed to predict the multiplier, whereas the Bagnold–Bruk equation based on stream power gave results that are acceptable, in view of the variability of data obtained from nature.

Table 8.3.

Estuary	Date	Typical grain sizes (mm)		Grain densities (t/m^3)		$1/z_s$	$1/x_s$	$n = \frac{\text{ratio of time scale for solid movement}}{\text{ratio of time scale for water movement}}$			
		(estuary)	(model)	(estuary)	(model)			Einstein–Brown equation 8.25	Kalinske equation 8.26	Bagnold–Bruk equation 8.27	Measured in model
Wyre[a]	1949	0·18[b]	1·0	2·7	1·2	70	1000	42	143	10	1*
Forth[a]	1949	0·1	0·5	2·7	1·2	144	1800	260	550	45	1
Mersey[a]	1956	0·18	0·13	2·7	2·7	60	550	2·7	5·9	0·64	0·33
Tees[a]	1958	0·2	0·55	2·7	1·1	75	750	495	440	44	65
Dee[a]	1965	0·24	0·43	2·7	1·4	60	1500	522	527	21	50
Loire[c]	1958	1·5	0·58	2·7	1·18	100	667	930	365	55	23–37
Gironde[d]	1958	0·2	0·7	2·7	1·18	100	667	22	40	6	5

Sources of data:

[a] Discussion of paper 7206s. *Inst.C.E.* (1969), by M. A. Kendrick and W. A. Price.

[b] The long-term effects of training walls, reclamation and dredging on estuaries, Sir C. C. Inglis and F. J. T. Kestner, *Proc. ICE*, *9*, March 1958.

[c] Étude sur modèle réduit de la Loire Maritime: *Lab. Nat. d'Hydraulique*, Chatou, Série A, Oct. 1958.

[d] Estuaire de la Gironde: *Lab. Nat. d'Hydraulique*, Chatou, Série A, March 1958.

* This was controlled by addition of material to the model at the required rate.

Wave action at the mouth of an estuary can have a dominant effect on sediment transport. With the aid of waves, sediment can be moved great distances by weak tidal currents. If models of the mouths of estuaries have to reproduce the effect of waves, it is necessary to select suitable scales. Wave action can induce sediment transport and circulation of water. This is the case in locations exposed to long-period waves in shallow water, but it is nearly always significant in the breaker zone, regardless of wave period. Sometimes the action of long-period or shallow-water waves is merely to move material on the bed to and fro, so that it can be transported by tidal streams. It is then only necessary to reproduce waves in the model as a means of initiating sediment movement, the direction and period of waves being less important than correct reproduction of tides. At other times, waves influence the flow pattern to such an extent that they must be reproduced properly in model, with regard to period and direction.

Wave height poses a special problem. If the vertical scale of a model is distorted, wave steepness must be increased if geometrical similarity is to be preserved. On the other hand, waves must break at the correct location so that wave-induced currents and sediment transport are properly modelled. Because waves break at a steepness of about 1/7, their height must be reproduced at the horizontal scale if they are to break at the correct point. Compromise is clearly needed. Small distortion of scale is necessary and a balance must be found between the effect of waves and of tide. This balance is obtained by regulating the duration of wave action during the tidal cycles by a process of trial. Fortunately bed materials can be selected that would give a high value of amplification factor n, and which would respond well to the reduced wave height in a distorted model, but would nevertheless not respond to the weak tidal currents in the outer, coastal regions of most estuaries. Comprehensive studies of this problem have been made by Bijker[25] and Kamphuis[26] and these provide bases for preliminary selection of scales and bed material for modelling the combined action of waves and tides.

8.10 MAINTENANCE OF MODEL CALIBRATION

The calibration of a model by comparison with field data is an essential part of its operation. Reproduction of tidal rise and fall, and transport of sediment, can only be ensured if good field data are available, and the results given by the model can never be better than the data, taken as a whole. Nevertheless individual items of field information may be wrong and, in many cases, model results have drawn attention to errors in the datum of tide gauges.

Successful calibration of a model does not necessarily ensure that the model will continue to reproduce tides and flows correctly under changed conditions. Changes may affect channel resistance to such an extent that roughness based on initial calibration is no longer suitable. This is not often a serious problem if it is borne in mind that parts of an estuary having similar conditions of depth and bed material are likely to have similar roughness. Consistent adjustment of roughness will result in roughness elements having similar size and spacing over regions of similar depth (see figure 8.4). Introduction of shoals or deepened channels should be accompanied by introduction of an appropriate roughness for the new conditions. Mobile beds pose greater problems because the roughness is likely to change during an experiment as the bed forms develop. It is necessary to check calibration with regard to tide and flow after the bed has been allowed to develop.

A more serious problem is concerned with maintenance of correctly-scaled boundary conditions following major changes to the bed of the model. All models, whether physical or mathematical, are operated by imposing boundary conditions, e.g. of tidal rise and fall at the seaward end. Major changes which would affect tidal rise and fall in nature would be prevented from doing so by the model control apparatus. The effect is most serious in models of part of an estuary with tidal control apparatus at each end. Relatively small changes can modify tidal level or flow at the positions where these are monitored. There is no simple solution to this problem, though its effects can be reduced by moving the tidal generators so far from the region of experimentation that changes at the boundaries are expected to be small. Trials involving barrages or large reclamations that are certain to affect tidal propagation must be interpreted with care. The tidal generator should be placed so far seawards of the mouth of the estuary that disturbances within it will have negligible effect there. Unfortunately, tidal data is seldom available far enough offshore and tide curves have to be devised such that good reproduction is obtained at the most seaward tidal gauging station. This can be done with the aid of numerical methods[27]. One way of ensuring that tidal reproduction is correct after changes have been introduced in an estuary is to carry out pairs of tests with the tide generators in different locations. If the results are identical before the changes, but differ after them, one can be sure that the inner tidal gauge is affected by tidal reflections and is at an unsuitable position. This is relatively easy to check in the case of mathematical models but may be impossible in most physical models.

8.11 SUMMARY

1. Physical models are the only method of study of estuaries that can reproduce three-dimensional flows varying smoothly in both space and time.
2. Effects that can be reproduced easily are tidal rise and fall and flow patterns in plan in well-mixed shallow estuaries.
3. Other effects can be reproduced less easily, but nevertheless with acceptable accuracy. These include density currents due to variations in salinity; the Coriolis effect caused by the earth's rotation (but not simultaneously with density gradients); the effect of structures such as jetties and spurs.
4. Turbulent mixing of pollutants, temperature differences, erosion, transport and deposition of bed material can only be reproduced qualitatively. Useful guidance can be obtained from physical models, but scale effects limit the accuracy of results.
5. Care must be taken to ensure that changes introduced into a model would not alter the boundary conditions of tidal imput at the location of the tidal generator(s) in the prototype.

6. The greatest single merit of physical models is their capacity to reproduce, on an easily-observed scale, the intricate three-dimensional flow in a large estuary. They are, in consequence, an aid to thought and planning without equal. They can lead to substantial economy in planning field measurements by indicating the main features of flow and they can draw attention to unknown effects such as local eddies and flow disturbances. The latter can then be checked in nature with the aid of a few simple observations.

7. Compared with other predictive methods of study, model experiments are slow and expensive in labour. Even a simple study will take a few months to complete. However, the cost is small in relation to field studies, which must extend over at least a year and must involve very large resources of men and equipment.

NOTATION

The notation is essentially the same as for Chapter 2 with the addition of symbols defined in the text. Some important ones recur in several equations and are also defined below:

suffix b indicates a property of the bed material
suffix m indicates a value in a physical model
suffix p indicates a corresponding value in its prototype
suffix s indicates a scale ratio, e.g. $U_m/U_p = U_s$
$n = (q_b/q)_s$
q = volume of water flowing through unit width per second
q_b = volume of solids flowing through unit width per second
R = hydraulic radius; cross-sectional area/wetted perimeter
$\mathbf{R}$ = Reynolds number, UH/ν or UR/ν
$S_b = \rho_b/\rho$
w_f = fall velocity of solid particles in water
$\beta = R_s/z_s$
ρ = density of water
ρ_b = density of solid grains

REFERENCES

[1] Osborne Reynolds, On model estuaries, *British Association Reports* (1889), (1890), (1891)

[2] Osborne Reynolds, On certain laws relating to the regime of rivers and estuaries and on the possibility of experiments on a small scale, *British Association Report*, **55** (1887). (See summary of references[1] and [2] in ref. [3])

[3] *Osborne Reynolds and engineering science today*, Manchester University Press (1970)

[4] Allen, J., *Scale models in hydraulic engineering*, Longmans, Green (1947)

[5] Ball, D. J. and Barney, G. C., Application of multivariable control theory to hydraulic models, *Proc. I.E.E.*, **122**, 2, Feb. (1975)

[6] Yalin, S. M., *Theory of hydraulic models*, Macmillan (1971)

[7] Birkhoff, G., *Hydrodynamics*, Ch. IV, Oxford University Press (1960)
[8] Langhaar, H. L., *Dimensional analysis and theory of models*, John Wiley (1960)
[9] Laboratoire National d'Hydraulique, Chatou, *Reports ref.* RB/CD–B 572/92 (1958), and RB/AW/CD-B 572/92 (1958).
[10] Engelund, F., Hydraulic resistance of alluvial streams, *Proc. ASCE*, **92**, HY2, March, 315–326 (1966) and Closure of discussion, *Proc. ASCE*, **93**, HY4, July, 287–296 (1967)
[11] Raudkivi, A. J., *Loose boundary hydraulics*, Ch. 13, Pergamon Press (1967)
[12] Simons, D. B. and Richardson, E., Resistance to flow in alluvial channels, *US Geo. Survey*, Professional Paper no. **422-J** (1966)
[13] Einstein, H. A., Formulae for the transportation of bed load, *Trans. ASCE*, **107**, 561–577 (1942)
[14] Kalinske, A. A., Movement of sediment as bed in rivers, *Trans. Am. Geophys. Un.*, **28**, 4, 615–620 (1947)
[15] Bruk, S., Basic bed load functions, *IAHR Congress*, New Delhi (1973)
[16] Bagnold, R. A., An approach to the sediment transport problem from general physics, *US Geol. Surv.*, Professional Paper no. 1 **422-I** (1966)
[17] Engelund, F. and Hansen, E., *A monograph on sediment transport in alluvial streams* Teknisk Forlag, Copenhagen (1967)
[18] Komura, S., Similarity and design methods of river models with movable beds, *Trans. Jap. Soc. C.E.* **80** (April), 31–41 (1962)
[19] McDowell, D. M., Scale effects in models of shallow tidal waters, *Proc. ICE*, Paper No. 7209s (1969)
[20] Knight, D. W., *Proc. Inst. Civil Engrs.* (Discussion), Paper 7209s (1970)
[21] Knight, D. W., *Ph.D. Thesis*, Aberdeen University (1969)
[22] Zwamborn, J. A., Reproductability in hydraulic models of prototype river morphology, *Houille Blanche* **3**, 291–297 (1966)
[23] Bijker, E. W., Determination of scales of moveable bed models, *Delft Hyd. Lab*, Pub. **35** (1965)
[24] Einstein, H. A. and Chien Ning, Similarity of distorted river models with movable beds, *Trans. ASCE*, **121**, 1956
[25] Bijker, E. W., Some considerations about scales for coastal models with movable bed, *Delft Hyd. Lab.*, Pub. No. **50** (1967)
[26] Kamphuis, J. W., *Scale selection for wave models*, C.E. Research Report No. **71**, Queen's University, Kingston, Ontario (1972)
[27] Pinless, S. J., The reduction of artificial boundary reflexions in numerically modelled estuaries, *Proc. Inst. Civil Engrs*, Part 2, **59**, 255–264 (1975)

9

Control of estuaries

The behaviour of estuaries is governed by many interacting effects and the control of any one of them is likely to be followed by changes in others. For example, stabilisation of channels by means of training works may be followed by siltation outside the channel and an overall loss in tidal capacity. Deepening of channels through the region of marked vertical saline gradients is usually accompanied by landward penetration of saline water. These two examples are well-known cases in which the immediate consequences of changes can be forseen. Although the immediate effects of control work can usually be foretold, the long term consequences can only be visualised within broad limits. The further ahead one tries to look, the more indistinct does the vision become. Any attempt to control an estuary will therefore be, at best, a temporary process. The long-term consequences can become very complicated and success in forecasting them requires wide experience of estuarial systems. Some of that experience is presented in Chapter 10, but there is no substitute for close study of actual systems. It follows that the effect of control works must always be carefully observed by field measurements. It will often be necessary to do further work to ensure that objectives are maintained.

It is not possible to describe control measures that could deal with the infinite variety of situations that can arise in real estuaries. There are, however, some general principles of behaviour that must always be taken into consideration and these are discussed here. There are also some well-established methods of control including training works and dredging operations. These are also described in this chapter.

9.1 GENERAL PRINCIPLES OF BEHAVIOUR

1. Fluid will continue to flow with unchanged momentum in a straight line unless a pressure gradient or gradient of shear stress acts on it. All training works act by providing such a gradient. Guide banks nearly tangential to the flow prevent the rectilinear motion of water and cause a build-up of head that provides the necessary pressure gradient with a component perpendicular to the bank. Spurs constructed at a large angle to the flow retard it

locally, causing kinetic energy to be converted to potential. Permeable groynes dissipate some energy through turbulent eddying, but the added resistance to flow causes some of the kinetic energy to be converted to potential; their power to deflect water is less than that of solid spurs of similar size and location, but the basic method of deflection is the same.

2. A stream has a certain capacity to transport sand and gravel, depending on the speed and depth of flow and on the properties of the material available for transport, as described in Chapter 4. If sand is available for erosion, any local increase in speed will be accompanied by erosion of the bed. Conversely, any decrease will be accompanied by deposition. Deepening a channel in sand can be accomplished by increasing flow speed locally, by forcing the total flow through a smaller cross-section. Any sand scoured from the bed as a result of local training works will be deposited when the current has left their zone of influence.

3. The power of a stream to transport fine sediment was discussed in Chapter 4. It does not depend on flow speed in any simple manner. Clay-sized particles tend to consolidate on the bed in flocculated form. The flocs are broken up and the solids brought into suspension when a critical shear stress is exceeded. The material remains in suspension at a lower shear stress, but when stress falls below a limiting value flocs form and gradually settle onto the bed and slowly consolidate. At low speeds of flow, material will concentrate near the bed and may move as a layer having relatively high density. Local increase of flow speed does not necessarily result in increase of local depth in consolidated silts and clays. When erosion does occur, the eroded material may be added to the wash-load of material moving to and fro in the channel and dispersed over a wide area. It seldom causes shoaling immediately downstream of the training works that cause erosion.

4. All flows are three-dimensional. In estuaries, the path followed by a flood current often differs from that followed by the ebb current as shown in figure 1.13. Local changes that accompany flood and ebb currents have a vital effect in maintaining the equilibrium of the system. In particular, the channels that carry the main ebb current at low tide are usually dominated by the ebb, with net movement of sediment on the bed in a seaward direction. They are terminated by a shallow bar at the seaward end of the shoals that form outside an estuary. Other channels are usually terminated by a shoal at the landward end. These are flood-dominated channels. These ebb- and flood-dominated channels can be readily identified on charts. A considerable volume of sediment can be circulated between the ebb-dominant and flood-dominant channels. In their natural state, they continually re-adjust their courses by erosion and deposition and, in this process, redistribute material over the estuary. Any realignment of channels affects the natural circulation of sediment and may cause deposition in parts of the system that were previously in dynamic equilibrium. This effect must be carefully studied whenever training works are planned.

5. Reduction in inter-tidal volume of any part of an estuary reduces the tidal influx through every cross-section seawards of that point. If the reduction in volume is small enough to have a negligible effect on water levels, the reduction in flow is equal at all cross-sections and the effect is therefore greatest

immediately downstream of the reduced part. In practice, a large reclamation or barrage affects tidal range so that changes in tidal influx vary from section to section.

In any estuary in which the alluvial material is in dynamic equilibrium, reduction of inter-tidal volume has serious consequences. The weakening of currents seawards of the work causes deposition of sediment. The effect is greatest when the reduction is made near the upstream tidal limit. It has resulted in serious siltation in many estuaries; a process that is self-perpetuating as long as there is enough sediment moved by the tidal current to bring about appreciable intertidal deposition.

6. In many estuaries, there are considerable shoals, embayments and inlets that contribute little to the cross-sectional area available for conveying water in the main channel, but which add to the inter-tidal volume.

This inter-tidal volume outside the main conveying channel contributes to the mixing processes in the estuary and is one of the factors responsible for the landward penetration of saline water. This was described in Chapters 1 and 3. Closure of such embayments and inlets reduces mixing and, therefore, saline penetration; but it also reduces inter-tidal volume and tidal currents at all points seawards.

7. Deepening a channel in a region of large salinity gradient reduces the ability of fresh water to hold back saline water due to reduction of shear stresses and of horizontal pressure gradients. Penetration of saline water inland may be greatly increased by such deepening.

8. The speed of tidal propagation depends on the square root of the hydraulic mean depth, to a first degree of approximation.

It was shown in Chapters 1 and 2 that movement of a tidal wave through shallow water results in retardation of the low-water part of the wave relative to the high water part. The wave front steepens progressively as the wave advances, the time of rise becoming shorter than the time of fall of tide. The consequent increase in flood velocities over ebb velocities causes a strong landward bias in the direction of sand transport by the tidal streams. If the hydraulic mean depth of the tidal channel can be deepened for a considerable length, tidal propagation can be made more rapid at all states of the tide, the effect being greatest near low water. The landward bias of sediment transport due to tidal flow can then be reduced in the affected reaches of the estuary. At the same time, reduced frictional losses and changes in the reflection of tidal energy back towards the sea can lead to increased tidal range in parts of the estuary.

9.2 METHODS OF CONTROL

Measures taken to control estuaries can be passive or active. Passive control consists of defence against an undesirable situation such as bank erosion or local siltation. In such cases, problems are local and it is tempting to seek local solutions on grounds of economy. In many situations local action is ineffective and may prove to be costly and dangerous. Active control has the object of bringing about a change of regime. An example is replacement of a meandering low-water channel which has shallow crossings with a deeper, stable channel.

The difficulty with active control is that the immediate objectives may be achieved at reasonable cost, but the long-term consequences in the estuary as a whole may be difficult to predict.

Methods of active control include the following.

(1) Changing the watercourse in plan or cross-section by introduction of structures or by dredging.

(2) Changing the local direction of transport of solids by controlling secondary flows.

(3) Controlling the inflow of fresh water and solids by changing the river hydrograph.

(4) Controlling the local flux of solids by modifying tidal propagation or by dredging.

9.3 CONTROL OF THE WATERCOURSE

Many methods have been used to control watercourses. Some are passive, consisting of revetments for bank protection or blankets of coarse material to control depth by prevention of scour; some are active, consisting of training banks or groynes to control the local mean speed of flow, spurs to direct the main flow where required, devices to control secondary currents or alteration of banks and bed by dredging.

Revetments

These consist of masonry or stone facings of banks for prevention of erosion. Erosion near the waterline can be caused by wave action due to wind or the wash of passing vessels. A bank of alluvial material may be weakened by the alternate rising and falling of water setting up pressure gradients, and seepage, normal to its face. Erosion can also occur due to attack by currents, particularly round the outer bank of curved reaches.

A feature of flow in curved reaches is that the highest shear stress at the bed occurs near the outer bank. There is also a tendency for the retarded layer close to the bed to be deflected towards the inner bank. The water close to the outer bank is able to erode available material. Shoals within an estuary can be rapidly eroded in this way, the face of erosion – sometimes called a 'fret' – being nearly vertical.

Figure 9.1 shows a typical eroding face in the Hooghly estuary, at Barrackpore. Note the alternating hard and soft layers – a common feature of many estuaries that carry clay-sized particles as well as coarser silt and sand. The layers containing cohesive materials resist erosion but may be undercut by removal of sand and silt from layers beneath them.

If a shore line needs protection by means of a revetment, it is essential that the bank should be clad to as low a level as possible and that undercutting should be prevented, for example, by means of a sheet-pile wall along the toe (figure 9.2). Random stone can be used; it can settle into scour holes and easily be topped up. But revetments are expensive and must be properly maintained. Moreover, the revetment prevents material from being taken into suspension in the stream, so aggressive erosion can occur immediately downstream unless the revetment is carried beyond the region of curvature.

Figure 9.1. Eroding bank of tidal River Hooghly at Barrackpore, 30 km upstream of Calcutta. The top of the bank is about 4 m above water level

The point of greatest attack will usually be different during flood and ebb; this is another reason why revetments have to be longer than the region under immediate attack.

Training banks

These are built more or less parallel to the direction of flow and are used to guide the current in a desired direction or to concentrate the flow at a particular point. Sometimes, training banks have been built up to high water level so that the whole flow can be forced into a chosen path. They can also prevent water from entering a channel from adjacent shoals and thus prevent a major source of entry of sediment. Training banks used in this way are

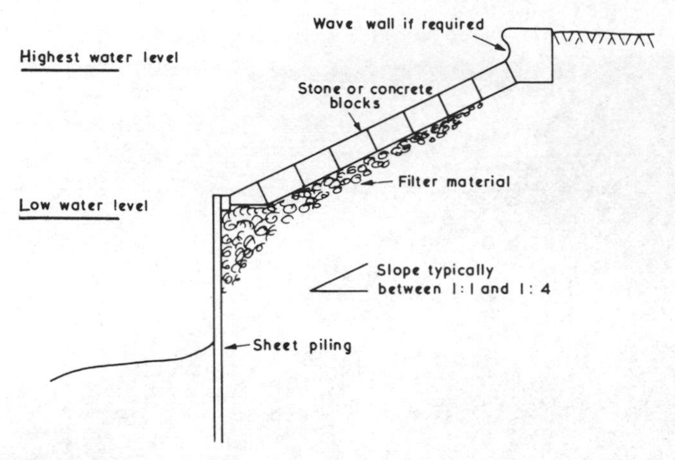

Figure 9.2. Cross-section of a revetment with sheet-pile toe wall

specifically intended to prevent the natural circulation in estuaries which consists of ebb-dominant flow in some channels, and flood-dominant flow in others.

It is more common for training banks to be built with crest level near low water level. The intention is to stabilise the channel – most frequently the ebb-dominant channel – and to concentrate the ebb into it, without interfering unduly with circulation in the adjacent channels and shoals. Nevertheless it is invariably found that training banks of this type have an effect on natural patterns of sediment circulation. They are usually accompaned by considerable accretion of sediment outside the trained channels[1]. Figure 9.3 shows accretion that followed training of the Ribble Estuary. If such accretion is not objectionable, training banks can achieve a useful purpose. Unfortunately they are often built to meet a particular situation that requires urgent action. They are seldom built in the best location to achieve their long-term objectives: and once built, they are very expensive to alter[2].

Training banks cause flow through unit width of channel to be increased locally. The rate of sediment transport is increased and erosion is forced to occur. When water leaves a trained section it thus carries more sediment than it would if the flow were untrained. Whereas fine silt and clay are transported away by the stream, sand in movement must be deposited until the proper balance between flow and sand transport is established. Unless training banks extend into deep water, shallow bars must form just beyond their ends. Once a reach of a river is trained for any purpose, it is quite likely that training banks will have to be extended progressively into deep water if their effectiveness is to be preserved. Moreover, if low water training banks lead to accretion outside the trained channel and eventually overtop, as in figure 9.3, it will be necessary either to dredge or progressively raise the banks.

Training banks must be planned with great care, usually with the aid of model experiments. The minimum amount of work that will achieve the immediate objective should be done, further extensions being planned in

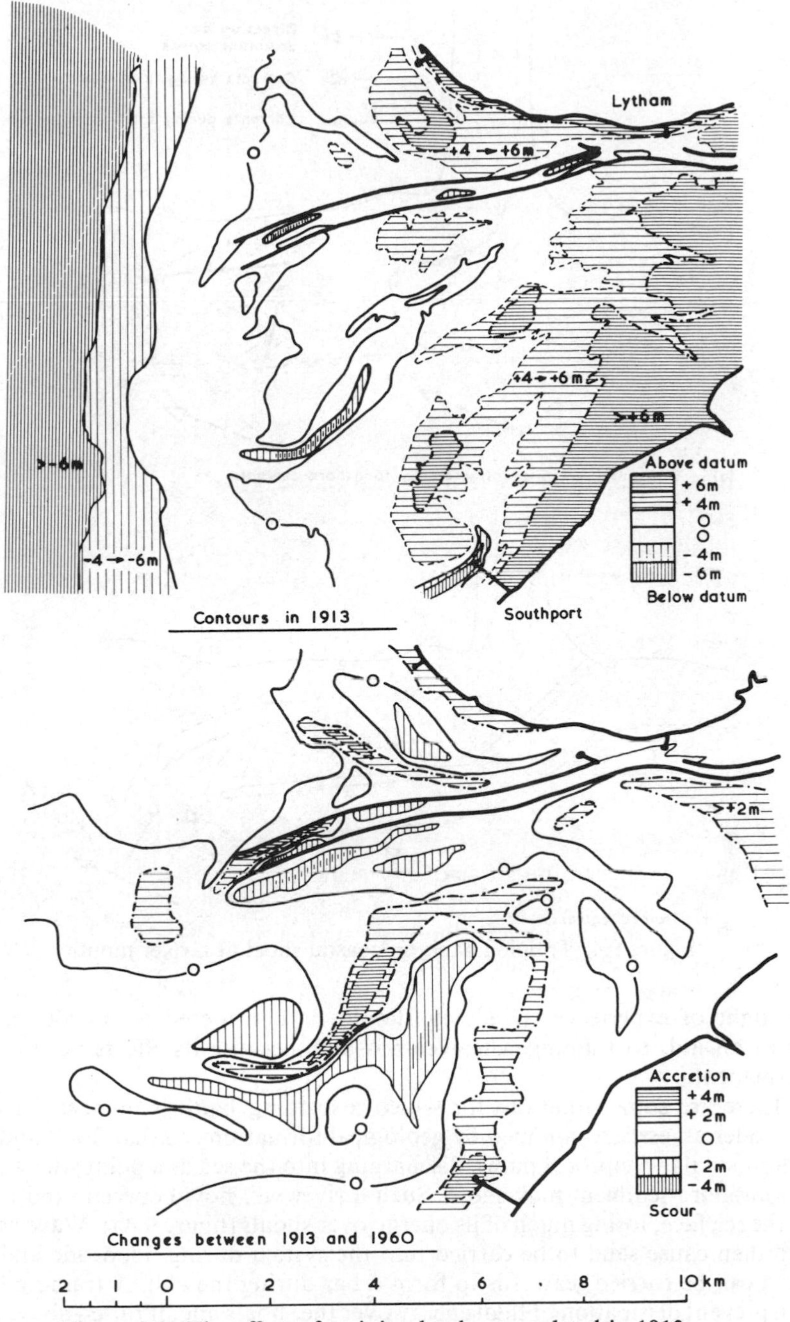

Figure 9.3. The effect of training banks, completed in 1913, on the Ribble estuary

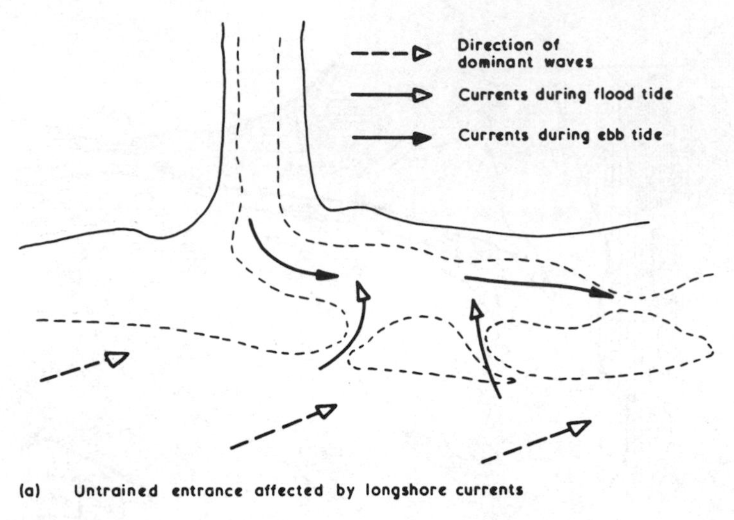

(a) Untrained entrance affected by longshore currents

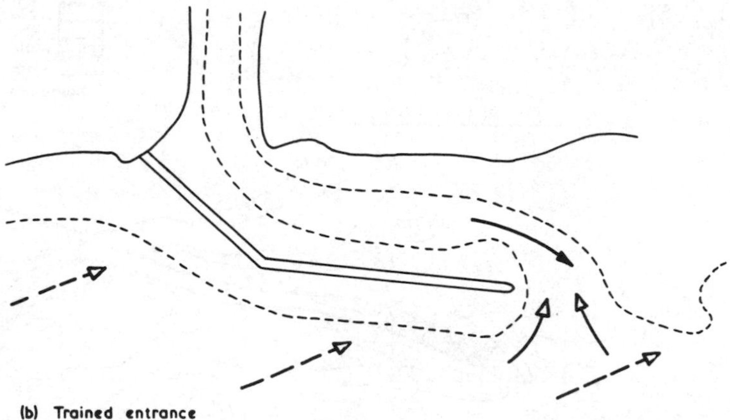

(b) Trained entrance

Figure 9.4. Training bank on coastal shoal at a river mouth

the light of experience. Dredging should be considered as an alternative, or as an aid, to training wherever possible, because its effects need not be permanent.

There are some situations for which a training bank is an ideal answer. Consider an estuary confined by geological formations so that flood and ebb follow similar reciprocal paths, discharging into the sea at a point where there is longshore sediment movement. Such a river will flow between sand banks at the sea face, losing much of its energy over shoals (figure 9.4a). Wave action can then cause sand to be carried into the system during flood tide and this sand can be carried seawards to form a bar during the ebb. A training bank can prevent dissipation of tidal energy over the shoals and, if raised above high water level, can act as a breakwater to prevent ingress of wave-induced sand motion (figure 9.4b).

It is relatively easy to construct a random stone training bank with crest level

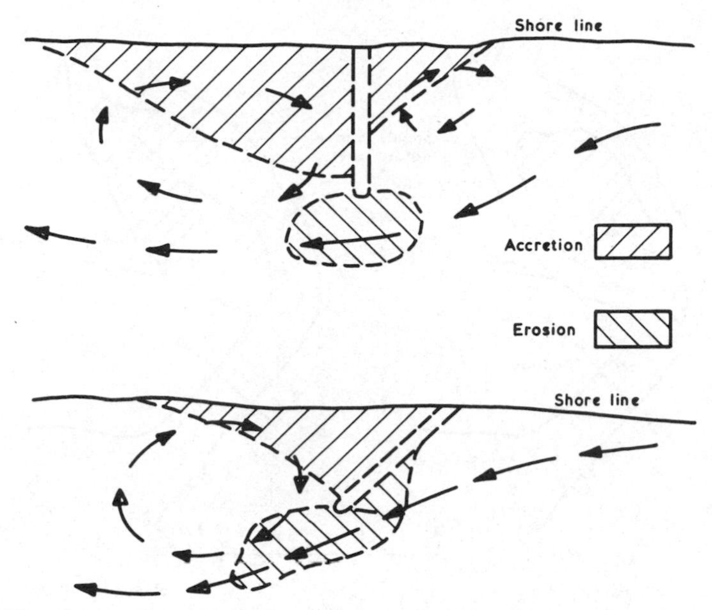

Figure 9.5. The action of spurs perpendicular to the stream (top) and inclined away from the stream (bottom)

near low water if the tidal rise is enough to allow loaded barges to operate over its crest for much of the time. A higher crest level will restrict or prevent the use of barges and special plant may be needed to raise the bank to its full height. A crest level above high water can be achieved quite easily if the training bank is connected to dry land, but in all other cases construction requires special plant, or facilities that are only justifiable if major work is being done.

Spurs or groynes

These are constructed with their crests forming a large angle – between say 45° and 135° – to the flow. They may be built from the shore, from an island, or from a training bank. They may be solid structures impervious to the flow or they may be made permeable. Solid spurs prevent flow normal to their sides and cause kinetic energy to be converted to potential (figure 9.5). The region of elevated surface causes flow to be deflected away. Spurs directed at an angle of 90° or less towards the oncoming flow have the greatest effect in deflecting the flow. Spurs directed away from the flow have a smaller effect and the flow at the tip may run nearly parallel with the spur[3]. The former have been called repelling spurs.

Single spurs can often work well when only local, passive effects are required. A good example was a spur constructed in the River Rupnarain in 1944 to prevent the imminent danger of erosion of the shore and destruction of a railway bridge[4]. Several schemes were tried in an hydraulic model. The single spur was immediately effective in diverting the main stream away from

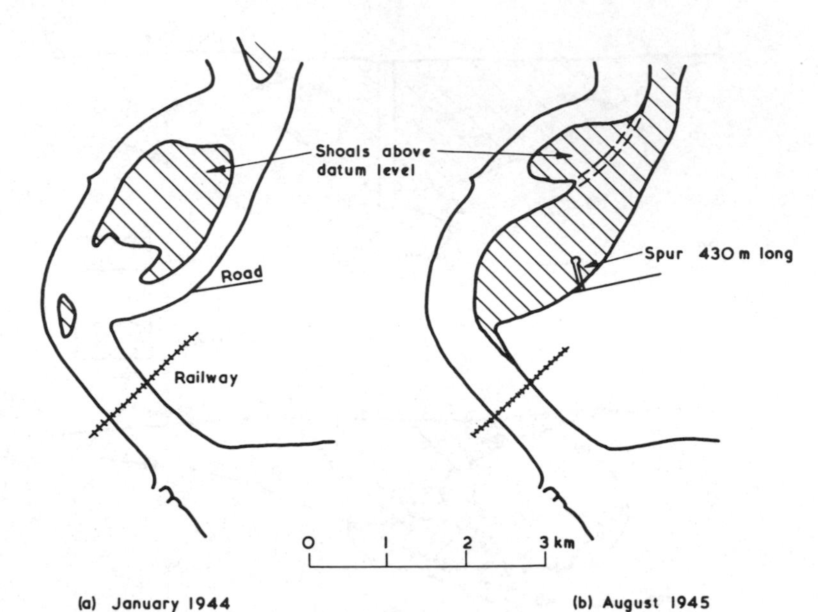

Figure 9.6. Action of a single spur in the tidal River Rupnarain (based on figure 9 and 12 of reference[4], courtesy the Institution of Civil Engineers)

the bridge (figure 9.6). Siltation on either side of the spur caused the shore line to re-form at its tip and within a year or so the spur had been totally assimilated into the new shore. There has been no recurrence of bank erosion that would threaten this bridge.

This example illustrates an important point. Spurs operate by deflecting the flow. They are most effective when they work with natural forces rather than against them. A revetment could not have been used to prevent bank erosion in this case. It would have been a very expensive structure built at the point of attack and for some distance on either side of it. A revetment would have taken longer to build than the spur and would not have been effective until nearly complete. At any time, bank erosion could have out-flanked the work during construction or after completion. It is most unlikely that a revetment could have been built with the available resources. The spur, on the other hand, was built upstream of the point of severe bank erosion and it began to be effective as soon as it projected into the stream. It could not be easily outflanked. Its effect could be seen and controlled at any time by adjusting its final length.

Single spurs can cause large-scale eddying and consequent energy loss; moreover, their effect is local and can be drastic, with severe erosion at the tip and heavy accretion in the eddying zones. The worst effects of spurs can be almost eliminated by using a series of spurs, so spaced that they interact to prevent gross eddying (figure 9.7).

In practice it has been found that, if spurs have length L, they are most

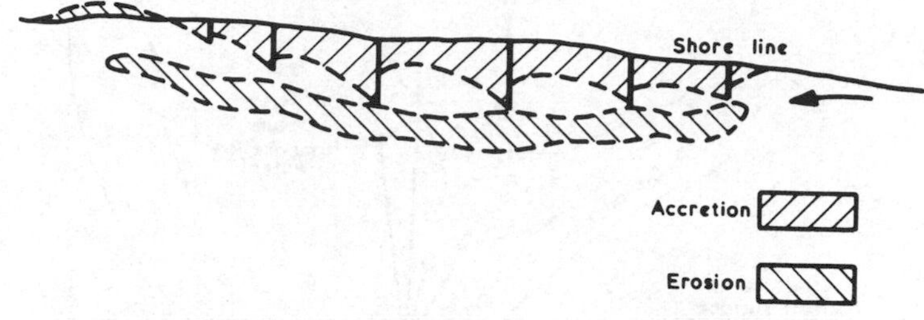

Figure 9.7. Action of a series of spurs

effective if the spacing is not more than L on outer banks and 2 to $2\frac{1}{2}L$ on inner banks of curved reaches[3].

Spurs can also be used, like training banks, for control of flow channels[5]. A series of spurs can stabilise a channel and direct the flow into a narrow path over shoals and crossings, thereby increasing the natural depth. They are usually built with crest height at, or below, low water level at the tip, but rising gradually towards high water level at the base. The object is to provide a large cross-section for flow near high water, but to concentrate the flow at low water.

Spurs have one particular disadvantage compared with training works. They offer greater resistance to the flow and the resulting energy loss may cause rapid siltation and therefore loss of tidal capacity outside the main channel. This effect can be reduced by construction of permeable spurs. On the other hand, spurs have two decided advantages. Construction can proceed from the shore and is usually easier than construction of a training bank. The most important advantage of spurs, however, is flexibility. The effect of a spur can be seen to develop as it is built. Its crest height or length can be modified and construction stopped as soon as the desired effect has been achieved. Modification can be done without removing and re-building the whole structure. The basic design can be varied to suit local materials of construction and it can be made permeable to any chosen degree.

Any structure in an estuary, such as a jetty or berth, acts as a spur and can modify the local pattern of flow and water surface level. The latter can have an effect on the behaviour of ships alongside a jetty; the British Transport Docks Board Research Laboratory have demonstrated that a transverse surface gradient set up by a jetty at Immingham is sufficient to prevent ships from being forced against the jetty by the current.

9.4 CONTROL OF TRANSVERSE CIRCULATION

When water enters a curved reach of a river, the momentum of flow causes a concentration of shear stress near the outer, concave bank. Near the inner bank, velocities are weak and shear stress is low. In sharp bends, there may even be a separation of flow at the inner bank with reversal of flow close to the shore. The result is that bed material is transported rapidly round the outside of the bend but only slowly round the inside. Depths round the outside are greatest,

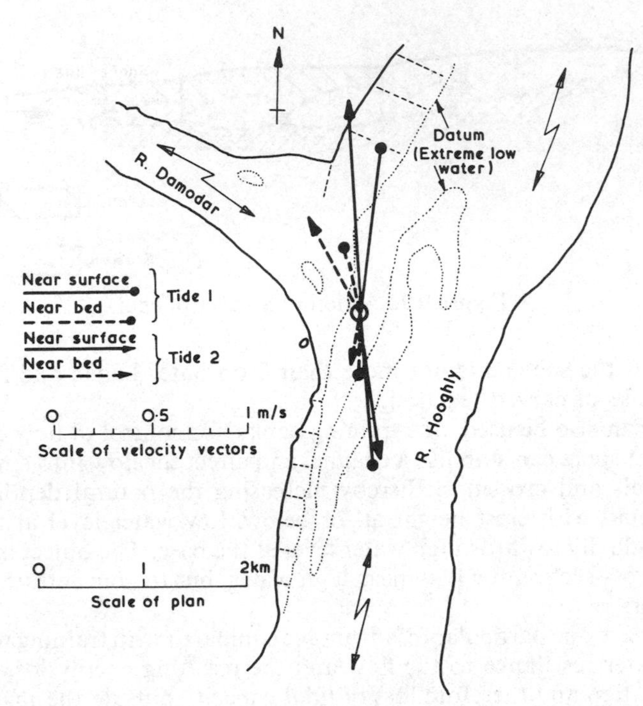

Figure 9.8. Observations in the River Hooghly showing effects of curvature of flow (courtesy the Calcutta Port Trust)

and granular materials remaining on the bed are coarse. Round the inside of a bend, on the other hand, the shoals consist of fine sand, silt or clay-sized particles.

In addition to the variation of shear stress over a cross-section, flow direction at a point may vary with depth. The effect can be quite marked in curved reaches; some measurements obtained at a station in a curved reach of the River Hooghly are shown in figure 9.8. The variation of velocity is caused by lateral pressure gradients brought about as the water is deflected round a bend. These pressure gradients produce radial accelerations which cause larger deflections of the slower moving water near the bed, than of the faster moving water near the surface[6, 7]. In real estuaries the depth of water is very much smaller than the width. This, coupled with the high level of turbulence, usually prevents formation of such a simple pattern with flow towards the outer bank at the surface, and away from it at the bed, at all points in a cross-section. The effect is often confined to a layer close to the bed and is not consistent over the whole width of a stream. Moreover, in reversing tidal streams the sense of curvature is often locally different between ebb and flood flows.

Variation of shear stress at the bed and transverse circulation due to flow curvature should be taken into account when designing works in rivers and estuaries. Intakes, locks and jetties sited round the inside of a bend suffer

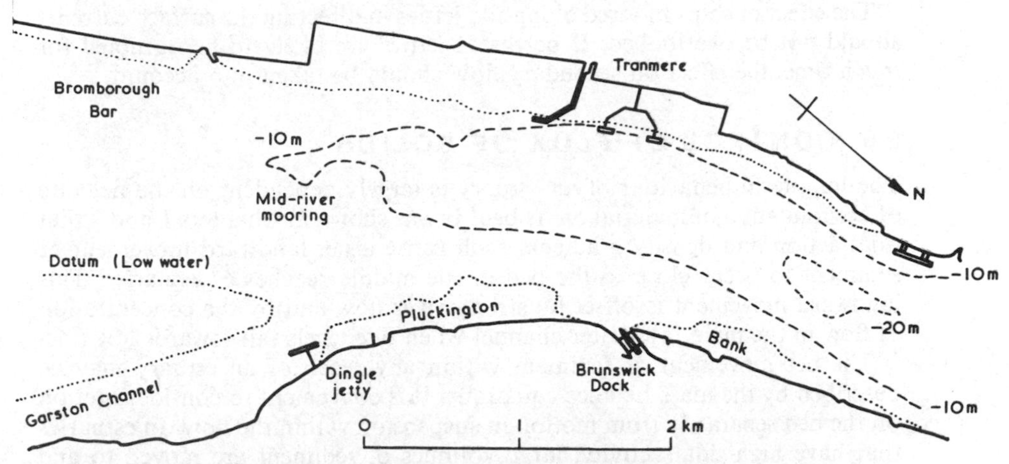

Figure 9.9. Brunswick dock entrance and Tranmere oil jetty

much heavier siltation than similar works sited round the outer side. Maintenance dredging costs can be very high in such cases. Engineering works have often been sited without proper regard for such hydraulic effects because other considerations have been thought to be over-riding. The consequence has been high cost of maintenance dredging throughout the life of the structure. One example was the entrance to the Brunswick Dock system at Liverpool. The entrance channel was dredged through the Pluckington Bank (figure 9.9) to a depth of 1 m below datum. There is a tidal rise between 8 m and 10 m above datum at this point. Annual maintenance dredging of this entrance averaged about $1{\cdot}5 \times 10^6$ hopper tons, though in some years it was much higher. On the opposite side of the river, the Tranmere jetties are maintained at depths of 12 m and 13 m below datum with dredging of only $2{\cdot}3 \times 10^5$ hopper tons a year.

Advantage can be taken of secondary current effects by careful location of installations. Attempts have been made to regulate these effects in rivers[8] and proposals have been made to control them in estuaries by means of vanes inclined to the flow. One technique, developed at the *Laboratoire de l'Electricité de France*[9], is to fix vertical panels at the bed of the stream, inclined to the flow at angles between 10° and 45°, and extending only a short distance above the bottom. They have been used to control secondary flow over a crossing and into a branch of a river. In the former, depths over the crossing were increased, and in the latter, bed-load was diverted into the branch, bringing about its closure during periods of low discharge. In each case, fixed panels were attached to piles in the bed of the river, the overall effect being similar to the system known as 'bandalling' in India[10].

In reversing flows, it is not possible to provide fixed panels on the bed of a channel that would operate in the same sense on both flood and ebb currents. An alternative is to control surface flow by means of floating guide vanes pivotted about one end. So far, these have been tried in laboratory channels with some success, but none had been tried in practice up to 1975.

The effect of ships moored alongside jetties in directing the surface currents should not be overlooked. If berths at jetties are likely to be occupied for much time, the effect on secondary flow should be taken into account.

9.5 CONTROL OF FLUX OF SOLIDS

The long-term behaviour of an estuary is largely dependent on the net rate of accumulation of material on its bed. It was shown in Chapters 1 and 2 that tidal action and density gradients each cause a net landward movement of sediment to occur close to the bed in the middle reaches of estuaries. This landward movement is offset by strong river flow and by the concentration of flow in the main low-water channel when tide levels fall towards low tide.

The net movement of sediment within any reach of an estuary may be described by the mass-balance equations. It is convenient to consider motion on the bed separately from motion in suspension within the flow. In estuaries that have high tidal activity, large volumes of sediment are moved to and fro with each tide. The material balance equation contains large terms and net movement is represented by the difference between two large numbers. For example, it has been estimated that the mean annual influx of solids in suspension into the Mersey Estuary during rising tides is in the order of $140 \times 10^6 t$. During the nineteenth century there were considerable fluctuations in the volume of water below high tide level, but no trend of net accretion or erosion over the century as a whole. Evidently the natural forces were in long-term equilibrium. From 1911 to 1960, following various engineering works in the tidal Mersey and its sea approaches, there was an average deterioration of $1{\cdot}1 \times 10^6 t$ of solids a year. This is less than 1 % of the weight of solids moving in suspension during flood tides at any time. A more detailed analysis of sediment movement in the Mersey Estuary is given in Chapter 10.

9.5.1 Bed load transport

Sediment in motion in an estuary moves to and fro under tidal action, but sand moving close to the bed moves a comparatively short distance during each spring tide and does not move at all until the critical shear stress at the bed is exceeded locally. During neap tides there may be little or no movement, but the rate of transport increases very rapidly with increase of shear stress. In tidal flows the shear stress depends on the acceleration as well as the velocity. However, near the times of maximum flood or ebb current, the acceleration is small and shear stress is nearly proportional to the square of the flow speed U in a well-mixed estuary. Bed-load transport will then be approximately proportional to $(U - U_c)^5$, where U_c is the velocity needed to initiate bed movement. Control of the flux of bed material in any given situation therefore depends strongly on control of flow speed.

Control of flux of solids can be achieved in several ways. One is by control of the hydrograph of fresh water discharge, by construction of dams and barrages. These are usually built to enable peak flows in rivers to be reduced, the stored water being made available in times of low discharge. Because sand transport varies more or less with $(U - U_c)^5$, removal of peak flows leads to a considerable reduction in transport of solids seawards in those parts of a tidal system in which fresh water flow has a dominant influence. In a typical estuary, this would occur in the upper reaches and could also occur in the low water channels if these have a much smaller sectional area than the high water channels. An exception is the case of stratified flow, described in Chapter 3, Section 5, in which a moderate increase of flow of fresh water would also increase net landward movement close to the bed, and a reduction would decrease the rate of net landward movement. The situation is more complicated in reality because flow might be stratified in an estuary during neap tides and moderate river discharges, but mixed during spring tides.

Large river discharges displace a saline wedge seawards while increasing the intensity of landward movement in the lower layer. In many natural systems that display long-term stability, occasional discharges that flush the saline water out of the mouth are a necessary part of that stability. Control of peak discharges to provide useable water and to control floods can seriously disturb the equilibrium of such an estuary. When planning river basin development, the consequences of removal of peak flows should be examined and the possibility of restoring equilibrium by occasional flushing with high discharges should be considered.

Flow speed may also be controlled by changing the cross-section. Local increase of depths over shoals can be achieved by means of training works or dredging. Although the discharge per unit width may be increased locally by training a reach, the total discharge of solids over the whole width will eventually be the same as it was in the untrained river. Local dredging, on the other hand, results in an increase in cross-section and in reduction in the total rate of sediment discharge for a given rate of flow of water. Local deepening has little effect on an estuary as a whole, so the rate of sediment transport in adjacent reaches is almost unchanged. Sediment is brought to the dredged reach at the same rate as formerly and some is deposited in the dredged reach. Regular dredging is needed to maintain depths in the channel. Knowledge of the rate of sediment transport before and after deepening enables the rate of dredging to be estimated using the equations described in Chapter 4.

9.5.2 Sediment in suspension

The behaviour of fine solids travelling in suspension is described in Chapters 1 and 4. The material travels a considerable distance to and fro with each tide. The characteristics of the fine material and the salt content of the water each have a bearing on its behaviour. Silt particles behave like sand so far as initiation and cessation of movement from the bed is concerned, but once brought into motion they move with the water and only settle slowly out of suspension when the level of turbulence is reduced. Clay-sized particles form flocs in saline water, but they settle rapidly in still water or in weak currents to form a loose, unconsolidated layer when they first reach the bed. This layer has the properties of a liquid with very high solid content; it may be thixotropic, requiring a small but finite shear stress to bring it into movement but behaving as a viscous liquid once in motion.

Flocculated mud settles out of a turbulent stream only at very low speeds of flow over the bed. On the other hand, it requires a greater shear stress and flow speed to bring it back into suspension. Thus, suspended solids may be carried landwards by the rising tide, but if they are deposited along the margins of an estuary, they may remain there during the falling tide. Similarly, material carried up river during neap tides may accumulate near the bed to form a layer of liquid mud, but this may disperse during the following cycle of spring tides. In any case, the quantity of mud moving in suspension at any time depends more on the availability of material for erosion than on the intensity of shear stress, once the critical shear stress has been exceeded.

A mud layer will have a density and critical shear stress that gradually increases with distance below its surface. As flow over the mud gradually speeds up, the critical shear at the bed will be exceeded for successive layers. The availability of material for movement is therefore limited by the gradually increasing critical shear stress as the surface layers are removed. The key to the problem of fine solids is the amount of unconsolidated material that is available for erosion. Control can be achieved by reducing it, either by removal from the system or by avoiding disturbance of the bed unless absolutely necessary.

Removal of fine solids from the system can be done by dumping solids ashore. Dumping spoil at sea, even at a considerable distance offshore, will not guarantee that the material will not return. In most offshore situations there is a net landward movement at the bed induced by density effects, towards any estuary that discharges much fresh water. This is discussed in the next section, on dredging.

Disturbance of the bed may be caused by the passage of ships, but the most frequent cause is through the action of dredging. One of the most effective methods of control of intractible problems caused by fine sediment has been to modify dredging techniques and to minimise dredging.

9.6 CONTROL OF FLUX OF SOLIDS BY DREDGING

Before discussing this subject, it is first necessary to describe the operation of dredgers as tools of maintenance of depth. Bucket ladder and grab dredgers are used for maintenance dredging in special situations, such as alongside quays or in docks, but the main tools used for channel maintenance are suction hopper dredgers. They consist, basically, of a pump which sucks sand or mud from the bed and discharges it into a hopper within the ship. The hopper is fitted with doors or valves through which the spoil can be discharged at a disposal site.

The first suction dredgers were stationary dredgers that were anchored at the dredging site and sucked sand from their immediate vicinity. The result was a deep hole around the suction pipe. Stationary suction dredgers have no place in maintenance of channels that have a mobile bed. All modern dredgers intended for this purpose are designed to move over the bed while dredging. The suction pipe is fitted with a specially designed head that is dragged along the bottom behind its point of entry to the ship. Such dredgers are known as trailing suction dredgers (figure 9.10).

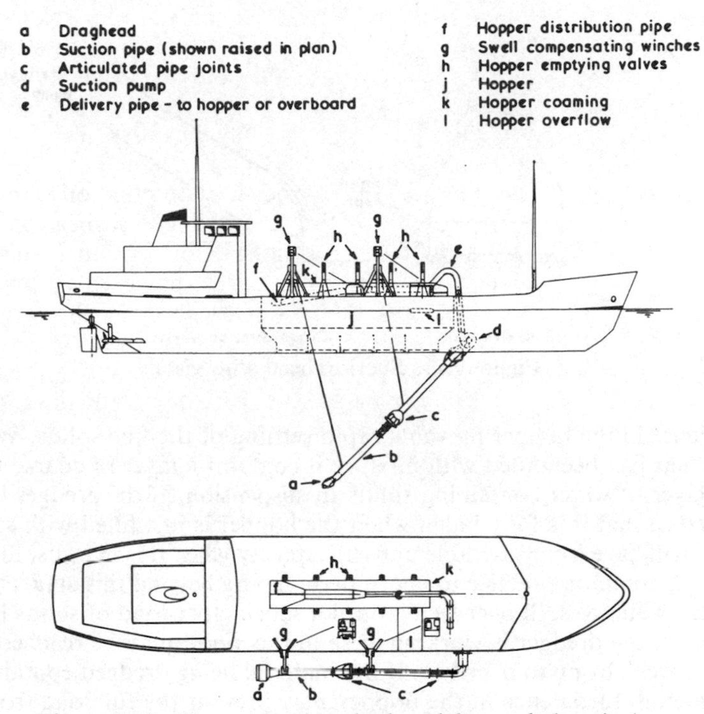

Figure 9.10. Hopper suction dredger (plan and elevation). (Based on Stantrail dredger of IHC, Holland)

Recent research into the design of suction heads has led to great improvements in their effectiveness [11, 12]. Modern heads may be equipped with visors that automatically adjust the face of the head to the slope of the ground (figure 9.11); with blades to break up consolidated material; and water jets to break up the ground and to assist the flow of solids into the suction pipe, at high concentration, without choking. These suction heads rest on the bottom and are drawn over it at relatively high speeds, of order 1–2 knots, or over 0·5 m/s. Trailing suction dredgers can make long, shallow cuts in the bed, from end to end of most bars and crossings. The efficiency of such dredgers depends largely on the power required to propel the ship forward while dredging; this has been found to be least, with many bed materials, when jets are used while dredging with a modern suction head.

Modern suction pipes are articulated and supported by winch-controlled wire ropes so that they can withstand some vertical movement due to waves and irregularities of the bed without damage. The winches can be fitted with swell compensating gear that can maintain nearly constant tension in the supporting wires when working in waves or over an undulating bottom.

A suction hopper dredger removes a mixture of water and solids from the bed. Their ratio depends on the nature of material being dredged, on the design of suction head and pumping system and on the method of operation. The mixture is pumped into the hopper where coarse solids settle quite rapidly.

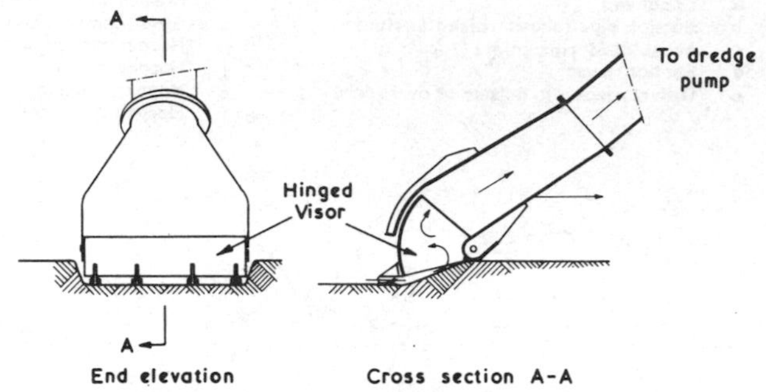

Figure 9.11. Suction head with visor

Turbulence in the hopper prevents rapid settling of the fine solids. When the hopper has just been filled with mixture it contains a layer of coarse material and a layer of water containing solids in suspension. If the dredger has been designed so that it is fully laden when the hopper is just filled with saturated sand, it will have a considerable unused capacity when it is first just filled with spoil. It is common practice to continue pumping beyond this stage, allowing silt-laden water to spill over weirs, until a satisfactory load of solids has been taken in. If the dredger is working in sand this point may be reached shortly after the weirs begin to overflow. If the material being dredged contains much fine material, turbulence in the hopper may prevent the full load from being reached even if pumping is continued indefinitely. It is uneconomic to continue dredging once the rate of loading falls below a certain value. A typical loading curve is shown in figure 9.12.

The practice of allowing silt-laden water to return to the estuary has serious consequences. In addition to silt in the overflow from the hopper, the suction head disturbs the bed, with the result that fine material is brought into suspension. The dredging operation then consists of removal of relatively coarse material in the hopper and return of most of the fine material to the estuary. The fine material, which was formerly consolidated on the bed or mixed with coarser material, is put into suspension and added to the wash-load in the estuary.

The process of removal of the coarser fraction of the bed material, while allowing the fine material to be returned to the system from hopper overflow and disturbance by the drag-head and ship's propellors, is compounded by spoil-disposal methods in common use. If some material can return from disposal sites by combined currents and wave action, it is inevitable that the fine material will be returned most readily.

There are exceptional situations in which fine material may be allowed to be disturbed or discharged in large quantities. They arise when currents clearly carry material away from the area of dredging so that it cannot return in a less-consolidated form. One such case is dredging of the mouth of the River Orinoco[13], where material from the channel has simply been cast to one side of a dredger through a long boom, to be carried away by the littoral current. Another is side-cast dredging in the River Plate[14]. The River Plate

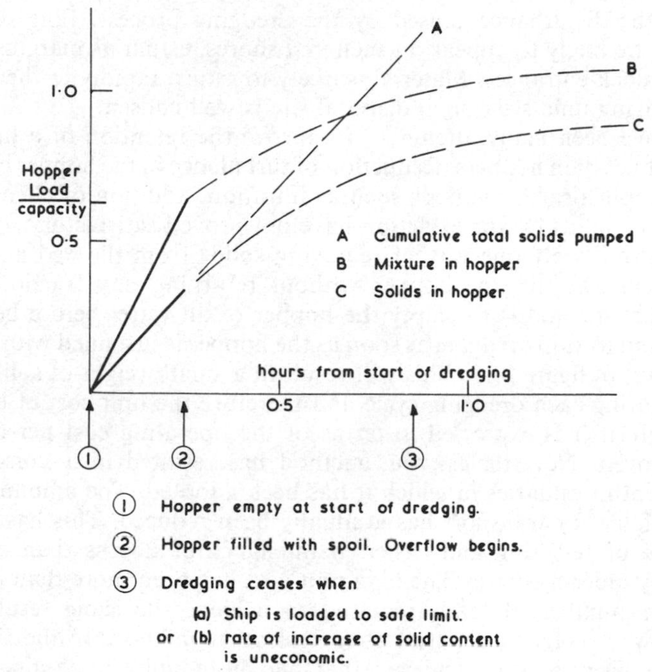

Figure 9.12. Typical loading curves for hopper suction dredger

discharges into a large fresh water embayment before finally reaching the sea. Circulating currents within this embayment carry spoil away from the site of dredging. Because the water in the embayment is fresh, there are no saline density currents and the material on the bed is not flocculated. Tests have shown that little material returns directly to the dredged cut and the dredging operation changes neither the total quantity of silt nor its grain size distribution. Such cases, however, are exceptional and use of side-cast or agitation dredging in most estuaries is harmful.

9.7 THE DREDGING CYCLE

The complete cycle of dredging consists of:

1. Removal of solids from the bed through the suction pipe.
2. Accumulation of solids in the hopper of the dredger.
3. Return of fine solids to the estuary via the overflow from the hopper.
4. Disturbance of the bed by the drag head and propellors of the dredger, resulting in the transport of fine solids by the current.
5. Removal of spoil within the hopper to a disposal site.
6. Dumping of spoil; coarse solids settle rapidly to the bed, but fine solids may be transported away from the site by tidal currents before they can settle.
7. Dispersal of some – or all – of the solids from the disposal site by waves and tidal streams.

8. Siltation at several places as a direct result of the disposal of dredged spoil and the disturbance caused by the dredging process. Fine solids, in particular, are likely to appear on sheltered shores, estuarial margins and in docks and dock entrances. Material is likely to return rapidly to the original site of dredging unless the spoil disposal site is well chosen.

There have been many attempts to improve the retention of a high proportion of fine silt in hoppers. Reduction of turbulence in the hopper can help, but more sophisticated methods such as filtration, addition of chemicals or separation of solids by vortex action have not proved satisfactory so far. At present, there is only one way of removing solids from the bed and transporting them away in the hopper without returning any fraction to the estuary. That method is to empty the hopper of all water before beginning to dredge and to stop dredging as soon as the hopper is just filled with mixture – at point (2) in figure 9.12. This will result in a small weight of solids being removed during each dredging cycle and therefore the unit cost of dredging will be high, if it is expressed in terms of the operating cost per tonne of solids removed. Nevertheless, this method has resulted in a considerable improvement in estuaries in which it has been adopted. The amount of fine solids available for transport has gradually been reduced. This has led to a slower rate of re-silting and easier maintenance of depths than could be achieved by older methods. The high unit cost has been more than offset by the smaller number of loads required to achieve the same results. This technique was evolved after trial of several other methods in the Delaware estuary[15]. Depths were achieved that had been quite unobtainable previously, despite intensive operations in which hopper overflow took place; moreover, the total cost of dredging was reduced despite the higher unit cost of dredging.

9.8 SPOIL DISPOSAL

The disposal of dredged spoil should be at a site from which there is no chance that the material will return to the dredged cut or to other undesirable places. This ideal is seldom attainable in practice and a choice has to be made between several possible compromises. Spoil may be placed, in decreasing order of effectiveness:

1. Ashore.
2. Offshore, at a site from which it cannot return to the estuary.
3. At a site from which only a manageable quantity returns to the dredged cut.
4. Far enough from the site of dredging that there will be a useful delay before it returns.

Choice of a spoil disposal site is clearly an essential part of a dredging operation and a full economic study of the alternatives is necessary. The best choice from the point of view of the contractor, in economic terms, is not likely to be the best choice for the port management unless the dredging contract is drawn up in such a way that the overall effect of dredging can be taken into account.

In many estuaries, the distance between the site of dredging and a spoil dumping ground at sea is small enough for economic dredging operations to be practical. In long tidal systems, however, dumping at sea from some dredging sites is quite impossible. Spoil must then either be placed ashore or returned to the river. For example, in the River Hooghly, much of the dredging is done to control the location and depth of crossings during seasonal changes. The tidal river is so long that it is impractical to carry the spoil out to sea and strong currents and intensive land use make it impossible to place the spoil ashore. Moreover, the amount of fresh sediment brought down by rivers during the monsoon is much greater than the amount dredged annually. Spoil is deposited in the river at locations from which it is unlikely that it would return quickly to the site of dredging. In other cases, such as the River Delaware, the annual volume of dredging has exceeded the annual supply of fresh material. It has been found that the disadvantages of placing spoil ashore are outweighed by the improvement of the estuary as a whole when this is done.

It has already been shown that dredging causes sorting of fine from coarse particles. This process may be taken further by the process of spoil disposal. Considerable dilution occurs when the spoil leaves the hopper. Large particles settle first, while fine particles settle slowly. As they settle they may be carried by tidal currents away from the dumping site, generally shorewards in the direction of net drift. Working of the bed by wave action can cause further separation of fine material, which may then be carried shorewards by wave-induced flow and by density currents. Spoil dumped in the Irish Sea from the Mersey Estuary has been shown to behave in this way [16]. Regular surveys and analysis of the mass balance equations described in Chapter 4, have shown that about 25 % of dredged material remains at the dumping site, while most of the remainder, consisting of silt and clay, returns quite rapidly to the system.

9.9 QUANTITATIVE ESTIMATION OF DREDGING PERFORMANCE

Quantitative estimates of dredging performance can be made by use of the mass balance equations. Consider the case of the Delaware Estuary in the United States as an example of use of equation 4.21. This partially-mixed estuary has a classic trumpet shape in plan and has no tortuous channels or inter-tidal flats. The bed material in the saline intrusion zone is mud which enters the estuary exclusively in freshwater runoff (S_R)[15] and amounts to some 4·6 Mm^3 per year of *in situ* material. The estuary navigation channel has been progressively deepened and is maintained by dredging. Prior to 1915 the dredging was performed by agitation or by hopper dredgers that dumped freely within the estuarine system. After 1915 the dredged material was dumped into two areas at the margin of the estuary. A stationary cutter pipeline dredge pumped sediment ashore from these areas. Progressive deepening of the navigation channel to some 12·2 m below low level was attempted in 1940. Rapid shoaling resulted and by 1950 some 22 Mm^3 of material was dredged annually. In spite of this, considerable difficulty was found in loading the dredges and navigation depths could not be increased above about 11 m.

Equation 4.21 can be applied using data from Wicker's paper[15] and constants found for the Mersey Estuary. It shows that a considerable quantity of the dredged material was returning to the navigation channel. Field observations indicated that ΔV, S_A, $S_O = S_E$, S_{LB} and S_P were small or zero and that D averaged some 12·3 Mm³ between 1943 and 1951. The quantity of returning dredged material (S_D) is then given by equation 4.21 as 7·7 Mm³, i.e. only some 37 % of the amount deposited at the spoil ground (12·3 Mm³) was permanently placed ashore, and the rest returned to the estuarine shoals.

The moral of the example is clear. Sediment should not be deposited within an estuary such as the Delaware in areas where it can be moved by tidal currents, since the residual density circulation will return it to established shoaling zones. In the case of the Delaware, the re-handling areas were abandoned and sediment was subsequently pumped directly from the dredgers into the hold of a converted ship from which it was pumped ashore. By 1963, projected navigation channel depths were reached (12·2 m) and annual dredging was reduced to 5·4 Mm³.

Solution of equation 4.21 depends on knowledge of the amount of spoil removed by dredging. Dredging records have often been expressed in such a crude manner that quantitative estimates could not be made; for example in hopper tonnes, based on the number of loads removed each year, with no indication of the density of each load. It is now usual to record the amount of solids removed, based on simple measurements.

If the hopper of a dredger contains a known volume of a mixture of water and solids, the total mass in the hopper is

$$\rho V = \rho_w V_w + \rho_s V_s \tag{9.1}$$

where ρ is the density of the mixture,

V is its volume

suffix $_w$ indicates the corresponding values for water

suffix $_s$ indicates corresponding values for solids in the mixture

The total mass of the mixture in the hopper can be found from the change in the volume displaced by the ship during the loading operation, while the total volume in the hopper can be measured. In each case, care must be taken to allow for the heel and trim of the ship. In practice, pneumatic draught gauges are widely used to indicate these quantities in the control room of the dredger. They can compensate for the trim of the ship and can indicate the total load in the hopper.

The total mass of the mixture in the hopper is:

$$W = \rho V \tag{9.2}$$

and its volume

$$V = V_w + V_s \tag{9.3}$$

if the mixture is saturated with water.

The density of water ρ_w will depend on the salinity and will be between 1·00 and 1·024 t/m³. The density of solids ρ_s in the hopper will usually average about 2·65 t/m³ but may vary somewhat if there is much organic matter present.

The overall density of mixture in a hopper includes settled solids, suspended solids and water. If the hopper is just filled with clean sand with a negligible volume of free water above the saturated mixture, the density in the hopper is likely to be nearly 2·0 t/m^3. At this density, the mass of solids in 1 m^3 of mixture would be about 1·6 t. In practice it may not be possible to obtain such a high density; a dredger designed to work well in silt requires a hopper with a relatively large volume. Such a hopper filled with sand would overload the ship. General purpose dredgers are often designed to be fully laden when the hopper is filled with a mixture of density 1·33 t/m^3, which would have a solid content of about 0·5 t/m^3.

The density of a silt mixture is very variable. Stiff clay cannot be dredged efficiently with a trailing dredger; a cutter suction dredger or a bucket dredger would be needed. Silty clay and highly consolidated silt can be dredged with the aid of fixed cutting vanes on the underside of the suction head; but these materials are not likely to be encountered in the course of maintenance dredging unless a channel has been neglected for several months or years. Moderately concentrated silt can often be dredged with little or no dilution. The soft silt can then be discharged into the hopper with a paste-like consistency, so that a comparatively high solid content can be achieved. The density of the mixture in the hopper could be about 1·6 t/m^3, with a solid content of about 0·9 t/m^3. At the other extreme, the maximum concentration obtainable in the hopper while dredging soft silt from docks, or entrances, may be less than 1·2 t/m^3, with a solid content of less than 0·25 t/m^3.

9.10 CONTROL OF TIDAL PROPAGATION

Training works or dredging may be used specifically to modify tidal propagation, with the object of reducing landward movement of sand or increasing the tidal range available for movement of ships. It is sometimes advocated that large tidal rivers should be trained with this in mind, so that removal of shoals and deepening of channels would modify tidal propagation.

The effect of deepening can be studied quite well by using one of the computational methods described in Chapter 6, or by construction of an hydraulic model. For example, in the Lune estuary (UK) removal of shoals in 1847–51 resulted in deepening of the main channel by about 1 m over a length of over 7 km. The duration of flood tide near the mouth increased from 3 h 20 min to 3 h 45 min[1] immediately after completion of the work. The ebb tide was shortened by a corresponding amount.

The effect is greatest when there is a large ratio of high water depth to low water depth; that is, when low-water channels are shallow. Extensive deepening can then have a measurable effect on tidal propagation, leading to a reduction of maximum flood-tide velocities and an increase in ebb-tide velocities. However, the changes are relatively small and any advantage could be offset by increased penetration of saline water. In the case of the Lune estuary siltation occurred outside the deepened channel as a result of construction of training works and caused the duration of rising tide to revert to 3 h 20 min over a long period (see Chapter 10).

9.11 CONCLUSIONS

The management of an estuary is a subtle process. It consists of first understanding and then guiding the forces of nature; but natural forces cannot be predicted in detail and the consequences of guiding them can only be forecast in broad terms. Management is a continuing process of guidance of the system, observing its response and adjusting the guidance as required. Careful observation of the system and use of the minimum effort to achieve a goal are the principles to be followed; to these might be added the avoidance of permanent works unless they are absolutely necessary.

These ideals are practical in many cases, but it often happens that major surgery is needed, as, for example, in the Rhine Delta, the mouths of which were closed by barrages following the catastrophic flooding due to the storm surge of 1953. The resulting system is then bound to be quite different from its original state. This need not be bad; heavy siltation of the Lune estuary following its training in the mid nineteenth century has resulted in a splendid habitat for wading birds, while the original objective of channel stabilisation has been achieved, as described in Chapter 10. In fact, nature has a great power to adapt to new conditions. The important point is that all the effects of changes should be forseen as far ahead as possible. In the case of the Rhine Delta, a careful and intensive study followed the 1953 catastrophe, and the resulting consequences of barrage construction on ecology were considered in detail. In the case of the Lune and Ribble estuaries in England, the effects of change were not foreseen, or were interpreted wrongly.

The management of estuaries will continue to be an art, but the scientific principles outlined in this book, coupled with experience of the behaviour of various systems, should prevent ignorance of the major consequences of engineering works. In the final analysis, experience of other systems provides the best indication of things that could go wrong as a result of management decisions. A small sample of such experience is given in this book, particularly in the final chapter.

REFERENCES

[1] Inglis, Sir C. C. and Kestner, F. J., The long-term effects of training walls, reclamation and dredging on estuaries, *Proc. Inst. Civil Eng.*, **9**, 193–216 (1958)

[2] Agar, M. and McDowell, D. M., The sea approaches to the port of Liverpool, *Proc. Inst. Civil Eng.*, **49**, 145–156 (1971)

[3] River Training and Bank Protection, *United Nations Economic Commission for Asia and the Far East*, UN Sales No. 1953, **11**, F 6, Bangkok (1953)

[4] Inglis, Sir C. C., Training works constructed in the Rupnarain River in Bengal, *Inst. Civil Eng.*, Maritime Paper No. 3 (1946)

[5] Franco, J. J., Research for river regulation dike design. *Proc. Am. Soc. Civil Eng.*, **92**, WW 2, Aug. (1967)

[6] Fox, J. A. and Ball, D. J., The analysis of secondary flow in bends in open channels, *Proc. Inst. Civil Eng.*, Paper No. 7087, **39** (1968)

[7] Francis, D. R. J. and Asfari, A. F., Velocity distribution in wide, curved open channel flows, *J. Hyd. Res.*, **9**, no. 1 (1971)
[8] Rousselot, M. and Chabert, M. J., Paper S1–5, *XX Congress of the Permanent International Association of Navigation Congresses* (*PIANC*), Baltimore, (1961) (published Brussels)
[9] Remillieux, M., First experimental closure of a secondary channel... by means of bottom panels, *PIANC*, *Bulletin No. 22*, IV, (1966)
[10] River training and Bank Protection (*ibid.*), p. 30
[11] Drag Heads, *Ports and Dredging*, No. **78**, IHC Holland, Rotterdam (1973)
[12] Hadjidakis, A., Increasing the output of trailing dredgers when working in fine compacted sand, *Ports and Dredging*, No. **60**, IHC Holland (1968)
[13] Hayward, H. G. A., Sidecast (boom) dredging – *PIANC Bulletin No. 4*, II, 35–64, Brussels (1962)
[14] Thorn, M. F. C., Loading and consolidation of dredged silt in a trailer suction hopper dredger, *1st Int. Symposium on Dredging Technology*, Canterbury, Paper B 1, British Hydromechanics Research Association (1975)
[15] Wicker, C. F., Maintenance of Delaware Estuary ship channel, *XXI International Navigation Congress*, Stockholm (1965), Section II, Subject 3, 209–226, PIANC, Brussels
[16] Halliwell, A. R. and O'Connor, B. A., Quantifying spoil disposal practices, *Proc. 14th Coastal Engineering Conference*, ch. 153, vol. III, June (1964)

10

Discussion of case histories

Many examples of estuarine behaviour have been used to illustrate particular topics in earlier chapters. Estuarine studies, however, cannot be done on the basis of a few isolated examples. To conclude the book, we have returned to the theme of Chapter 1, namely, that estuaries are regions in which complex interactions take place, which cannot be understood unless a global view of a system is obtained. Case histories, necessarily brief, are used to show how various strands have been brought together in the study of a few estuaries, each having a different character.

10.1 NAVIGATION CHANNEL TRAINING SCHEME—LIVERPOOL BAY (UK)

The following case history illustrates the consequences of undertaking large-scale dredging operations and training wall construction in the SE corner of the Irish Sea. The study area is known as Liverpool Bay and consists of some 180 sq km of inter-tidal sand banks ($d_{50} \approx$ 200–400 μm) and shallow water channels. The bed configuration prior to engineering activities, and the general location of Liverpool Bay is shown in figure 10.1. The relationship of Liverpool Bay to the Mersey Estuary is shown in figure 1.9, which also gives tidal data, while some typical tide curves in the estuary are shown in figure 1.4.

The large tidal rise and fall produces high velocities, exceeding 3 m/s on spring tides in Crosby Channel and the Narrows. Tidal influx into the Narrows averages nearly 2000 m^3/s during a spring tide, the maximum rate exceeding 3000 m^3/s, whereas fresh water flow into the estuary averages only 56 m^3/s, with peak flows exceeding 1200 m^3/s. Flow into Liverpool Bay as a whole includes the fresh water discharges from the estuaries of the Dee, Mersey and Ribble and averages about 160 m^3/s.

Wave action in the Bay is limited by the size of the Irish Sea, which prevents formation of long period waves. The mean annual wave height is only 0·8 m, but in winter months significant wave heights of 5 m, with periods between 6 and 8 seconds have been recorded at the Bar.

The main navigation route to the Port of Liverpool in 1890 (figure 10.1) was via the Queen's and Crosby channels, although smaller ships also made

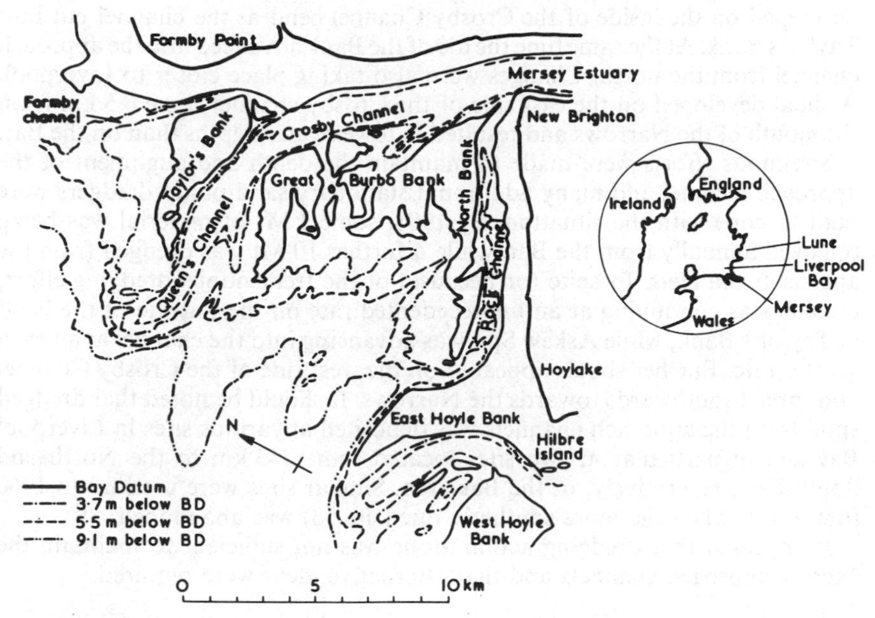

Figure 10.1. Sea bed contours in Liverpool Bay in 1890 (based on figure 2 of reference[2], courtesy the Institution of Civil Engineers)

use of the Formby and Rock channels. Main navigation channel depths were more than adequate for nineteenth-century ships, except at the seaward end of the Queen's Channel where the Bar existed with a depth of some 3–4 m below LBD (Liverpool Bay Datum: approximately the level of low water on a 9·15 m range spring tide in the Mersey Narrows, or 4·432 m below Ordnance Datum at Newlyn). Competition for the Atlantic trade in the 1880s was so fierce, however, that the local Port Authority (the Mersey Dock and Harbour Board) decided to dredge a channel through the Bar.

10.2 DREDGING OPERATIONS AND THEIR EFFECT

Dredging operations started on the Bar in 1890 with the employment of two 500-tonne capacity stationary sand pump dredgers [1]. A narrow channel with a depth of – 6·4 m LBD was dredged through the Bar and this was maintained with little difficulty. The success of this first dredging operation led to the introduction in 1893 of a 3000-tonne capacity vessel. By 1894, three dredgers were removing some 4 M hopper tonnes (weight of material dredged based on the average density of material in the hold of the dredgers) annually, and were starting to experience some difficulty in loading on the seaward side of the Bar. Agitation dredging, using water jets, was then introduced in 1895 in order to remove the silty material and 'allow the ebb tide to carry it away'. Little difficulty was experienced in maintaining a depth of some 9 m below LBD at the Bar, but it proved impossible to maintain the alignment of the navigation channels. A rapidly expanding shoal (Askew Spit, see figure 1.9) developed on the inside of the Crosby Channel bend as the channel cut into Taylor's Bank. At the same time the toe of the Bank advanced into the approach channel from the north. Changes were also taking place closer to Liverpool. A shoal developed on the east side of the Crosby channel some 6·5 km from the mouth of the Narrows and resulted in lower water depths than on the Bar.

Strenuous efforts were made to maintain the depth and alignment of the approach channel and many additional stationary sand pump dredgers were built to cope with the situation. By 1908, some 1 Mt of material was being removed annually from the Bar, while a further 10 Mt was dredged from the approach channels. In spite (or because) of the tremendous dredging effort, erosion was continuing at an unprecedented rate on the outside of the bend at Taylor's Bank, while Askew Spit was advancing into the channel at an even greater rate. Further shoals appeared on the west side of the Crosby Channel and spread southwards towards the Narrows. It should be noted that dredged spoil from the approach channels was deposited at various sites in Liverpool Bay and in particular at two sites located some 4–5 km to the North and South-East, respectively, of the Bar area. Similar sites were used up to 1960 (figure 1.9) when the more southerly one (site 53) was abandoned.

It appeared that dredging action alone was not sufficient to maintain the Mersey approach channels and that alternative ideas were required.

10.3 REMEDIAL ACTION

In 1909–10 a 3·6 km length of training wall was constructed along the face of Taylor's Bank on the outside of the Crosby Channel bend. The intention was to prevent the continued northward movement of the channel and also to prevent a partially-formed channel from breaking through Taylor's Bank and establishing a new system of channels, as occurred frequently prior to 1890.

The training wall was constructed from limestone blocks, of some 10–200 kg in weight, to a height of some 1·5 m above LBD. In addition, a very large stationary sand pump dredger, the Leviathan (10000 t capacity) was specially built to work on the approach channels. Between 1910 and 1917, following completion of the initial section of training wall, dredgers removed some 13–17 Mt of material annually from the approach channels, as well as between 1–3 Mt from the Bar.

The initial training wall succeeded in stabilising the position of the Crosby Channel bend, but did not prevent considerable changes taking place in channel width and depth. The channel cross section at Askew Spit continued to narrow and deepen, and in fact, only stabilised in 1930 when widths of some 300 m and maximum depths of some 21–22 m below LBD were reached: a depth controlled by the exposure of the hard underlying boulder-clay sub-stratum. The early training work had little or no effect upon the shoals at the southern end of the Crosby channel and much dredging effort was employed to control them, but with little success. Consequently in 1912, the Port Authority engineers made proposals to extend Taylor's Bank training wall to the west and to build training walls on both the east and west sides of the Crosby Channel in an effort to deal with the shoal areas. Work began in 1914 on the westward extension of Taylor's Bank wall, and helped to prevent a

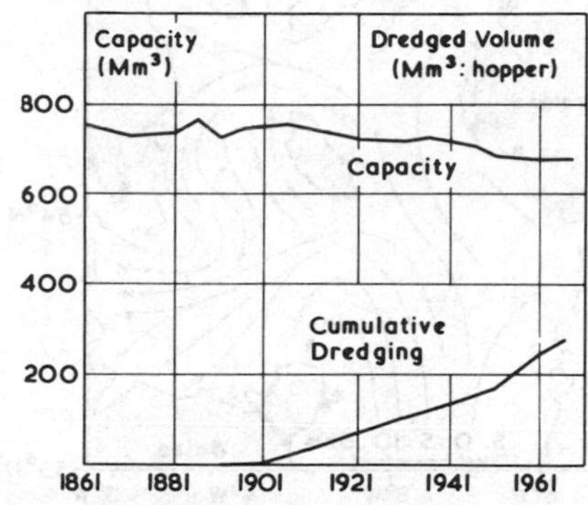

Figure 10.2. Capacity and cumulative dredging figures for the Mersey estuary upstream of Rock Light (based on figure 6 of reference[2], courtesy the Institution of Civil Engineers)

channel establishing itself through Taylor's Bank at the seaward end of the earlier wall. Some 430 m of wall were constructed in 1914 before work ceased due to the start of the First World War.

Between 1918 and 1935, various training schemes were proposed and tested in hydraulic models. The final training scheme shown in figure 1.9 involved the dredging back of a portion of Taylor's Bank which intruded over the proposed line of the channel. The major part of the training walls were completed by the start of the Second World War and enabled a minimum depth of 7·3 m below LBD to be maintained during the war years with only minimal maintenance dredging (< 1 Mt).

Between 1945 and 1957, Taylor's Spit was again dredged back to its 1935 position, the training walls were extended in the seawards direction, while gaps and settlement were repaired in sections already constructed. Annual maintenance dredging at the end of the period was fluctuating between 5 and 9 Mt. The training scheme had successfully stabilised the approach channel and sufficient depth of water was available for the large volume of shipping traffic which used the Port of Liverpool.

10.4 EFFECTS OF TRAINING AND DREDGING ACTIVITIES

The construction of the training walls coupled with intense dredging activity and local spoil disposal practices produced a redistribution of sediment in Liverpool Bay which had two major consequences. First, sediment accreted on banks and channels located outside the main trained section, and in particular the Rock and Formby channels, filled in at an accelerated rate and largely disappeared by 1957 (figure 1.9). The main increase in height of the sand banks occurred on Great Burbo Bank and was such that the training wall was over-topped at both the northern and southern end of the bank. This trend continued at the southern end even when the walls were raised locally by about 2 m. The second effect was that a large volume of sediment entered the Mersey Estuary. Figure 10.2 shows the decrease in fluid volume which

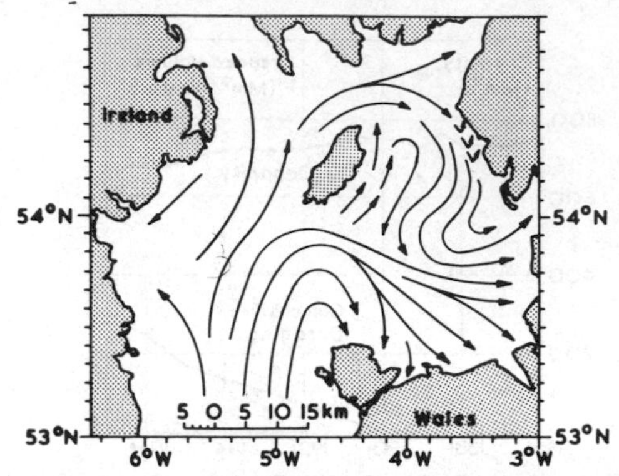

Figure 10.3. Near-bed tidal-residual circulation pattern in winter for the Irish Sea, based on sea-bed drifter results (based on a figure from reference[3], courtesy the Controller of Her Majesty's Stationery Office)

occurred upstream of New Brighton (Rock Light) between the bed and the high water level of a standard tide of 9·45 m high water level in the Narrows. The fluid volume, that is, the capacity of the estuary, decreased by some 75 Mm^3 between 1911 and 1961, despite the removal of some 200 Mm^3 of material by dredging plant[2].

10.5 INTERPRETATION OF EVENTS

The stability of a particular tidal area is controlled by its long-term tidal-average water movement pattern (see Chapter 4). Liverpool Bay has a very complicated long-term pattern which has four interacting components: a salinity–density component; a geostrophic component (due to the Earth's rotation); a wind/wave component; and a tidal inertia component controlled mainly by the configuration of the flow channels. The geostrophic, salinity–density and wind-induced components are important in the offshore zone, while the wave mass-transport and the tidal-inertia components are more important in the near-shore zone. Offshore field observations using sea bed drifters[3] show long term motion in the offshore direction (almost due North) near the sea surface and onshore motion near the sea bed (see figure 10.3). Calculations by Heaps[4] show that the observed motion is consistent with a geostrophically-modified salinity–density current system, and that bed velocities are normal to isohalines (lines of equal salinity).

The long-term pattern in the near-shore zone is largely controlled by tidal inertia and consequently it can be illustrated by hydraulic model tests using fresh water as the flow medium. Figure 10.4 shows model results[5] of near-bed tidal-average drifts (tidal-average velocity × tidal period) for conditions in 1911 just after the start of the training scheme. Actual conditions along the offshore boundaries of the grid in Fig. 10.4 are probably more consistent with figure 10.3, since both salinity–density current action and the mass-transporting action of waves will produce onshore near-bed drift; the dominant wave direction is from the northwest.

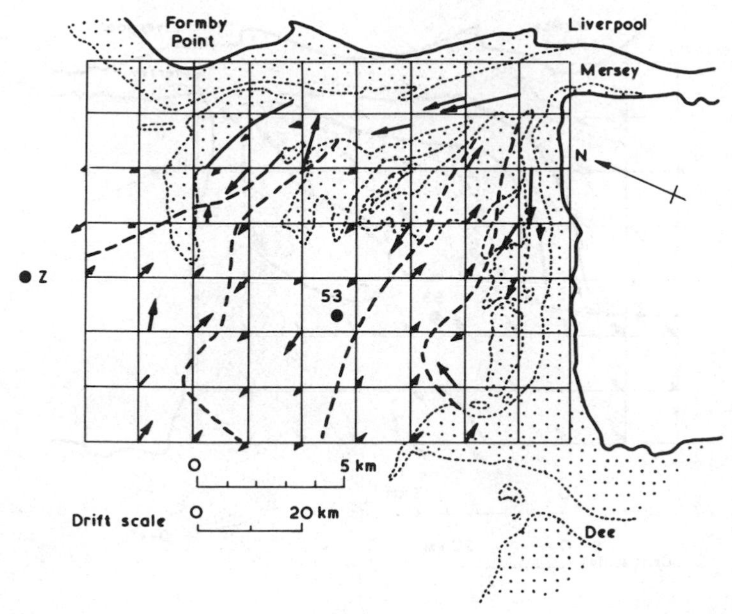

Figure 10.4. Near-bed tidal-residual drifts for Liverpool Bay, based on fixed-bed hydraulic model results with the bed moulded to 1911 contours (based on figure 8 of reference[5], courtesy the Institution of Civil Engineers)

The reasons for the development of shoals in Crosby channel and the growth of Askew Spit is partly explained by figure 10.4. Sediment dumped at the deposit sites is able to reach both the Crosby channel and Askew Spit via two concentrated flood-dominated transport paths. The large spatial variation of tidal-average drift near Askew Spit also suggests the presence of major accretion zones, while the onshore movement of Taylor's Spit is consistent with the local direction of the long term drift pattern.

Sedimentation at Askew Spit and in the navigation channel was also encouraged by the close proximity of the deposit sites to the shoal areas (the excursion of water during the flood tide is in the order of 14 km); by the supply of suspended material from dredging operations in Crosby Channel and on the Bar; by the high flow resistance of the irregular holes produced by the stationary sand pump dredgers and by the tidal flow pattern at Crosby Channel bend, which moved eroded material across the channel in both upstream and downstream directions towards the inside of the bend. Clearly, the initial growth of Askew Spit was the key factor in the problem, since erosion of Taylor's Bank and the bed of the navigation channel, together with sediment disturbances during training wall construction, would supply large amounts of fresh material to accelerate the growth of the shoal areas.

The chain of events following training wall construction was the result of two factors; firstly, modification in the long-term pattern of water movement and secondly, the continued use of the old dredging fleet and deposit grounds.

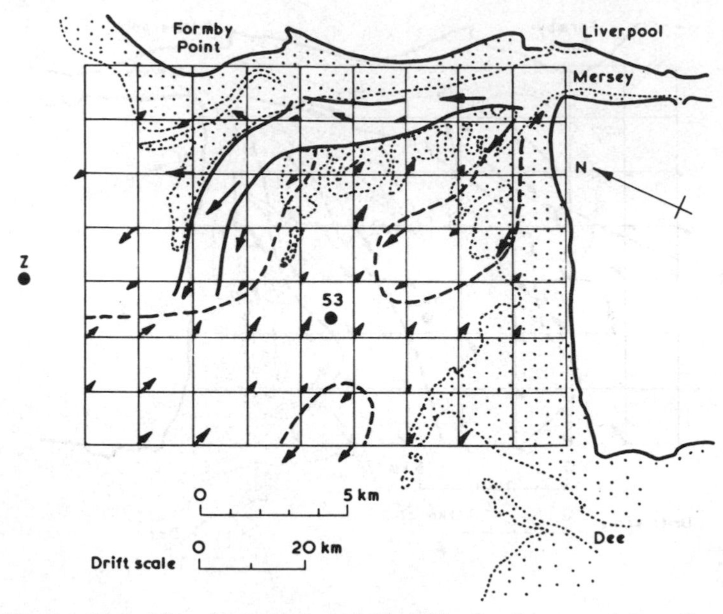

Figure 10.5. Near-bed tidal-residual drifts for Liverpool Bay, based on fixed-bed hydraulic model results with the bed moulded to 1957 contours (based on figure 9 of reference[5], courtesy the Institution of Civil Engineers)

The pattern of water movement following training wall construction is shown in figure 10.5[5]. The training walls succeeded in increasing the ebb flow in the navigation channel (maximum ebb velocities were increased on average by some 18 % over the increased flood velocities). Consequently, the ebb flow reduced in the Rock Channel and the flood-dominated zone extended and moved inshore. The reduction of ebb flow over Great Burbo Bank gave the strengthened flood tide a longer time to move sediment inshore and the Bank increased in height where the net drift pattern was zero and overtopped the training bank locally (figure 10.5). The tide and wave action in the shallow water on Burbo Bank then increased the supply of sediment to the mouth of the Mersey where density currents could progressively move it into the inner estuary. The supply of sediment was also increased, due to scour in the navigation channel, following construction of training banks and by the dumping of dredged spoil at Site 53. However, sediment entering the navigation channel over the West Crosby training bank at the northern end of Burbo Bank would tend to contribute more to shoaling in the channel than in the inner estuary.

The extension of the flood-dominant zone in Liverpool Bay and the reduction in ebb flow in the Rock channel were also responsible for the considerable changes that took place in the configuration of the inter-tidal sand banks behind the West Crosby bank, so that by 1969 (figure 1.9) little evidence remained of the channel's existence.

The Formby channel suffered a similar fate to the Rock channel. Sediment from Site Z also contributed to shoaling in the Formby channel area since field observations with current-meters and sea bed drifters[9] show a long-term drift towards Taylor's Bank and Formby Point.

10.6 POST-TRAINING ACTIVITIES

As as result of the hydraulic model studies of Price and Kendrick[5], deposit Site 53 was abandoned, stationary sand-pump dredgers were withdrawn and a single trailer suction dredger was introduced in 1961. The effect was quite dramatic; the quantity of sand dredged from the navigation channels was reduced by some 50 % in a very short time due to the reduction in flow resistance, less bed disturbance, and a reduction in sediment supply from Site 53. The quantity of fine silty material dredged has reduced more slowly than the sandy material, probably because the deposition of fine material is relatively unaffected by changes in bed resistance and, because it can travel large distances in suspension, it can still return quite rapidly from Site Z.

The success of trailer suction dredging also led the Port Authority to deepen the navigation channel to −8·5 m LBD in 1966/67. Maintenance dredging of the channels rose to a peak of 2·6 Mt per quarter year following deepening, but then declined to about the same value as prior to deepening – some 1·3 Mt per quarter year, compared with the 1936 peak dredging of some 6 Mt per quarter. There appear to be no adverse effects from the deepening scheme on the Mersey Estuary and little difficulty in maintaining navigation channel depths.

10.7 DREDGING OF EASTHAM CHANNEL – MERSEY ESTUARY (UK)

Eastham Channel is the main approach route in the Mersey Estuary for ships using the Queen Elizabeth II (Q.E. 2) oil dock, and the Port of Manchester via the Manchester Ship Canal (see figure 1.9). The approach channel was deepened in July 1953 in order to provide access for 32000 ton tankers to the Q.E. 2 oil dock at Eastham. Nearly 1 Mt of sediment was removed by bucket dredgers from the downstream end of the channel (Bromborough Bar) between July 1953 and January 1954 with the result that the depths were increased, on average, by some 1·2 m. Attempts were then made to obtain greater depths by increasing the dredging rate to some 3 Mt per year, which is six times the pre-1953 rate. Unfortunately, channel conditions went from bad to worse and by early 1962 water depths were lower than ever before at −1·5 m LBD.

As a result of hydraulic model experiments undertaken by Price and Kendrick[5] at the Hydraulic Research Station at Wallingford (UK), the Manchester Ship Canal Company decided to reduce the dredging rate by 30–40 % and switched from bucket dredgers to a trailer suction dredger. Approach channel depths increased almost immediately and by February 1963 minimum depths of −2·75 m LBD were obtained at Bromborough Bar. Over the next few years, dredging rates were further reduced, until by the late 1960s dredging was stopped altogether. Channel depths were then some −5 m LBD. Some deterioration occurred in 1970, but only a minimum amount of dredging was required in order to maintain adequate depths.

10.8 DISCUSSION OF EVENTS

In order to understand the sequence of events produced by dredging activity, it is necessary to consider both the physical characteristics of the estuary, and its tidal-average sediment circulation pattern.

The Mersey Estuary consists of a wide tidal basin with extensive inter-tidal sand ($d_{50} \approx 150\ \mu m$) and mud banks connected to Liverpool Bay via an almost inerodable channel (the Narrows) with depths of some −18 m LBD. Fresh water discharges are moderate (56 m^3/s mean flow) and the majority enter the estuary over Howley weir (52 %) at Warrington and the Weaver Sluices (18 %) opposite Hale Head. The estuary can therefore be regarded as nearly well-mixed (see figure 2.7). Almost the whole inner estuary between Dingle and Warrington dries out at low water due to the very steep bed slope upstream of Eastham and the large tidal range. The middle reaches of the estuary between Dingle and Stanlow Points are characterised by the Eastham, Middle Deep and Garston low water channels which generally show only small changes of position. Upstream of Eastham, the low water channel is narrow (50–60 m) and readily changes its position. It discharges eventually into either Garston or Eastham channels, although between 1929 and 1964 it showed little tendency to meander and discharged continuously into Eastham Channel (see figure 10.6).

The tidal-average sediment circulation pattern which existed in the Mersey Estuary during the 1950s and early 1960s has been demonstrated by Price and Kendrick's hydraulic model tests. Figure 10.7 shows the movement of fluorescent tracer particles on the bed of the saline mobile-bed hydraulic model during average fresh water flow rates, and a succession of identical large spring tides (9·24 range). Sediment entering the estuary through the Narrows would thus move progressively landwards up the Middle Deep channel and then seawards down the Eastham and Garston channels to accumulate in the vicinity of Bromborough and Garston Bars; both of which mark positions of zero longitudinal tidal-average movement. Sediment contained in fresh water flows or set in motion by low water channel movement upstream of Eastham would also tend to accumulate near Garston or Bromborough Bars, although the finer fractions tend to deposit in low tidal velocity areas along the estuary margins such as docks and dock entrances and the top of inter-tidal mud banks.

The presence of persistent shoaling areas in the Mersey Estuary is thought to be the result of the tidal-average flow pattern produced by the mixing of salt and fresh water. Indeed, the middle reaches of the estuary correspond to the point where the tidal-average longitudinal density gradient is equal to the tidal-average water-surface slope (see Chapter 4). A single shoal area does not exist, however, because the residual tidal motion is also modified by the wide curved plan-shape of the estuary. The heavier near-bed fluid tends to be deflected by centrifugal forces towards the centre of the estuary and the Middle

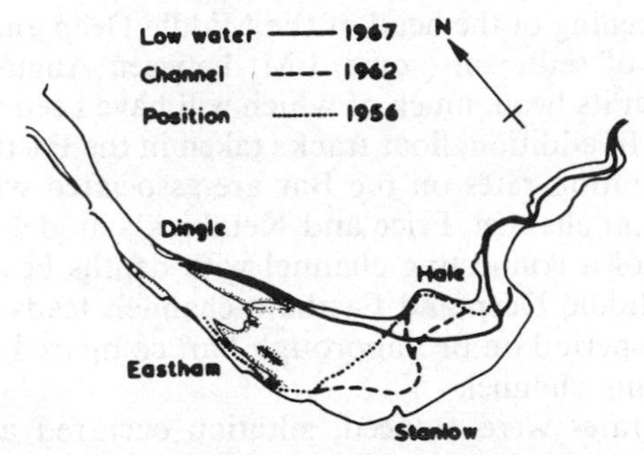

Figure 10.6. Position of Mersey Estuary low-water channel at various times between 1956 and 1967 (based on figure B1 of reference[6], courtesy the Institution of Civil Engineers)

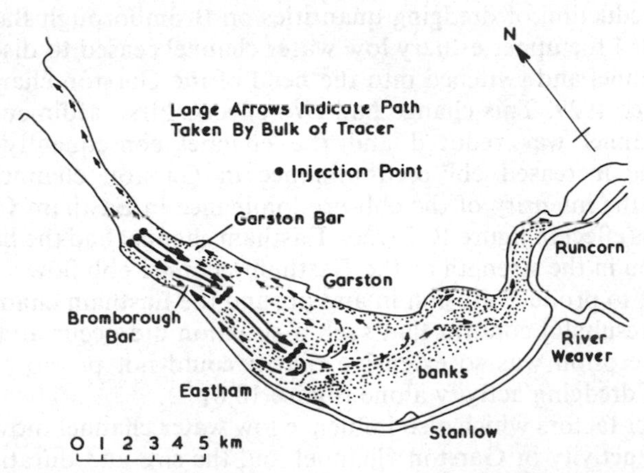

Figure 10.7. Tidal-average paths followed by fluorescent bed particles in a mobile-bed hydraulic model of the Mersey Estuary (based on figure 14 of reference[5], courtesy the Institution of Civil Engineers)

Deep channel (see figure 1.9). A further effect of the bed slope of the upper estuary is that the salinity shoaling point is moved downstream to a position of quite large tidal-average salinity (50 % of ocean values).

The reason for the lack of success of the dredging programme is readily explained by reference to the residual sediment circulation pattern. Deepening of Bromborough Bar would initially increase the tidal-average flow of water in both the Eastham and Middle Deep channels (cf. the analogous effect of increasing U_f in table 3.2). The increased flow would then lead to erosion of the upstream end of the Middle Deep and Eastham channels and to transport of this material predominantly down Eastham channel to the Bar area; sediment being deposited in large quantities as the ebb flow decelerates into the deep water downstream of the Bar. This hypothesis is supported by field and hydraulic model results. From 1953 onwards field surveys showed a widening and deepening of the head of the Middle Deep and also the erosion of large amounts of sediment (some 3 Mt between August 1958 and May 1959) from sands at its head, much of which will have been transported down Eastham channel. In addition, float tracks taken in the Eastham channel area show that high siltation rates on the Bar are associated with an increase in ebb flow in Eastham channel. Price and Kendrick's model tests also showed that the presence of a connecting channel with depths below LBD between the head of the Middle Deep and Eastham channels leads to twice as much siltation in a given period on Bromborough Bar, compared with the situation without a connecting channel.

Once dredging rates were reduced, siltation occurred and both residual flows and sediment supply were also reduced. The more efficient trailer dredger then allowed depths to be improved gradually. Field surveys again tend to confirm these views since the connecting channel between the Middle Deep and Eastham channels disappeared following the reduction in dredged quantities. Several further factors, must, however, be considered in the subsequent reduction of dredging quantities on Bromborough Bar.

In May 1964 the upper estuary low water channel ceased to discharge into Eastham channel and switched into the head of the Garston channel for the first time since 1929. This change had two effects. First, sediment supply to Eastham channel was reduced and the channel consequently deepened. Secondly, the increased ebb-predominance in Garston channel led to a reduction in the intensity of the ebb-predominance in Eastham Channel (cf. the analogous effect in figure 10.7 when Eastham channel had the higher flow). Any reduction in the strength of the Eastham channel ebb flow would allow the flood tide to produce erosion in and around the Eastham channel. Again, field survey results[6] confirm that such a situation did occur and show that the scale of erosion was so extensive that it could not possibly have been produced by dredging activity alone (figure 10.8).

Two further factors which also influence low water channel movements are the dredging activity in Garston Channel and the size and duration of river flows. It is interesting to note that dredging rates were increased in Garston Channel during the period when dredging rates were reducing in Eastham Channel. It is possible, therefore, for the Garston dredging to have increased the ebb-predominance in Garston Channel and the flood attack on upstream inter-tidal banks. It seems more likely, however, that natural changes in river flow rate will have had a greater effect on upstream banks since salinity penetration increases with reducing flow rate. An examination of rainfall and river discharge data for the estuary shows that the years 1962–4 were exceptionally dry (77 % of long term values) while the period 1954–60 was, on average, wetter than usual (by 6 % with a peak in 1958 of 24 %) and these periods agree with periods of large and small siltation at Bromborough. Extreme river flow rates also have a more direct effect on channel movement in the upper estuary. The low water channel is found to meander away from Hale Head (figure 1.9) during low river flow rates and to remain close to it during high flows. An examination of the lateral position of the upper estuary low water channel between 1956 and 1966 shows that the low water channel remained close to Hale Head between mid-1957 and early 1960, but started to meander

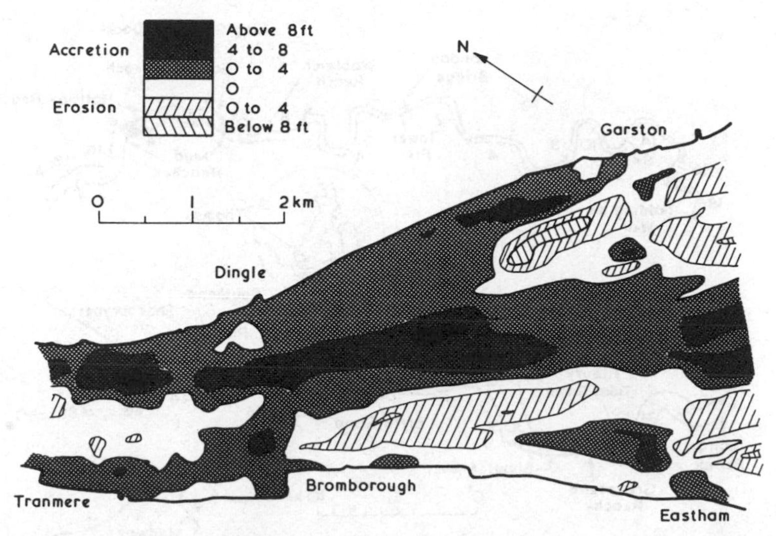

Figure 10.8. Accretion–erosion chart for the Mersey Estuary for the period 1961 to 1966 (based on figure B2 of reference[6], courtesy the Institution of Civil Engineers)

thereafter. By mid-1962 a small channel existed across the mouth of Dungeon Bight and by late 1963 had developed into the main channel (see Fig. 10.6). The downstream change to Garston channel was completed in May 1964, and proceeded to cause siltation problems in the vicinity of Garston Bar. Indeed, siltation and high channel velocities near low water in the late 1960s resulted in the abandonment of an oil jetty at Dingle and the provision of a mid-river mooring system, shown in figure 9.8, with sea-bed pipelines to the existing shore facilities. Little or no dredging has been required in mid-river, as befits the flood-dominant channel.

In 1971, the upper estuary channel again moved to Eastham and caused some siltation but by late 1973, the channel had moved back to Garston although the upstream configuration remained similar to that in the mid-1960s. It seems, therefore, that the reduction in engineering activities in the Mersey Estuary and Liverpool Bay has allowed the estuary to return to a more natural condition in which the estuary capacity changes only slowly (see figure 10.2) and the low water channel shows a greater frequency of movement as occurred prior to 1890.

10.9 DREDGING OF THE MUD REACHES – THAMES ESTUARY

The Thames estuary has a simple, almost 'ideal' plan shape and is tidal up to Teddington Weir, some 10 km upstream of Southend (figure 10.9). Tides are semi-diurnal with a mean spring and neap range of some 5 m and 3·3 m respectively. Some 84 % of the fresh water flow into the estuary is over Teddington Weir and averages 70 cumecs, although typical winter discharges

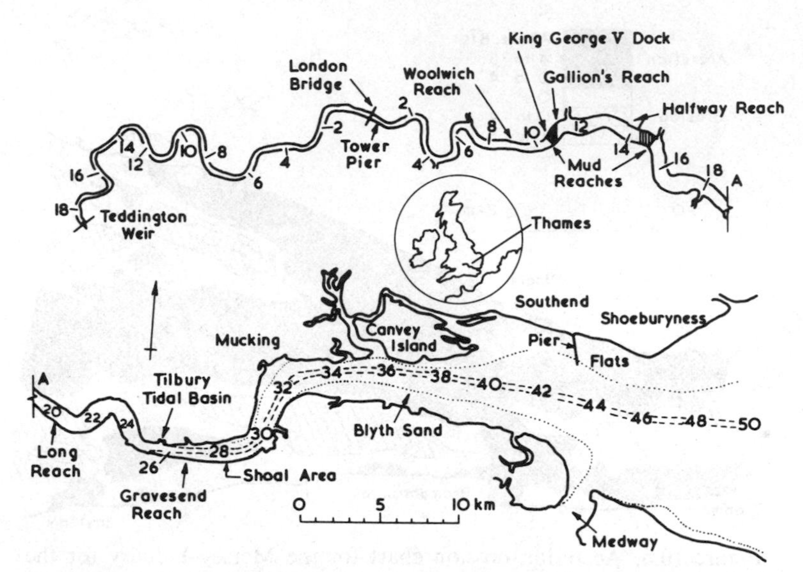

Figure 10.9. Location plan for the Thames Estuary (based on figure 1 of reference[7], courtesy the Institution of Civil Engineers)

can reach 3–4 times this value. The estuary is, therefore, 'well mixed', with surface and bed salinities differing by only 1–2‰.

The estuary has a hard bed between Teddington and Mucking and is composed of gravel, stones, clay or chalk, but fine easily-eroded sand ($d_{50} \approx 150\ \mu$m) is found in Sea Reach near the seaward end. Fine black mud ($d_{50} \approx 35\ \mu$m) containing some 20–35 % organic matter forms the bed material between Gallions Reach and Halfway Reach (the Mud Reaches), and for a distance of some 1·6 km at the seaward end of Gravesend Reach. Black mud is also found mixed with fine sand at Southend Flats and Blyth Sands.

Ships bound for the Port of London use a dredged navigation channel which has typical widths and mean-tide depths of 300–180 m, and 13–11 m, respectively, from the sea up to King George V Dock. Upstream geometry reduces progressively and causes considerable distortion to the tidal wave (see figures 6.1 and 6.2). The navigation channel was deepened between 1909 and 1928 and has been almost self-maintaining over the greater part of its length. There were, however, two areas (Lower Gravesend Reach and the Mud Reaches) where persistent shoaling used to occur and continuous dredging was required to maintain adequate navigation channel depths. Indeed, by the end of 1956, conditions in Gravesend Reach were little better than in the 1830s and some 3 Mt (hopper measure) of material were dredged on average each year from the estuary, including the navigation channel, the impounded dock system, and Tilbury Tidal Basin [7]. The shoaling problem was so severe that the Port of London Authority (PLA) commissioned a field and hydraulic model study of the estuary. The work was undertaken in 1955 by Inglis and Allen of the Hydraulic Research Station (Wallingford). Much of the following discussion is based on Inglis and Allen's work [7].

10.10 DISCUSSION OF THE SHOALING PROBLEM

Discussion of any siltation problem must first be concerned with the source of the sediment causing the problem. Sediment enters the estuary from both landward and seaward sources. Inglis and Allen found that the River Thames and other tributary streams supplied silt and clay to the estuary amounting to some 15 % of an average year's dredging quantities (based on pre-1956 figures) while sewage sludge dumped near the north shore at Mucking provided a further 8 %. Domestic and industrial effluents, storm water drains and small amounts of dredged material dumped in the estuary made up a further 22–27 %. Consequently, some 50–55 % of the dredged material was supplied from seaward sources and was, therefore, either eroded from coastal and offshore areas or from the navigation channel spoil disposal site at the Black Deep, which was situated some 48 km seawards of Southend. An examination of the spoil ground itself showed that the sea bed consisted of clean sand and clearly suggested that sediment was returning to the estuary, particularly as it was normally dumped shortly before the start of the flood tide.

An examination of the tidal average residual drift in the estuary confirms the view that sediment returned to shoal areas from the Black Deep. The field and hydraulic model study by Inglis and Allen[7] showed that the tidal-average drift in Woolwich Reach at mile 9 (figure 10.9) was almost zero for a large proportion of small-range tides, but seawards for the largest tides. A landward drift was found on all tides opposite Southend (near mile 43), in the Yantlet Channel off Shoeburyness, and at the Black Deep spoil disposal ground.

Further confirmation that the Mud Reaches corresponded to a salinity-induced shoal area was provided by field and model tracer studies. Radioactive soda-glass particles with a similar size-distribution to the inorganic part of the estuary mud were released onto the estuary bed at mile 26, just after high water level. The tracer moved seawards on the ebb tide and some deposited on the estuary bed between Gravesend Reach and Canvey Island. Successive tides then moved tracer material upstream, allowing some to deposit on the way, until the Mud Reaches were reached about a week after the start of the test. No trace material was ever found upstream of the Mud Reaches. A parallel experiment was also conducted in Inglis and Allen's saline model, in which nearly neutrally-buoyant celluloid balls were released from the same place as the radioactive tracer. The balls moved along with the tide near the model bed and eventually accumulated in the Mud Reaches, where their flood and ebb tidal excursions were equal. Figure 10.10 summarises the two experiments.

The Mud Reaches seem, therefore, to correspond to a salinity-associated shoaling zone, with a reasonably high tidal-average salinity (≈ 12 ‰) due to bed slope hindering salt penetration. However, some analytical calculations (see Chapter 4) suggest that the Mud Reaches exist due to tidal inertia effects alone, while the Gravesend Reach shoal is due to salinity gradient effects[11]. Neither suggestion agrees with the hydraulic model tests, which reproduced the zero drift point in the Mud Reaches only, when operated with saline water

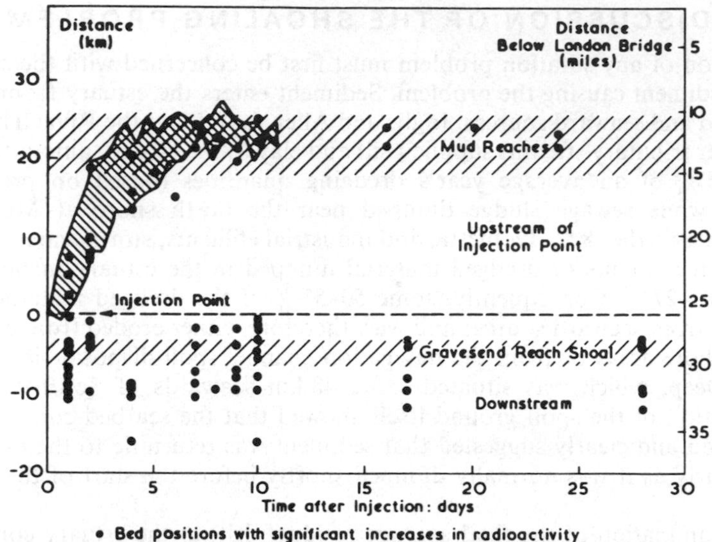

Figure 10.10. Comparison of results from a radioactive tracer experiment in the Thames Estuary and the movement of near-bed floats (celluloid balls) in a fixed-bed model of the estuary (based on figure 20 of reference[7], courtesy the Institution of Civil Engineers)

at the mouth. The reason for the discrepancies is most likely to be the simplifying assumptions needed to obtain analytical results.

The existence of the Gravesend Shoal is probably the result of local geometric effects, as suggested by Inglis and Allen[7]. Figure 10.9 shows that the width of the estuary almost doubles in Gravesend Reach between miles 26 and 28. In fact, the cross-sectional velocity between miles 26·22 and 28·5 decreases by about 30 %, so allowing sediment to deposit as in the case of Bromborough Bar on the Mersey Estuary. In addition, the dredged channel has a straight alignment but the natural flow in the reach is curved so that sediment tends to accumulate on the inside of the bend.

Subsequent to Inglis and Allen's field and model work, the Black Deep was abandoned as a spoil disposal site and material was pumped ashore, with the result that dredging quantities were reduced by some 50 %.

10.11 LOW WATER CHANNEL TRAINING SCHEME – LUNE ESTUARY

The Lune Estuary is situated on the NW coast of Lancashire (UK) and discharges into Morecambe Bay in the central part of the Irish Sea (see figures 10.1(a) and 10.12). The estuary is tidal up to Lancaster weir, some 14 km from the sea, and is subjected to large tides with a mean spring range of some 8 m near the sea. The estuary bed has a steep longitudinal gradient

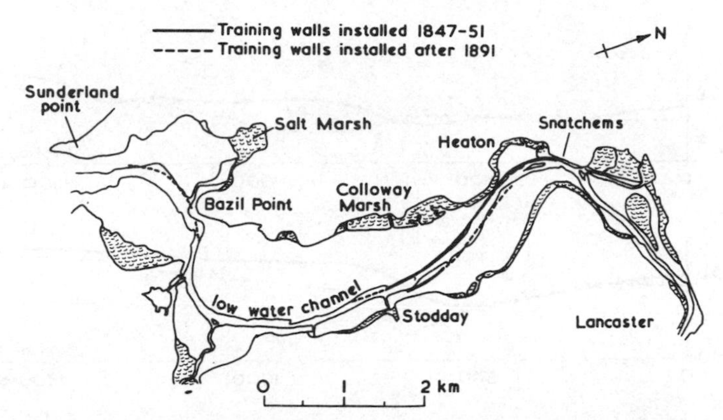

Figure 10.11. The Lune Estuary in 1838, showing various training schemes (based on figure 13, of reference[8], courtesy the Institution of Civil Engineers)

($\approx$ 1/1000) and consequently the greater part of the estuary dries out at low water on spring tides, except for a small low water channel which formerly meandered back and forth across the estuary through banks of silt and fine sand ($d_{50} \approx 150\ \mu$m). The steep bed slope causes considerable distortion of the tide curves, even near the sea at Chapel where the duration of the flood tide is only some 3·67 hours. River flows are small to moderate and the estuary can be classified as nearly well-mixed during spring tides.

Prior to 1847, the low water channel downstream of Heaton showed frequent large scale movements and consequently a scheme of river training was initiated under the guidance of Robert Stevenson and Son. The work started in 1847, with the dredging away of various hard areas on the channel bed in the upper reaches, and produced a gradual deepening of the channel by just over a metre. A low training wall, of dumped stone backed with gravel and clay, was then constructed to a height of almost a metre above the original low water level, from Snatchems to Stodday. A similar wall was planned on the eastern side of the channel just upstream of Stodday but only a short length was built before lack of money caused work to cease (figure 10.11).

The effect of the dredging and training scheme was felt immediately. The channel enlarged and deepened in the trained section and for some distance downstream, and as a consequence the duration of the flood tide was found to have increased by some 25 minutes. The success of the scheme was offset, however, by the rapid accumulation of silt and fine sand in areas behind the walls. Indeed by 1851, some 2·4 Mm^3 had accumulated between Heaton and Glasson. By 1856 further changes were occurring and, in particular, a flood channel was starting to develop behind the western training wall at Stodday (Fig. 10.11). Stevenson then recommended extension of the walls but lack of money held up the scheme and the works were not completed until the 1890s.

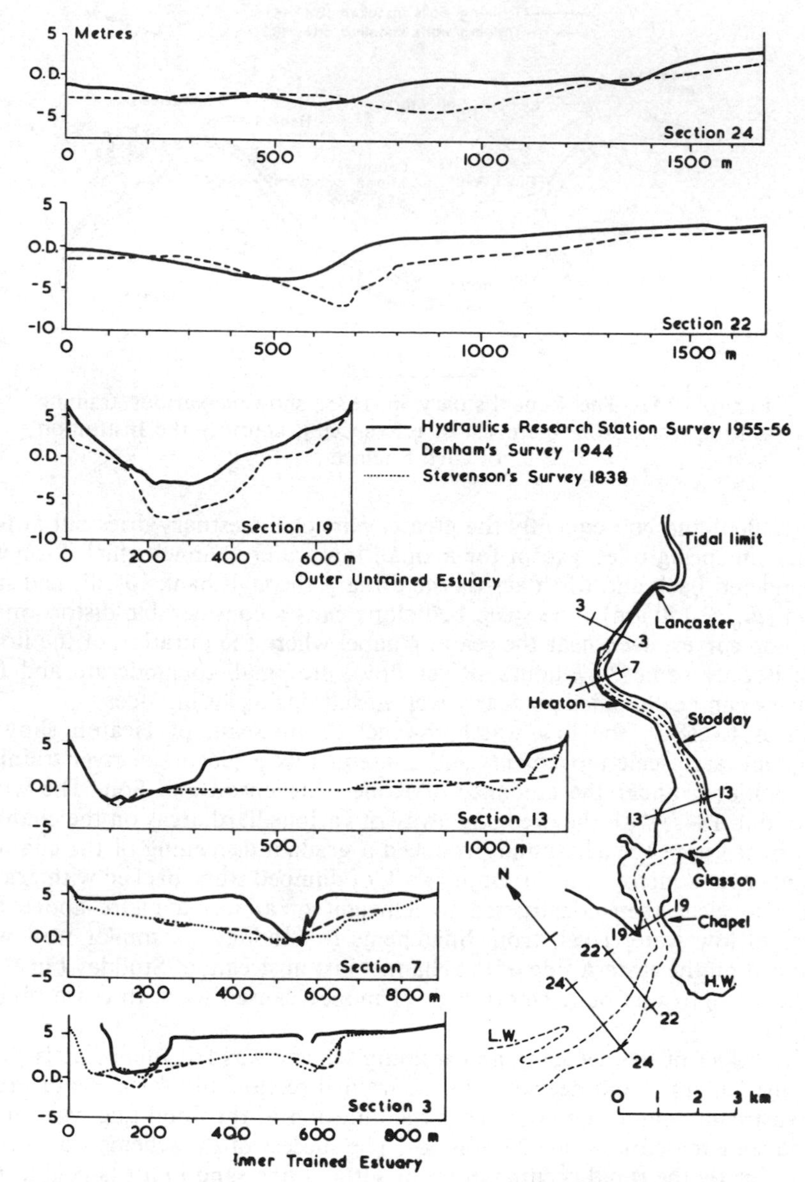

Figure 10.12. Effect of training schemes on bed levels in the Lune Estuary between 1838 and 1955 (based on figure 14 of reference[8], courtesy the Institution of Civil Engineers)

10.12 LONG TERM EFFECTS OF TRAINING

Following completion of the training scheme, the estuary was allowed to adjust its bed without interference from man. Indeed, very few records of what occurred in the estuary existed until the mid 1950s, when Hydraulic Research Station personnel surveyed the area [8].

The survey results showed that massive accretion (26·5 Mm^3) had occurred in the estuary and that the tidal volume on a spring tide had been reduced by some 47 %. In the middle and upstream reaches of the estuary the inter-tidal banks behind the training walls had risen to salt-marsh level (an accretion in places of nearly 5 m) and some narrowing of the low water channel had also occurred where the salt marsh over-topped the walls. However, the maximum depth in the low water channel was much the same as in 1838. Downstream of Glasson the accretion was less severe and was associated more with sandy material which had settled on the bottom of the channel (see figure 10.12) [8]. Seaward of Sunderland Point, the system of flood and ebb channels also changed. The main ebb channel of 1844 accreted by some 6 Mm^3 and a new, shallower, and narrower channel developed to the south along the line of a former flood channel. The northern flood channel (near section 24 of figure 10.12) also became shallower at its head as it retreated westward, but it deepened considerably at its seaward end.

A comparison of estuary tides for 1939 and 1956 by Inglis and Kestner showed that low water levels above Heaton had improved, as was observed in the early 1850s, and the level of mean high water above Glasson had risen by some 0·6 m, following a similar increase in the height of the salt marsh. The period of the flood tide at the estuary entrance was also found to have returned to its pre-training value of 3·67 hours. Tidal discharges, calculated from the observed tide curves by cubature (see Chapter 6) showed a large reduction over pre-training values at all tidal states, with peak flood and ebb discharges being reduced by some 47 % and 20 % respectively at Chapel, and with slightly smaller reductions at Heaton [8]. The ratio of peak flood to peak ebb discharge was also found to have reduced from some 1·5 (prior to training) to 1 (after training) at Chapel, some 1·8 to 0·85 at Stodday, and some 2 to 1·1 at Heaton.

10.13 DISCUSSION OF EVENTS

The changes which occurred in the estuary following the construction of the training walls were an inevitable result of suppressing lateral movement of the low water channel and of modifying tidal velocities in an estuary with a high suspended load. Construction of the training walls initially concentrated the ebb flow in the trained section and thus allowed deepening to occur. The formation of greater flow depths, which was also aided by dredging away of hard bed zones in the upper estuary, then produced a more rapid movement of the tidal wave ($\propto (gH)^{\frac{1}{2}}$) into the estuary and a consequent lengthening of the flood tidal period. The subsequent lowering of low water level increased tidal range and thus flow velocities on the ebb tide, leading to greater channel erosion. In turn, the increased ebb-predominance of the trained section

Table 10.1 *Stability velocities for the Lune Estuary*[a]

Section no. (figure 10.12)	1838/1844			1955			
	$P' \times 10^7$ (m^3)	$A_{mtl} \times 10^2$ (m^2)	U_{st} (mm/s)	$P' \times 10^7$ (m^3)	$A_{mtl} \times 10^2$ (m^2)	U_{st} (mm/s)	$\frac{U_{st}(1955)}{U_{st}(1838)}$
18	5·37	22·1[d]	545	2·85	10·6[d]	600	1·1
13	2·66[b]	22·9	199	1·24	7·78	357	1·8
7	0·44[c]	10·7	92	0·32	2·79	259	2·8

[a] Based on data from ref. [8]. [b] Interpolated value. [c] Extrapolated value. [d] Based on Section 19.

NOTE: $U_{st} = P'/(A_{mtl}.T)$; $T = 44700$ s. P', tidal prism; A_{mtl}, mean tide level cross-sectional area.

produced an increase in flood predominance behind the training walls, and allowed sediment to be eroded from downstream areas and transported upstream. Progressive siltation then developed from the head of the estuary and led to a decrease in tidal volume and ebb flow in the low water channel. Siltation of the channel then reduced the period of the flood tide and decreased ebb tidal flows, allowing further channel siltation to occur. Deposition of the finer fractions of the suspended load along estuary margins allowed the inter-tidal banks to increase in width and height, but at a reduced rate, as flows also reduced. Eventually, salt marsh level was reached, since sufficient fine material was available for deposition, and a stable V- or U-shaped channel was formed along the line of the training works. Erosion of the salt marsh was discouraged at high bank levels by the growth of marsh grasses and at lower levels by the cohesive nature of the clay fraction ($\approx 10\%$ of the total) which deposited along with the fine silt and sand.

A stable channel situation was finally reached when the upstream transport of fine silt and sand was balanced by downstream transport on the ebb tide (equation 4.15). Field observations by Kestner [10] confirm that such a condition was reached and also suggest that estuarine deposits will tend to become sandier in future, since a greater proportion of fine sand was found to enter the estuary on the flood tide than leaves it on the ebb. The reverse situation was found to be true for finer fractions. Progressive shoaling of the estuary due to salinity/density tidal-averaged flows does not, therefore, appear to occur in the Lune, probably because of the steep bed slope and because the cross-sectional areas at low tide are very much smaller than those at high tide.

A dynamically stable system of channels also appears to have existed prior to the start of training wall construction, as can be demonstrated by calculating the stability velocity for various cross-sections within the estuary (table 10.1). The stability velocity for the estuary entrance (cross-section 18) is seen to be identical to that found for similar bed material in Dutch estuaries (see table 10.1 and Chapter 4). A larger stability velocity is required after training since deposits are more cohesive and, therefore, require high erosion

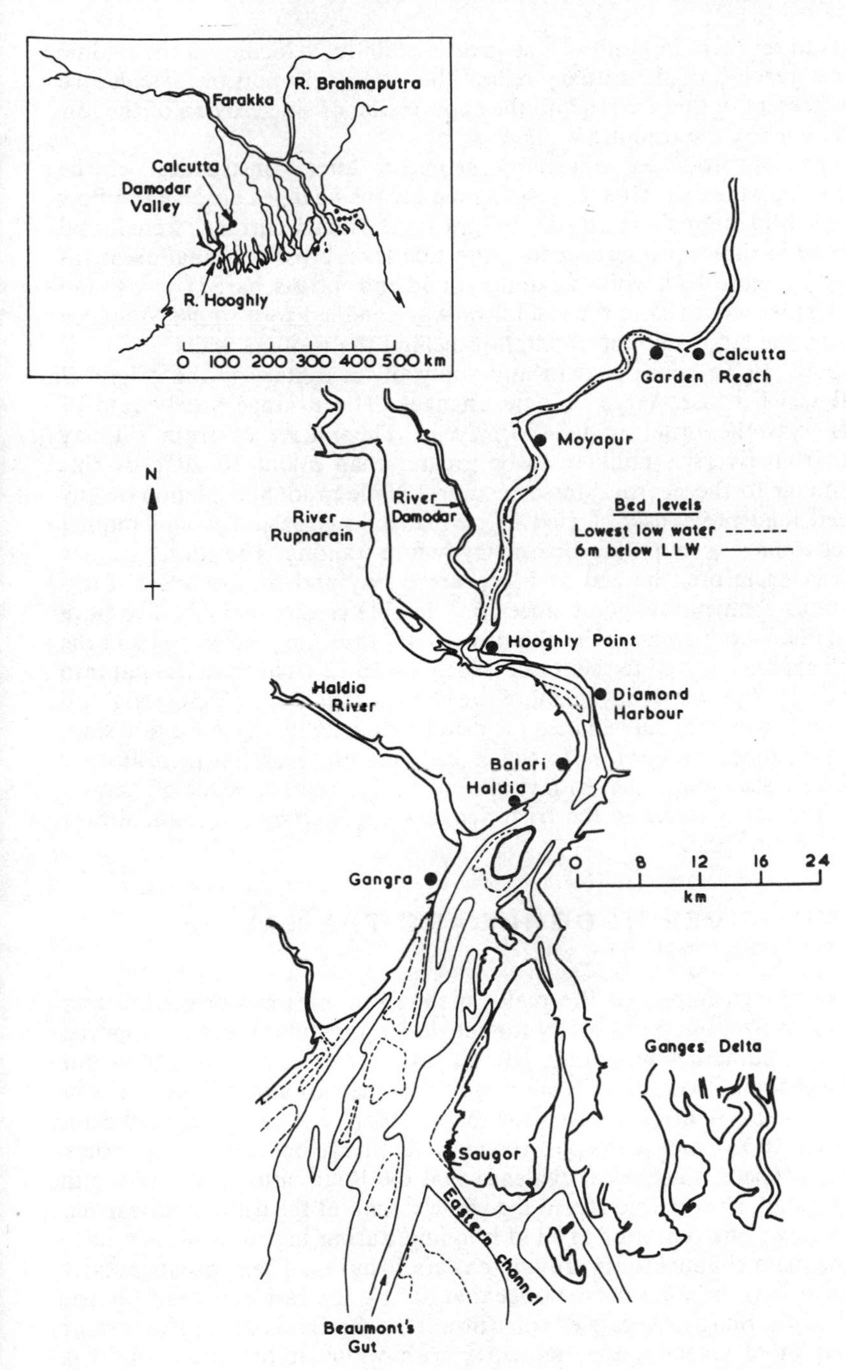

10.13. The River Hooghly system and the Ganga Delta

velocities to set them in motion. The greater stability velocities in the middle and upper reaches of the estuary reflect the greater proportion of cohesive material present in those areas and the suppression of side erosion of the low water channel by the training walls.

The changes produced in estuary geometry after construction of the training walls were also directly responsible for the changes observed in flow discharges. Maximum flood tide discharges at the estuary entrance were found to decrease in direct proportion to mean tide level cross-sectional areas, as suggested by figure 4.15, while maximum flood and ebb discharges were found to have similar values since the tidal flow was confined to a single V-shaped channel by the large amount of siltation behind the training walls.

One point to be considered in any study of an estuary is the origin of material which causes large volume changes. The average yearly rate of accretion was the order of 0·24 Mm^3/year. The inflow of fresh silt/clay material from rivers is unlikely to be greater than about 10–20% of this value (similar to the nearby Mersey estuary) while gradual depletion of any suspended load previously carried in permanent suspension would require the water to have a soup-like consistency before training. The source of new material is, therefore, the bed of Morecambe Bay and the Irish Sea. Field observations confirm this point since considerable erosion is known to have occurred in offshore areas as the ebb flow from the estuary weakened and the main ebb channel moved to the south (see figure 10.12). The material put into suspension by the lateral movement of the ebb channel and by increased flood scour would have been carried into the estuary directly by the flood tide since the offshore zone is very close to the estuary mouth. Material from further offshore can also reach the estuary due to the combined effect of density currents and wave action in the Irish Sea, as suggested by the residual flow pattern of figure 10.3.

10.14 THE RIVER HOOGHLY AND THE BAY OF BENGAL[13]

The River Ganga is one of the really great rivers of the world, not only because of its size, but particularly for the dense population that it supports. It drains the southern slopes of the Himalayas and carries perennial flow, but with a large annual variation, from about 4000 m^3/s to about 57000 m^3/s in an average year. It discharges to the Bay of Bengal through a vast delta (inset, figure 10.13), and in the past the main discharge has followed a variety of channels. There is historical evidence that the Bhagirathi (non-tidal), and Hooghly (tidal), rivers, which form the western limb of the delta, once carried the main flow, but centuries of land building during monsoon floods have forced the main channel to its present, eastern location. The consequences of this change have become most noticeable during the last 100 years or so; before then, the Bhagirathi carried some flow from the Ganga throughout most years, but more recently, the frequency with which it has dried out has increased until, from 1918 onwards, it has closed during each dry season. The Bhagirathi–Hooghly system was then only fed by spill from the Ganga during the monsoons (June to September) and for a short time thereafter.

The River Hooghly provides access to the port of Calcutta and the new port of Haldia (figure 10.13). It carries large flows of fresh water during the monsoon season, most of which spill from the main channel of the Ganga. These spill flows average between 1400 and 2300 m^3/s over the monsoons as a whole, but peak discharges reach between 2500 and 3500 m^3/s. For the rest of the year, natural flow from the Ganga falls rapidly after the end of the monsoon and, from about November until June, the only significant fresh water flow into the head water has been groundwater seepage. A barrage across the Ganga at Farakka was completed in 1970[14] so that perennial fresh water flow into the Hooghly can now be provided.

There is also run-off from local hills to the west, discharging into the River Hooghly through a few small tributaries and one major tributary. The latter drains the Damodar Valley and discharges into the Hooghly via the River Rupnarain at Hooghly Point (figure 10.13). The mean fresh water flow is much smaller than that from the Ganga, but cyclones that occur during the monsoon periods have caused catastrophic flooding, with recorded discharges of order 20000 m^3/s for a few hours. During the 1950s flow from the Damodar Valley was regulated by construction of several major reservoirs, resulting in considerable reduction in peak discharges. The River Rupnarain has a broad, shallow estuary in which the tidal influx is about 75 % of that in the River Hooghly. Table 1.2 gives some typical dimensions and tidal flows in the River Hooghly, while figures 1.2 and 1.3 show tidal behaviour.

It will be seen from figure 1.2 that the rising tide becomes very steep-fronted up river from Hooghly Point; in fact, a bore having a height of up to a metre in deep water and more than 3 metres over shoals travels from a point just upstream of Hooghly Point right through the port of Calcutta. The frequency of occurrence of the bore has varied over the years. Before 1940, it occurred during equinoctial spring tides. In the 1960s there was a bore tide during every spring tide. The increased frequency of bores was a direct result of siltation in the upper reaches of the tidal river and was, in turn, a contributory factor to it. The primary cause, however, was the gradual reduction in fresh water flow from the Ganga.

During the dry season, saline water gradually penetrated to over 25 km north of Calcutta, while monsoon flows reduced salinity to a few parts per thousand at the mouth[15]. The estuaries flow through deep alluvium and have undergone major changes of path. In fact, the seasonal changes in alignment of the shallow crossings between bends determine the tracks used for navigation. Because of these frequent changes, regular surveys must be made and a dredging fleet is used to optimise the available depths. A River Survey Department monitors the depths available for navigation and conducts a complete survey of the river from a point upstream of Calcutta to the sea twice a year. Sensitive crossings are checked daily and more complete local surveys are done once or twice a month.

Seawards of Hooghly Point, the system widens progressively and the estuary can be divided clearly into ebb-dominant and flood-dominant channels. At Saugor Island, the width of the main estuary is about 22 km. Seawards of Saugor there are extensive tongues of sand extending into the Bay of Bengal to about 80 km from the shore. Beyond this limit, known as Sandheads,

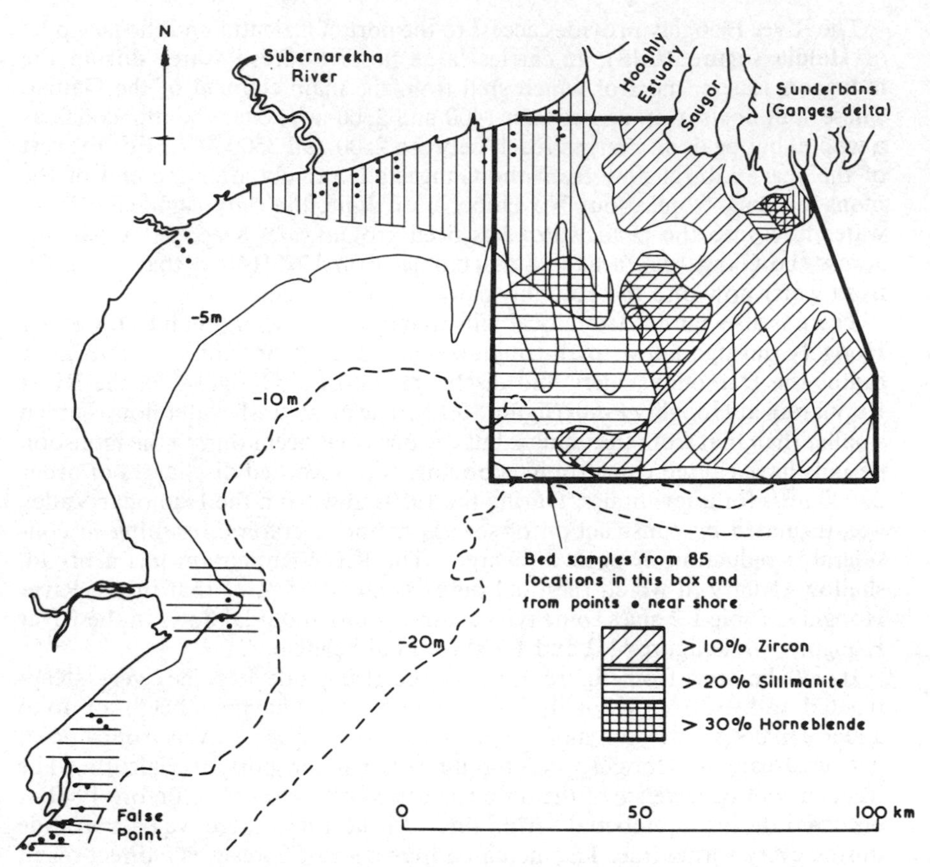

Figure 10.14. The Sandheads of the River Hooghly and part of the Bay of Bengal, showing distributions of some heavy minerals (courtesy Calcutta Port Trust)

depths increase rapidly to over 100 m; but within the channels through the sands, depths only exceed 10 m below lowest water level for short distances, while they may fall to less than 6 m on shoals and bars. The sand formations illustrated in figure 10.14 have migrated slowly eastwards and, depending on their locations, various channels through the shoals have been used for navigation. In 1941 and 1942 and from 1950 to 1959, a central channel called Beaumont's Gut was used. By 1960, its terminal bar had moved south and shoaled so much that it was no longer usable. The Eastern Channel has been used ever since and is now being maintained by dredging.

Study of a large system like the Hooghly requires considerable resources and adoption of economical methods of working. The Hydraulic Study Department of the Commissioners for the Port of Calcutta was established in 1962 to examine methods of regulating the estuary to obtain the best economic depths for navigation to Calcutta and its satellite port at Haldia. The Department acquired extensive facilities for hydrological work in the estuary, including a launch; a survey ship capable of operating 80 km offshore in the Bay of Bengal during moderately rough monsoon conditions; a Decca Hi-Fix chain for accurate position location in the Bay; and bases in Calcutta and near Hooghly Point for data analysis and logistical support for field work. Hydraulic models had been constructed several years earlier at the Central Water and Power Research Station, Poona[16] to help study the general problem of the River Hooghly. One was a model of the whole tidal system built to a horizontal scale of 1:2000, and a vertical scale of 1:160. The other was a model of the vicinity of the Port of Calcutta with horizontal scale 1:300 and vertical scale 1:60.

The programme of work of the Hydraulic Study Department began with measurements intended to clarify the physical behaviour of the system. It included systematic measurement of flow, sediment transport, temperature and salinity at a few selected cross-sections, with simultaneous measurements at three or more points in any cross-section, in addition to measurements of tidal level and surveys of the bed which supplemented the routine surveys. Radioactive tracer tests were used to determine local rates of sand transport. Apart from the scale of the problem, the methods adopted were similar to those used by Inglis and Allen in the Thames[7] and Price and Kendrick[5] and Halliwell and O'Connor[12] in the Mersey Estuary. The long-term objective was to recommend methods of improvement of depths for navigation from Calcutta and Haldia to the sea, and these are being met by construction of extensive training works, designed to reduce seasonal channel changes and to concentrate ebb and flood currents into single channels over 'crossings'.

Two particular items illustrate aspects of estuarine studies not covered by other case histories. These were concerned with use of natural tracers to study movement of sediment from the estuary to the Bay of Bengal; and with control of one particular crossing in the estuary known as Balari Bar.

10.15 USE OF HEAVY MINERALS AS TRACERS

Frequent channel changes have taken place in the Sandheads region seawards of Saugor (see figure 10.14). The changes occur most rapidly during the monsoon periods and during cyclones. They are affected by tidal flow, river-borne sediment, and by wave action due to the monsoon winds. Sources of sediment include the various tributary rivers as well as possible movement from the east coast of India. Experience at east-coast ports indicated that there could be a flow of over 10^6 m^3 a year of sand towards the north. The area covered by the channels and shoals is about 6400 km^2. Because sand movement takes place mainly during the monsoon, when wave action on the shoals can be severe, direct observation was impossible. Surveys at 10-yearly intervals gave evidence of slow migration of the channels from west to east, but there was little evidence to indicate the source of the sediment. One hypothesis was that material brought down the River Hooghly was gradually re-worked by monsoon wave action and cyclonic storms. In an attempt to learn more about this, extensive sampling of the bed was done on a grid 80 km $\times$ 80 km and along the adjacent coast (figure 10.14). The samples were subjected to analysis of the heavy mineral[17] content. This technique has been used in several coastal and estuarine studies[18]. The results were contrary to expectation. Because the various sources of sediment had quite different geological characteristics, it was possible to identify several minerals that could be used as natural tracers of sediment transport. One of these was zircon, which was found in the sediments of the River Hooghly, but was absent from the coastal sediments. Another was hornblende, found in a local river that discharges into the Bay of Bengal 60 km west of the mouth of the River Hooghly. A third was sillimanite, which was present in the sediments at False Point and was typical of coastline material to the immediate south, but was not found in appreciable amounts in sediments from the other sources. An indication of the results is given in figure 10.14. The fact that zircon was only found in and around the eastern side of the Sandheads shows that sediment derived from the rivers Ganga and Domodar had not recently contributed to the whole system of channels and shoals. Material from the coast and from the local rivers discharging into the Bay had moved from west to east across the Sandheads and into the main delta area. This knowledge contributed to plans for deepening the approaches to the port of Haldia by dredging.

10.16 IMPROVEMENT OF BALARI BAR

Balari Bar (figure 10.15) is situated at the landward end of a flood-dominated channel in a broad reach in which ebb currents dominate the eastern side. The ebb channel carries large amounts of sand during the monsoon and its seaward end is notoriously unstable. The flood channel, on the other hand, has been comparatively stable since the eighteenth century; this stability led to its selection as the location of the port of Haldia, construction of which began in 1964. Depths of nearly 6 m below datum (approximately lowest low water) were attained on the bar before 1940, but after 1945 the reach widened by erosion of the eastern bank, and a shoal between the channels developed into a new island, Nayachara. Depths on the Bar were then about 4 m below datum. By 1960, the island was 8 km long by 3½ km wide and covered with vegetation. Balari Bar began to deteriorate rapidly and the navigation track had to be rotated clockwise. Depths fell to only 2·6 m below datum. This development was accompanied by erosion of the shore of the eastern side of the reach, as shown in figure 10.15.

Various proposals for regulating the Bar were considered, including intensive dredging; construction of training banks on the Bar and island; and spurs in the eastern channel that might deflect more of the ebb into the Haldia Channel while arresting erosion of the shore. Dredging using existing craft was unsuccessful. An intensive search was made for evidence of the past behaviour of the system, during which it was found that a 9 km long shoreline of the next reach upstream, shown in figure 10.16, had eroded by some 600 m since 1896, most of the erosion having taken place since 1936. Further search led to discovery of a chart made in the late eighteenth century by a Major Rennel which showed the shoreline about 1000 m further back than it was in 1960. The chart was small and its accuracy doubtful, but other features had been reproduced quite well. A flight over the region showed clear evidence of

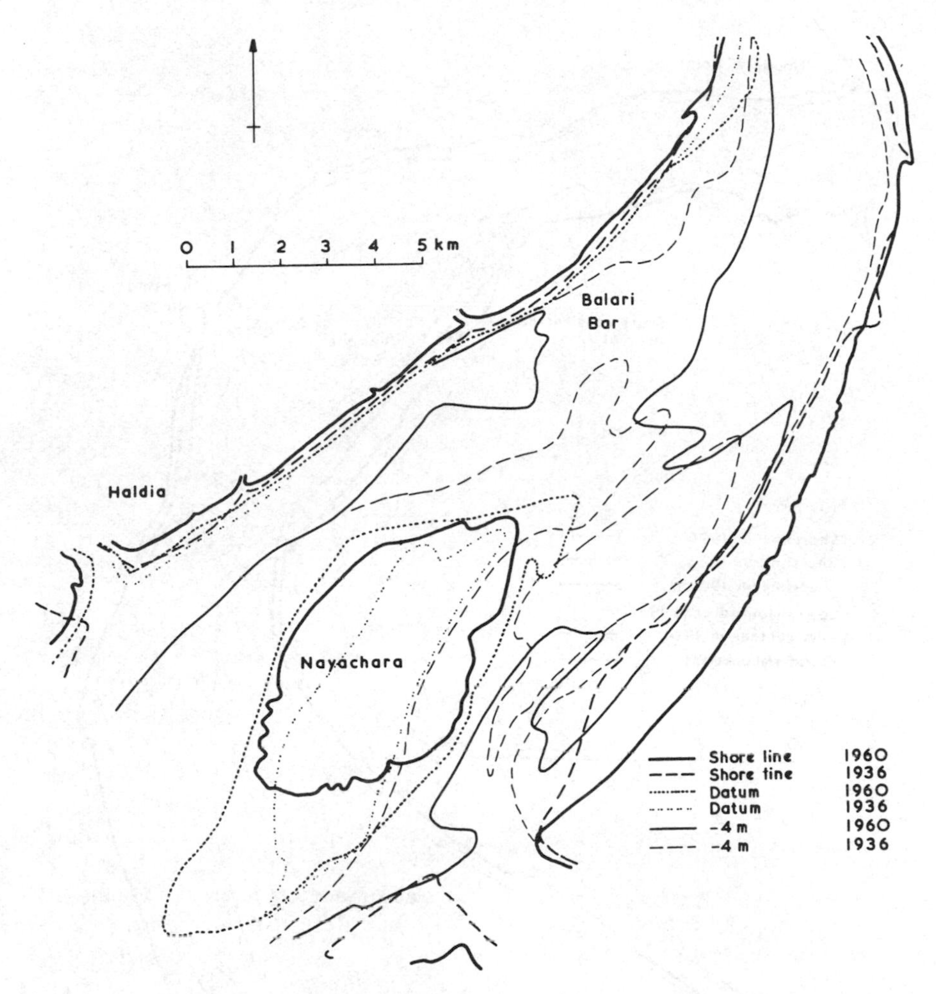

Figure 10.15. Balari Bar showing shoreline and contours in 1960 (before training) and shoreline and partial contours in 1936 (courtesy Calcutta Port Trust)

former shorelines, indicated by parallel lines of shrubs extending back to the approximate location of the shoreline shown on Rennel's chart. It was also found that the eastern shoreline opposite Nayachara had been further east of the 1960 shore in the late eighteenth and early nineteenth centuries.

Erosion of the Diamond Sand shoreline had changed the shoreline close to Balari Bar from a relatively smooth curve between 1896 and 1936 to a rather sharp curve in 1960, shown in figure 10.16. Erosion of the shore took place during the flood tide, but ebb currents could carry eroded material towards Balari Bar, where the abrupt change in direction of the shoreline would cause slack water and deposition.

The evidence suggested a natural cyclic change. Quantitative analysis was

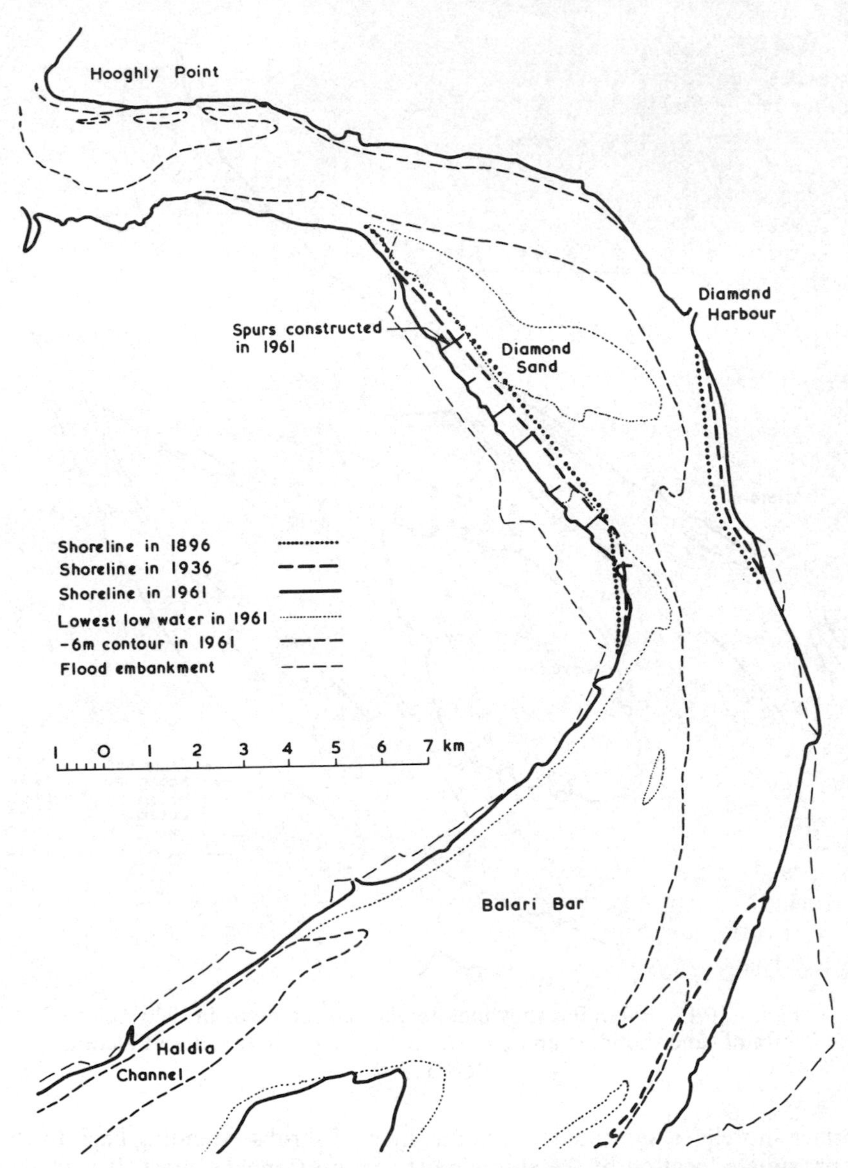

10.16. Diamond Harbour Reach and Balari Bar, showing shorelines between 1896 and 1961 and training works (spurs) as completed in 1961 (courtesy Calcutta Port Trust)

impossible, so the exact sequence of cause and effect could not be determined. However, it seemed certain that continued erosion of the Diamond Sand shore would be harmful, whereas restoration of the shoreline of 1889 would be beneficial.

Tests were done on the 1:2000 horizontal scale model at the Central Water and Power Research Station, Poona. The Haldia and Diamond Harbour reaches were moulded with mobile material to the 1960 contours. Operation of the model confirmed the unfavourable situation of Balari Bar. The Diamond Sand shore was then modelled to approximately the 1896 location, but the rest of the bed moulded to the 1960 contours. Operation of the model resulted in development of a new channel over Balari Bar, on a favourable alignment and with improved depths. Finally, tests were done with a series of spurs extending 600 m from the shore, with their crests near datum level (lowest low water). This also led to improvement of the Bar. It was decided to use spurs to stabilise the shore. Preparations were made during the final model tests so that when they were completed in December 1960, work could begin almost immediately. By the start of the monsoon in June 1961, one spur near the eastern end was nearly complete to its full length of 600 m, shorter spurs were completed at each end of the 9 km shore, and work on other spurs had begun. All the spurs were made from bricks made on the spot from baked estuary mud. Construction ceased temporarily with the onset of the monsoon. A cutter suction dredger then began to cut a channel on a new alignment close to that indicated by the model, but even as it worked, natural erosion occurred and it ceased operations after 6 weeks. By the end of the monsoon, the newly aligned channel was opened. Surveys revealed that over $3{\cdot}8 \times 10^6$ m^3 of accretion had occurred round the new spurs, while erosion of the shoreline had ceased.

Only the salient features of a complex situation are outlined here, but they serve to illustrate several points: the use of historical data to determine past development including evidence of cyclic changes; the interaction of neighbouring reaches of an estuary; the use of physical models as design tools; and finally the use of spurs to control erosion and to guide currents.

Control of one part of a complex system cannot be expected to result in control of the whole system. In the case of Balari Bar, a temporary respite was obtained – for several years, in fact – giving time for more complete study of the situation. Additional training works around Nayachara have since been constructed and the Bar is well under control.

NOTATION

A_{mtl} Cross-sectional area at mean tide level
P' Tidal prism
T Tidal period
U_{st} Stability velocity for an estuary

REFERENCES

[1] Cashin, J. A., Engineering works for the improvement of the estuary of the Mersey, *J. Inst. Civil Eng.*, **32** (1949)

[2] Agar, M. and McDowell, D. M., The sea approaches to the port of Liverpool, *Proc. Inst. Civil Eng.*, **49**, June (1971)

[3] Ramster, J. W., *Out of sight out of mind*, Appendix 6, vol. **2**, Dept. of the Environment, HMSO, London (1972)

[4] Heaps, N. S., Estimation of density currents in the Liverpool Bay area of the Irish Sea, *Geo. J. Roy. Ast. Soc.*, **30** (1972)
[5] Price, W. A. and Kendrick, M. P., Field and model investigation into the reasons for siltation in the Mersey Estuary. *Proc. Inst. Civil Eng.*, **24** May (1963)
[6] O'Connor, B. A., Discussion on Session B, *Dredging Symposium, Proc. Inst. Civil Eng.*, London (1968)
[7] Inglis, C. C. and Allen, F. H., The regimen of the Thames estuary as affected by currents, salinities and river flow, *Proc. Inst. Civil Eng.*, **7** (1957)
[8] Inglis, C. C. and Kestner, F. J. T., The long term effects of training walls, reclamation and dredging on an estuary, *Proc. Inst. Civil Eng.*, **9** (1958)
[9] Halliwell, A. R., Discussion on ref. 2, *Proc. Inst. Civil Eng.*, **51**, March (1972)
[10] Kestner, F. J. T., Short term changes in the distribution of fine sediments in estuaries, *Proc. Inst. Civil Eng.*, **19**, June (1961)
[11] McGregor, R. C., The influence of topography and pressure gradients on shoaling in a tidal estuary, *Geo. J. Roy. Ast. Soc.* **25** (1971)
[12] Halliwell, A. R. and O'Connor, B. A., Quantifying spoil disposal practices, *Proc. 14th Conf. on Coastal Engineering*, Ch. 153, Vol. III, June (1974)
[13] Bhattacharya, S. K., Deltaic activity of Bhagirathi–Hooghly river system, *Proc. ASCE*, WW1, **99** (1973)
[14] Mookerjea, D., Farakka Barrage project; a challenge to engineers, Paper 7725, *Proc. Inst. Civil Eng.*, **58** (1975)
[15] Gole, C. V. and Thakar, V. S., Progressive salinity intrusion during the dry season in the River Hooghly, Paper C-30, *13th Congress of IAHR*, Kyoto (1969)
[16] Joglekar, D. V., Ghotankar, S. T. and Chitale, S. V., Hydraulic model investigations of the Hooghly to improve its navigability, Paper A-4, *6th Congress of IAHR*, The Hague (1955)
[17] Shanmugam, A. T., Some aspects of sediment characteristics in tidal rivers and branches, *Proc. 10th Conf. on Coastal Engineering*, Sec. 4 (1966)
[18] Van Abel, Tj. H. and Pool. D. M., Sources of recent sediments in the Northern Gulf of Mexico, *J. Sed. Petrology*, **30**, no. 1, pp. 91–122 (1960)

Index

Abrasion 87, 118
Acceleration 32, 41, 43
  convective 43, 151, 163, 165, 168, 171
  due to gravity 66, 87
  flow *see* Flow acceleration
  spatial 50, 100
  temporal 49, 50, 100, 163
Accretion 117, 126, 267
Advection 54, 77, 178, 192
Aluminium salts 103
Amazon 7, 13
Amelsandse Gat 114
Ammoniacal nitrogen 179, 190
Amphidromic points 26, 27
Anti-dunes 91, 92
Arrested saline wedge 68–72

Bagnold–Bruk equation 217, 220
Balari Bar 274–277
Banks
  erosion 9, 84, 198, 227–229, 233, 234, 274, 275
  protection 228
  training 229–232, 235
Bar, underwater 251, 252, 257–261
Barrage 222, 238
  tidal 75, 152, 154
Bay of Bengal 269–274
Bay of Fundy 154
Bays 25, 60, 68, 250–256, 264, 269
Bed contours 251, 266
Bed forms 37
  anti-dunes 91, 92
  classification 91, 92, 93
  development time 94, 95, 212, 222
  dunes 90, 91, 212–214
  effect on bed-load transport 90, 93, 100
  effect on vertical sediment transfer 100
  formation 90
  Ribble estuary 92
  ripples 90–93, 112, 212–214
  size 93–95
  statistical distributions 93, 94
Bed friction
  due to bed forms 91–96, 171
    influence of waves 112
Bed irregularities 17, 49
  levels 96, 129, 142, 143
  load *see* Sediment
  material (in models) 213–221
  rigid 102
  sampling programme 118, 273
  shear stress 50, 89, 90, 105–109, 112
  slope 15, 44, 68, 71, 72, 111, 113, 149, 157, 258, 259, 263, 265, 268
Bends 21, 37, 40, 58, 252, 255
Bingham fluid 104
Biochemical Oxygen Demand 147, 179
  equations for distribution 179–181
Boundary conditions 45, 76, 97, 100, 102, 146, 161, 168, 169, 222
  finite-difference methods 162, 163, 166, 168
  method of characteristics 156, 158
Boundary layer 115, 116
Bristol Channel 151
Brouwershavense Gat (model) 207, 211
Brownian motion 17, 18, 109

Calcium carbonate 86
Calcutta 44, 270–273, *see also* Hooghly estuary and river
Capacity 253, 254
Case histories 250–277
Celerity, shallow water wave 5, *see also* Tidal wave characteristics
Centrifugal force 2, 3, 55, 61, 68, 113, 115, 149, 258
Channels
  braided 113
  control of 225
  curved 21, 65, 128, 236
  deepening of 225–227
  ebb-dominated 21, 226, 267, 269, 271, 274
  embayment 60
  flood-dominated 21, 22, 92, 226, 230, 255, 256, 261, 265, 267, 271, 274
  mobility of 113, 272, 273
  navigation 72, 75, 83, 112, 113, 117, 118, 245, 246, 250
  port approach 83, 96
  stability of 113, 116, 268
Chao Phya estuary 85
  minimum deposition shear stress 109
  sediment transport characteristics 110
Characteristic lines 155
  method for tidal analysis 155–159, 168
  velocity 155–157
Charts, survey 118, 126, 129, 274
Chesapeake Bay 110
Chezy's coefficient 37, 44, 114, 115, 169, 171, 201, 202
Chlorite 18, 103
Circulation patterns 147, 180
  Coriolis-induced 67
  induced by curvature 236
  Irish Sea 254
  Liverpool Bay 255, 256
  Mersey estuary, sediment 259
  salinity/density 115, 116
  salt wedge 70
  Tees estuary 71
Classification, estuaries 16, 44
Clay 17, 18, 20, 85–88, 103, 214, 215, 228, 230, 236, 239, 247, 265, 268
Clay-sized particles 17–20, 51, 214
Clyde estuary 158, 159
Coastal bays 60
Coastal currents 85
Coastal deposits 85
Coastal waters 13, 35, 42, 48, 111, 177, 178, 193
Coastal zone 60, 67, 75, 149, 154, 178
Coefficients
  decay rate, *see* Decay rate coefficients
  dynamic viscosity 56, 57
  eddy viscosity 56, 115, *see also* Diffusion coefficient, eddy
  energy loss 209
  entrainment 68–71
  exchange 188, 189, 191–193
  friction, *see* Friction coefficient
  heat-exchange 193
  kinematic viscosity 42, 67, 89
  molecular diffusion 53
Cohesive sediments 20, 103–112, 118, 228, 268, 269
Computers 27, 146, 156, 158, 168, 169, 172, 173, 178, 193
  analogue 125, 173
  digital 67, 172, 173, 177
  round-off errors 147, 166
Concentration, equilibrium, tidal 100, 101
  instantaneous 51, 103
  non-equilibrium 100, 101
  profiles, Chao Phya estuary 110
    equations for 97, 98
    Gladstone Lock 99
    Mersey Narrows 101
    non-uniform bed deposit 102
    stratified flow 102
    Upper Chesapeake Bay 108
  turbulent fluctuation 51, 56
  turbulent-mean 51, 56, 97–103
Conservative substance 53, 58, 60, 63, 178, 188
Continuity, conservation of 15
Continuity equation for fluid 33, 37, 39
  for other quantities 53–64, 70, 77, 96–98, 179–193
Control of estuaries 225–249
Control volume 53, 54, 117, 118
Convection definition 54
Convergence
  characteristic method 156
  finite-difference method 166, 167
Coriolis effects 25–27, 31–33, 43, 61, 67, 68, 113, 115, 125, 148, 149, 171, 187, 198, 206, 207, 222
Correlation technique 177

Correlation velocity/grain-size 87
C_i' method 101, 102
Cubature 160, 161, 267
applied to Hooghly estuary 44
applied to Mersey estuary 161
for friction coefficients 171
Current meter 49, 134, 257
Currents
coastal 85
Coriolis-induced 55, 68, 113
density, *see* Density currents
ocean 84
residual 72, 85, *see also* Tidal residual flow
tidal, *see* Tidal currents; Tidal velocities
wave-induced 22, 24, 111, 112
wind-induced 22, 24, 67
Curvature, channel 21, 65, 128, 236

Darcy–Weisbach friction coefficient 95, 96, 202, 210–212
Datum, Ordnance 251
Liverpool Bay 251
Decay rate coefficients 54, 64
algal growth 179
biochemical oxidation 179, 186
general pollutant 182, 184
photosynthesis 179
re-aeration 179
sedimentation and absorption on bed deposits 179, 186
Dee estuary 19, 83, 85, 220, 250
Deformation factor analysis 167
Delaware estuary 77, 84
cross-sectional area 184
dredging 244–246
physical model 184
tidal analysis 154
Deltaworks 115, 173
Demerara estuary 85
Density currents 68, 115, 243, 245, 254, 256, 268, 269
Density, fluid 13, 15, 24, 31–36, 39–45, 50, 51, 53, 65–67, 72, 128, 129, 177, 182, 184, 198
Density gauge, radioactive 118
Density gradients 13, 15, 17, 22, 39–45, 55–61, 65–68, 70–72, 99, 115, 238, 254, 258
Density, sediment 86, 87, 93, 120, 220, 246, 247
Density term in equation of motion 39–42, 44, 116, 148
Deposition of sediment 86, 87
Diffusion–advection equation 77, 177, 185, 186, 192
cross-sectional-average 61, *see also* One-dimensional models
depth-average 60, 61
engineering approximations 58–64
general form 58
lateral average 58–60
pollutant distribution 75, 179–193
slack-tide 63, 64, *see also* One-dimensional models
suspended sediment 97, 98
tidal-average 63
Diffusion coefficient, eddy 56, 58, 60, 65, 98, 111, 125, 183
coastal zone 67
molecular 53
thermal 54
momentum 65, 66
radial (in ocean) 67
salt 66
turbulent 20
variability of 64
Diffusion, numerical 167, 186
Diffusion processes 50, 53, 67, 77, 187
velocity 67, 75
Diffusive mixing 66
transport terms 56, 57, 59–61
Dimensional analysis 198, 202
Discharge
freshwater 63, 72, 148, 149, 156, 169, 180, 181
pollutant 58, 64
river 73
see also Freshwater discharges
Dispersion coefficients
cross-sectional-average (one-dimensional) 61, 69, 73, 74, 76, 179, 180, 183, 184, 186
depth-average 60, 61
equation for one-dimensional form 76
equivalence of exchange coefficients 189, 191, 192
field tests for 74, 183, 184
in homogeneous tidal zone 75, 76, 183, 184
lateral-average 59, 60
magnitude 65
pollution problems 177, 179–184

Dispersion coefficients (*cont.*)
slack-tide, one-dimensional 63, 64
steady-state, one-dimensional 64, 73
stratified situations 73
Tay estuary 73
tidal-average, one-dimensional 63, 64, 73, 74
Dispersion mechanism 61
Dissolved mineral salts 77, 84, 147
Dissolved oxygen content of water 147 179, 180
one-dimensional equation 179
steady-state 180, 182
saturation concentration 179
Thames estuary 190, 193
Distortion, of tide curves 265
Distortion scale, in models 197, 212, 215, 221
Diurnal components 3–5, 149, 154, 169
Diurnal inequality 49
Diurnal tides 3–5
Drag on cables/instruments 133, 134
Dredged channel 102, 140, 228, 251, 270
Dredged sediment 83, 252
equation for source 117, 118
records 118
volume correction factor 118
Dredgers
bucket 240, 257
cutter suction 245, 277
hopper 240–247
trailer suction 240–257, 260
stationary, sand pump 240, 251, 252, 255, 257
suction head, drag head 240–243, 247
Dredging 18, 19, 68, 115, 117, 228, 232, 239–247, 274, 277
agitation 243, 251
behaviour of fine sediments 19, 109, 157
maintenance 117, 237, 253, 257
spoil disposal and dumping 19, 117, 240, 243–246, 252, 255, 263
Dredging operations 104, 250
Eastham channel, Mersey estuary 258–261
Liverpool Bay 251–257
low water channel 265
Thames estuary 261–264
Dubai 16
Dunes 17, 90–96, 132, 135, 137, 139, 141, 142, 212, 213, 214
Dye tests 75, 180, 184, *see also* Tracer tests
Dynamic equilibrium 226, 227
Dynamic model 192
Dynamic viscosity 56, 67

Ebb predominance 21, 72, 92, 226, 230, 260, 267
definition 72
location of shoals 116
Savannah estuary 73
Echo sounder 19, 118, 129–132, 142
Eddies 67
entrainment of sediment 106
formation 49, 50, 67
motion 49, 87, 96
reduction process 50
size 49, 50, 67
Eddy diffusivity 56, 60, *see also* Coefficient of eddy viscosity; Diffusion coefficient, eddy
viscosity coefficient 115, *see also* Coefficient of eddy viscosity
Effluents 48, 84
Einstein bed-load formula 216
Einstein–Brown bed-load formula 216
scale for sediment transport 218, 219, 220
Einstein transport theory for sediment 100
Electromagnetic sensor 56
Energy loss 15, 164
coefficient of 209
due to bed resistance 91
mixing by wind/wave action 67
spread in a surge wave 157
transfer process 67, 171
Engineering problems 58, 61, 62, 111
Engineering schemes 48, 75, 117, 148
analysis by models 52, 146–223
effect on estuary stability 117, 118, 250–277
effect on flow structure 50–52, 77
equations for dissolved/suspended matter distribution 53–64, 96–98, 179–186, 188–189, 191
English Channel 26, 151
Entrainment 15, 66, 67, 71, 77
coefficient 68–71
of sediment 96, 100, 101, 103, 106
Equations
bed-load transport 96, 148

Equations (*cont.*)
continuity 33, 45, 148, 161, 173
cross-sectional-average 39, 148, 162, 164, 165
depth-average 37, 171
distribution of dissolved salts, pollutants 53–78, 177–193
ideal estuary 114
mass-balance 117, 118, 238, 246
momentum 30–33, 45, 148, 161, 173
cross-sectional-average 37, 38, 148, 162, 164, 165
depth-average 35, 36, 149, 171
salt balance 188
suspended sediment 96–98, 103, 148
tidal analysis 148–173
Equilibrium concentration 98, 100, 101
Equilibrium, daily (of an estuary) 115
Equinoxes 3
Erosion 84–90, 98, 100, 101, 104–107, 111, 112, 117, 226, 259, 260, 267, 268
bank 9, 84, 198, 227–229, 233, 234, 274, 275
due to low water channel movements 260, 261
of discrete sediments 87–90, 98, 100, 101
of offshore sediments 84, 252, 269
rate, mud 106, 107
time for equilibrium concentrations 101
velocity correlations 105
waves 112
Estuaries
case histories 250–278
classification 16
control of 225–249
entrance/mouth 2, 5, 9, 13, 18, 21, 22, 24, 44, 58, 68, 71–73, 77, 92, 111, 114, 115, 117, 149, 150, 267, 268
field measurements for 124–145
general behaviour 1–29
geometry 21, 65, 67, 114, 115, 149, 158, 160, 164, 166, 168, 169, 180, 181, 186, 188
constant area 180, 181, 186
depths 72, 150–152, 154, 183
effect on tidal-residual flow 71, 72
exponential width 9, 114, 150, 154, 185
length 76, 77, 114, 151, 154, 188
parallel-sided, flat bed 72, 149, 152
simple-shaped 62, 76, 149, 154
hydraulic models 197–224, *see also* Hydraulic models; Models, physical
ideal 114, 115, 152
internal flow structure 67–73, 147
landward/upper reaches 9, 70, 71, 73, 76, 85, 92, 94, 111, 168, 267, 268
mathematical tidal models 146–196
middle reaches 9, 13, 70, 71, 85, 258, 267, 268
mixing processes 48–82, *see also* Mixing
number 16, 76, 77
partially mixed 15, 58, 66, 70–74
seaward reach 71, 72, 77, 267
sediment movements 83–123
shoals 84
location 116, 118
stability, *see* Stability
stratified 14–16, 22, 64, 102, 148, 149, 193
tidal dynamics 30–47
water quality models 177–193
well-mixed 14, 15, 20, 21, 35, 40, 65, 66, 99–101, 107, 108, 193, 258, 265
Eulerian average 62, 116
Eulerian–Lagrangian approach 62
Exaggeration of scale, *see* Scale distortion

Fall diameter 93
Fall velocity 54, 93, 97–101, 106, 107, 111
Felspar sediments 86
Fick's Law 53, 56
Field data 76, 83, 89, 146, 173, 208, 217
Field measurements 5, 24, 30, 38, 39, 111, 124–125, 147, 217, 218, 225, 272, 273
Field observations 83, 96, 119, 187, 192, 254
Field surveys 259, 260
Finite-difference method 159–171, 185–187
Crank–Nicholson form 186, 191
errors in 157, 159, 166–168, 185–187
explicit form 161–164, 166, 168, 170, 171, 185
for mixed-segment models 191
for tidal analysis 159–171
for water-quality models 185–187, 193

Finite-difference method (*cont.*)
implicit form 161, 164–166, 168, 170, 171, 185–187
Lax–Wendroff form 168
leap-frog form 164
order of 159, 160, 167, 168, 185–187
stability of 163, 166, 168, 170, 171
Finite element techniques 172, 185, 186, 193
First law of thermodynamics 54
Float tracks 260
Floc 17, 18, 20, 86, 103, 118, 214, 226, 239
aggregate/group 103, 106
size 103, 107
Flocculated sediments 103–105, 109, 111
Flood, routing 157
Flow acceleration 50, 71, 90, 100
Flow deceleration 50, 71, 90, 100
Flow depth 98, 99, 107, 166, 187, 267
Flow field 55, 56, 149, 177
Flow kinetic energy 21, 38
Flow processes 48
Flow resistance 90, 91, 95, 118
due to bed form shape 91
due to grains 91, 95
effect of suspended sediment 51
in Liverpool Bay 255, 257
Flow structure 49–52, 148
Fluctuations
in density 66
in freshwater flows 115
in hydrodynamic force 87
in pressure 57
in velocity 34, 54, 66, 171
Fluid density 13, 15, 24, 31–36, 39–45, 51, 53, 128, 129, 177, 198
Fluid mud 109
Flume tests 24, 76, 107, 109
Forces
body 31–33
centrifugal 2, 3, 55, 61, 68, 113, 115, 149, 258
cohesive 20
Coriolis 31, 55, 61, 68, 113, 115, 149, 258
density 71–73, 116
drag 88, 89
electrostatic/electrochemical 104
gravitational 2, 3, 31, 35, 54
hydrodynamic 87
lift 88, 89
on sediment 87–89, 104
seepage 88, 90
turbulent 109
viscous 87, 89
Form drag 91
Forth estuary 220
Fourier components 166, 186
Freshets 18, 19, 91, 106
Freshwater 18, 243, 271
discharges 2, 63, 72, 85, 148, 149, 156, 169, 180, 181, 250, 258
flows 13–17, 23, 42, 49, 67, 68, 72, 85, 92, 111–118, 142–144, 258, 269, 271
effect on estuary stability 117
effect on tidal-residual flows 72, 117
in mixed-segment models 189, 191
Friction 25, 37, 38, 41, 42, 149, 150, 210, 212
coefficient of 91, 95, 164, 211, 213, *see also* Chezy's coefficient; Darcy–Weisbach's coefficient; Manning's friction coefficient
in hydraulic models 92, 199–202, 208–210, 212, 213
in mathematical models 92, 125, 150–171
Frijlink's equation, for sediment transport 219

Galerkin finite-element method 172, 185, 186
Ganga, river 269–274
Gaussian distribution 182, 184
Gironde estuary 85, 107, 211, 220
Grain drag 91
Grain size 86, 88, 89, 93, 100, 101, 212
distribution curve 87
effective 92, 107
friction factor 95
Gravel 2, 11, 17, 85–87, 226
Gravity 2, 3, 5, 6, 31, 32, 35, 36, 40
Green's law 154
Groynes 233
permeable 226
Guide banks 225

Harmonic constituents 3, 4
Harmonic method 154
Heat 49, 54, 57, 77

High water 43, 95, 149, 152, 169, 188, 254, 263, 267
  phase-lag 152
Hooghly estuary and river 6–11, 38, 43, 161, 269–274
  Balari Bar, improvement 274–277
  bank erosion 228, 229
  bed-load transport 9, 10
  channel migration 272
  curvature effects 128, 236
  freshwater flows 143, 271
  friction coefficients 38, 171, 210, 211
  heavy minerals as tracers 273, 274
  magnitude of terms in equations 43, 44
  mathematical modelling 170, 171
  physical models 273, 276, 277
  tidal discharge, ideal 114
  tides 6–11, 16, 271
  velocities 11, 170–172
  volume, seasonal changes 143
Horneblende 272, 274
Hudson estuary 77
Humber estuary 9, 84, 116
  Grimsby middle shoal 116
  Immingham 235
Hydraulic models 25, 27, 62, 76, 83, 96, 173, 178
  Brouwershavense Gat 207, 211
  Hooghly estuary and river 273, 276, 277
  in engineering schemes 52, 75, 119, 146
  Liverpool Bay training scheme 253, 257
  Loire estuary 220
  Mersey estuary 254, 257–259
  Thames estuary 263
  with mathematical models 75, 146
Hydraulic radius 37, 76, 148, 155, 163, 169, 202, 210, 211, 218, 227
Hydrodynamic source/sink 177

Ideal tidal-averaging period 70
Illite 103
Instruments, positioning and location 133–135
Interface 40, 41, 158
Intrusion length 24, 68–72
Intrusion, saline 23, 24
Ionic bond strength 104, 106
Irish Sea 13, 22, 26, 75, 85, 151, 245
  Liverpool Bay 22, 250–257
  Morecambe Bay 264, 269
  Near-bed tidal-residual flows 254
  wave action 250
Iron salts 103
Isohalines 43, 254
Iterative procedure 156, 157, 163, 166

Jamaica Bay 187
James estuary/river 66, 180
  BOD/DO distributions 181
Jet entrainment theory 177
Jetties 169, 222, 235, 236, 238

Kalinske bed-load formula 216
  scaling 218–220
Kaolinite 18, 103, 109

Laminar flow 42, 204
Laminar sub-layer 49, 89, 90, 106
Laser 56
Layers
  lower 66–72, 148
  salt 71
  upper 66–72, 148
Leibnitz's rule 59, 60
Liverpool 13, 237
Liverpool Bar 250–252
Liverpool Bay 19, 22, 42, 142, 250
  Askew Spit 251, 252, 255
  Crosby Channel 250, 252, 255, 256
  datum 251
  Formby Channel 251, 253, 257
  Great Burbo Bank 253, 256
  near-bed tidal-residual flows 254–256
  Queen's Channel 250
  Rock Channel 251, 253, 256
  spoil disposal sites 252, 256, 257
  Taylor's Bank 252, 253, 255
  Taylor's Spit 253
  training scheme 250–257
Loire Estuary model 220
Low water 7, 21, 24, 43, 88, 258, 261, 265
Low-water channel 250, 259, 260, 265, 267, 268
Low-water slack 19, 100, 192

Lune estuary 247, 248, 251, 264–269
  accretion following training 267
  ideal estuary relations 114, 115
  tidal changes 247, 265, 267
  training works 265–267

Madras 85
Magnesium salts 103
Magnetite 86
Manning's friction coefficient 37, 38, 76
Mare Island Strait 109
Mass continuity principle 53, 147, 177, 187, 188, 189, 193
Mass transport 67, 111, 112, 254
Mathematical models 52, 92, 111, 146, 173
  advantages/disadvantages 146, 173
  calibration 92, 147, 168–171
  comparison with physical models 147, 192
  for combined tidal/water quality situations 146, 186, 187
  for sediment transport 148
  for stratified estuaries 64, 148, 149, 187, 193
  for tidal situations 148–173, 186
  for water quality problems 177–193
  geometrical data 169
  mixed-segment 188, 189, 191–193
  particular examples 147, 148, 177, 178
  schematization 169, 186
Measurements 124–145
  simultaneous 126, 127
Median grain size 87, 250, 258, 265
Mersey estuary 12, 16, 19, 43, 45, 139, 246, 250
  Bromborough 86
  Bromborough Bar 257, 258, 260
  capacity 142, 253, 261
  case histories 238, 245, 250–264
  dredging figures 253
  Eastham Channel 12, 257–260
  fresh water flows 250, 258
  Garston Bar 161, 258
  Garston Channel 258, 260, 261
  hydraulic models of 22, 197, 220, 254, 257, 258, 260
  ideal estuary relationship 114, 115
  isohalines 43
  Middle Deep Channel 258, 259
  Narrows 11, 65, 66, 74, 100, 101, 161, 250, 252, 258
  position of low water channel 259
  residual sediment movement 258, 259
  silt content of water 9
  tidal curves 7, 9
  tidal data 12, 250
Meteorological effects on tides 5, 125
Minerals
  heavy 86, 272–274
  sediment 86, 103
Mississippi river 15, 16, 69
Mixing 16, 23, 24, 35, 125, 182, 188
  eddy size 49, 50, 67
  effects
    of neutrally buoyant particles 51
    of pollutant discharges 51
    of suspended sediment 51, 52
    of turbulent mean velocity gradients 55
    of vertical density gradients 50, 66
  engineering approximation for equations 58–64
  flow structure 49–52
  general equations for 53–56
  time 182, 183
  variability of diffusion/dispersion coefficients 64–74
  wind and wave action 49, 55, 56, 60, 65, 67, 68
Models, physical 22, 27, 125, 193, 197–224, 233, 257, 273, 276, 277, *see also* Hydraulic models; Mathematical models
  calibration 213, 217, 218
  fixed-bed (rigid-bed) 207, 214
  mobile-bed 207, 213–222, 223, 257, 273, 276, 277
Momentum 25, 30, 158, 225
  correction factor 57, 58, 98
  equations, *see* Equations, momentum
  exchange from tidal flats 164
  of velocity fluctuations 51, 56
  term 116
  transfer 52, 57, 64–66, 98
Monitoring 127, 144
Monsoon 269, 271, 273, 277
Montmorillonite 18, 103
Moody diagram 210
Mud 18, 37, 85, 107, 263
  flats 20
  flows 105, 106
  fluid 109

Mud (*cont.*)
layers 130, 131
pellets 106
physical characteristics of 86

Navigation channels 72, 75, 83, 112, 113, 117, 118, 250, 255, 257
deepening of 72, 75, 225–227, 257
stability of 113, 252, 253
Net movement 18, 19, 21, 22, 115
landward 19, 22, 42, 45, 115
seaward 22, 42, 45, 115
Newtonian behaviour 104
Non-conservative substance 54, 64, 178, 179
North Sea 93, 117
Null point, *see* Zero net movement; Zero tidal-average flow
Numbers
estuary 16, 76, 77
Froude 16, 90–92, 94
Reynolds 52, 68, 199, 210, 211
grain size 89
scale 211
Richardson 65, 66, 100
Richardson flux 66
stratification 16, 77
Strouhal 201
Numerical analysis 30, 45, 96, 100
for tidal problems 27, 125, 149, 155–173
for water quality problems 185–187
truncation error 157, 159, 162, 163, 166, 167

Oceans 3, 13
barge disposal of sludge 84
currents 84
diffusion coefficient 67, 180
dispersal 64
mixing 51
outfalls 58, 64, 177, 180
salinity 69, 77, 188
tides 2–5, 149–151
One-dimensional models 148–172, 177, 178, 182, 183, 185, 187
Oosterschelde 114
Ordnance Datum 251
Organic carbon 179
Organic clay 104
Organic material 84–86
Organic nitrogen 179
Orinoco river 242
Orthogonal co-ordinates 31, 53
Oscillating water tunnel 111
Outfall 58, 64, 177, 180, 188
Oxidised nitrogen 179, 190, 193

Particles
density of 86, 188
shape of 86, 87, 118
size of 54, 57, 86, 107, 108, 118
statistical distributions of 87
submerged weight of 96
Patuxent river 75
Phase lag 114, 116, 152
Phase velocity 149, 150, 154, 155–157
Photosynthesis 179, 180
Physical models 30, 111, *see also* Hydraulic models; Models, physical
Piled structures 50, 109
Plasticity index 104
Plate river 242
Pollutants 58, 118, 144, 146, 188, 189
continuous injection of 64, 178, 180–182, 184, 185, 188–193
discharges 48, 51, 64, 73, 75, 177, 180, 184, 193
dispersal 75
distributions in estuaries 146, 178, 179, 181, 187, 188, 190, 191
effect on vertical stratification 74, 193
effluents 84, 263
heavy density 51, 64
instantaneous injection 182–184
intermittent flows 178
low density 51, 64
mathematical models 147, 177–193
mixing times of 182, 183
neutrally buoyant 179
outfalls 58, 64, 177, 180
source of sediments 84–86, 103
vertical line source 67
Porosity 87, 103
Port approach channels 83, 96
Port of Liverpool 153, 250
Port of Manchester 257
Potomac estuary 77
Pressure 53
Pressure gradient 39, 41, 45
Pressure intensity 31, 32, 35, 128
Principal axes 57

Pseudo-advection 185, 193
Pseudo-dispersion 185, 193

Quartz 52, 86, 103
Quasi-steady state 191

Reclamation schemes 48, 68, 75, 222
  effect on mixing 50, 68
Reservoirs 117
Resistance bed forms 212, 213
Resistance–capacitance networks 172
Resistance flow, *see* Friction; Flow resistance
Reynolds analogy 56, 57
Reynolds stresses 34, 203
Revetments 228–230, 234
Rhine Delta
  analogue model 173
  physical model 199, 211, 248
Ribble estuary 84, 92, 94, 96, 230, 248, 250
Ripples 17, 90, 112, 212, *see also* Bed forms
  resistance in models 212
River 1, 93, 94, 98, 99, 112, 117, 118, 157, 179, 269
River currents 84, 147
River discharge 13, 18, 23, 49, 70, 72–75, 126, 142–144, 260, 265
River flow 1, 5, 13, 17, 18, 23, 45, 117, 168, 265, *see also* Freshwater flows
  extreme 117, 260
  low 13, 74, 75, 260
River gauging records 168
River sediment supply 84, 85, 117, 263, 269
Rotterdam Waterway 24, 117
Roughness, bed 37, 93, 95, 116, 199
Roughness coefficient 76, 91, 171
Rupnarain river 233, 234, 270, 271

Saint Lawrence river 171
Saline penetration 22, 23, 117, 225, 227, 263, 271
Salinity 8, 15, 18, 23, 43, 53, 59, 74, 146, 192
  bed 66, 178
  distribution in salt wedge 70
  effects
    on mean tide level 164
    sediment erosion 106
  equation for estuarine distribution 53
  gradients 43, 45, 66, 67, 71, 72, 98, 100, 116, 164, 198, 227, 263
  in models 198
  intrusion 23, 71, 72, 85, 111, 117, 245, 263
  shoal zones 111, 259, 263
  surface 66, 178
  used as tracer 74, 75
Salt 13, 53, 57, 66, 147
Salt marsh 20, 85, 267, 268
Salt water 13, 15, 17, 18, 22, 66, 188, 258
Salt wedge 67, 77
  circulation in 68–70
  effect of freshwater flows 72
  entrainment in 68–72
  interface 68
  length of 69, 70
  longitudinal salinity distribution 70
  sediment distribution in 102
Sand 17, 20–22, 24, 84–86, 88–90, 93, 94, 96, 97, 106, 110, 114, 118, 138, 226, 230, 247, 257, 265, 267, 268, 273
  sized particles 17, 20, 21, 24, 111, 236
San Francisco Bay 23, 85, 178
  mud 85, 104, 109
    limiting shear stress 109
Savannah estuary 72, 73, 84, 85
Scale 197–224
Scale distortion 197, 205, 209, 212
Scale effects 93, 147, 204–207
Scale models 197–224
Scour 126, 256
Sea-bed 25, 84, 254
  drifters 19, 254, 257
Seasonal effects 142–144
Seawater 53, 54, 55, 67
Secondary flows 41, 60, 228, 237
Sediment
  armouring effects 89
  bed load 20, 90, 96, 106, 112, 113, 118, 137, 148, 237, 238
  circulation patterns, tidal-average 258, 259
  concentration 8, 9, 18, 52, 97–103
  consolidation of 18, 88, 104, 105, 112, 118, 226
  distributions 98, 99, 101, 102, 169

Sediment (*cont.*)
entrainment 96, 100, 101, 103
established motion 90–103, 105–111
fine 17, 19, 21, 214, 226, 239
initial motion 87–90, 104, 105, 111
lightweight 86, 93
minerals 86, 103
movement 17–28, 83–118
littoral 85, 115, 118
weather effects 117, 118
ocean disposal 84, 117, 252
partially consolidated 104–106
properties 17, 51, 86, 87, 92, 103–105
saltation 90
sampling 118, 253
sources 83–85, 113, 117, 118, 246, 263, 270, 273, 274
suspended 20, 51, 137, 138, 148, 214, 215, 226, 239, 240
concentrations 68, 98, 100, 101
density gradients 67
equation for concentrations 54, 97, 98
equation for transport rate 103
load 90, 106, 113, 115, 117, 118, 267–269
rates 96, 113
scale 213–221
total load 96
transport 9, 11, 20, 21, 27, 125, 137, 138, 144, 148, 198, 213–221, 227, 230, 238, 273
discrete particles 86–103
flocculated particles 103–112
mathematical models 148
unconsolidated 104–112, 239, 240
wash load 18, 226
Sedimentation 24, 255
Severn estuary 19, 75, 85, 86, 188, 211
Portishead 86
Uskmouth 19
Sewage, sludge 84, 263
Shallow seas 147, 148, 171, 178
Shear force 55, 69, 149
Shear strength of deposits 106
Shear stress 20, 30–32, 34–36, 40, 41, 112–114, 135, 136, 138, 139, 144
bed 20, 50, 89, 90, 93, 107, 108, 116, 118, 150
critical erosion 86, 89, 215
effect of suspended-sediment/temperature 90
interfacial 149
limiting deposition value 108, 118
minimum value for deposition 109
similarity 200–206, 209, 210
stability 115
vertical distribution 56, 98, 99
viscous 49, 56
Shear vane 105, 106
Shear velocity 65, 93, 98, 99, 100, 101, 109, 112, 183
due to waves/currents 112
Shields' curve 89, 100, 111
entrainment function 217
Shoaling zones
salinity intrusion 72, 116, 263
tidal inertia and salinity effects 258
tidal inertia, effects of 116, 256, 263, 264
Shoals 9, 18, 22, 23, 37, 232, 251, 252, 255, 263
Silica 86, 103
Sillimanite 272, 274
Silt 17–20, 37, 51, 84–86, 88, 107, 109–112, 230, 239, 247, 251, 257, 268
Siltation 83, 96, 117, 261, 268, 269
Siltation rates 72, 251–253, 265, 267
Silt sized particles 17, 20, 21, 24, 236
Simulation methods for models
electrical analogue 173
hydraulic (physical) models 172, 173, *see also* Hydraulic models; Models, physical
mixed-segment approach 188–193
tidal prism concept 187, 188
Single-tidal-average approach 62, 70
Slack tide 18, 20, 24, 64, 94, 104, 105, 107, 108, 110
high water 101, 152, 192
low water 100, 192
Slack water 20, 178, 184
Solent 75
Sonar, side-scan 132, 144
Sounding 19, 126, 129–131
echo 19, 130–132, 144
line 130
Southampton Water 75
Specific weights
of fluid 216, 217
of sediments 96
Spoil
disposal of 19, 83, 85, 253, 264
dredged 83, 84, 252, 256
grounds/sites 83, 85, 252, 255, 263, 264

Spurs 222, 225, 233–235, 277
Stability
of estuarine channels 113, 116, 268
of finite-difference schemes 163, 166, 168, 170, 171
Stability velocity 115, 268, 269
Standard deviation
of longitudinal concentration curves 184
of velocity fluctuations 50, 98
Statistical functions for pollution models 189
Steady-state approach 63, 64, 100, 178, 180–182, 187, 188, 189, 193
Stochastic process 177
Storm surge 147
Stratification 16, 66, 73, 77, 107, 178, 180
Stream 164
Stream power 93, 216, 217, 226
Surveys, bed level 129–132, 142, 143, 271, 273, 277
Suspension exponent 98, 99, 101

Tay estuary 73–75, 219
Taylor series 159
Tees estuary 58–60, 66, 70, 71, 75, 220
Temperature 53, 147, 177, 179
effects
on bed forms 94
on particle fall velocity 99
on turbulent-transport rates 56, 98, 99
equation for distribution 54
simulation models for 190, 193
Thames estuary 9, 43, 44, 75, 84, 189, 192, 211
analytical tidal model 151–153
Black Deep 262, 263
distribution of dissolved quantities 190, 192
dredging 102, 261–264
effect of salinity on mean tide level 164
erosion constant for mud 107
Gravesend Reach 262, 263
hydraulic model of 263
ideal estuary equations 114, 152
Mud Reaches 116, 262, 263
radioactive tracer tests 263
sediment sources 85, 263
stability criterion 116
Teddington Weir 151, 261, 262
tidal excursion 188
Thermal conductivity 54
Thixotropic behaviour 104
Three-dimensional effects 21, 22, 30, 197, 198, 226
Three-dimensional flows 21, 22, 58, 67, 77
Three-dimensional mathematical models 148
Three-dimensional process 49
Three-dimensional sediment waves 90
Three-dimensional spreading action of eddies 77
Thyborøn Inlet 114
Tidal amplitude 3, 4, 26, 114, 154
Tidal analysis
by analytical models 149–154
by characteristics 155–158, 168
by finite difference methods 159–171
by finite element techniques 172
simulation methods 172, 173, *see also* Hydraulic models; Models, physical; Mathematical models
types of model 148, 149
Tidal-averaging 62, 63, 113, 116, 178
Tidal bore 43, 157, 158, 164, 168, 271
Tidal celerity 5, 115, *see also* Tidal wave characteristics
Tidal constituent 149, 151, 154, 169
Tidal currents 1, 2, 5, 18, 21, 22, 24, 26, 71, 85, 91, 92, 96, 107, 112, 190, 273
Tidal cycle 17, 43, 58, 62, 72, 76, 94, 100, 104, 110, 113, 150, 178, 183
Tidal discharge 15, 114, 115, 154, 160, 161, 163, 267, 269
Tidal elevation 115, 116, 168
Tidal excursion 63, 107, 188, 255, 263
Tidal flats 60, 62, 112, 158, 164, 171
Tidal flow 24, 41, 45, 49, 56, 60, 91, 101, 113, 118, 146, 161, 178
residual 71, 72, 96, 115–117, 245–256, 258, 260
turbulent structure of 50, 52
Tidal force
lunar 2, 3
solar 3
Tidal inlet 27, 115, 158
Tidal level (period-mean) 16, 37, 72, 114, 149, 157
Tidal limit 1, 6, 28, 188
Tidal period 2, 3, 5, 16, 21, 34, 59, 62, 76, 87, 115, 149

Tidal period *(cont.)*
  ebb 6, 9, 115
  flood 5, 9, 115, 265, 267
Tidal predictions 4
Tidal prism 16, 148, 182, 188, 267, 268
Tidal propagation 5, 7, 26, 227, 228, 247
Tidal range 3, 5–7, 9, 16, 19, 26, 59, 91, 100, 113, 152, 169, 267
Tidal river 1, 9, 21, 55, 56, 63, 64, 68, 76, 83, 193
Tidal velocities 11, 13, 16, 68, 72, 77, 114–116, 149–164, 171, 184, 267, *see also* Tidal currents
Tidal volume 7, 15, 16, 115, 158, 267, 268, *see also* Tidal prism
Tidal wave characteristics 7, 26, 114, 116, 149, 150, 152, 154–156, 267
Tide curve 5, 9, 168, 169, 250, 265, 267
Tide, diurnal component 3, 4, 149, 154, 169
Tide, ebb 6, 9, 11, 20–27, 89–91, 94, 110, 247, 263, 267, 268
Tide, equinoctial 3
Tide, flood 6, 9, 11, 20–27, 89–92, 94, 100, 110, 157, 255, 263, 267, 268
Tide gauge 36, 38, 128, 129, 144, 160, 168, 222
Tide, generating force 3–5, 26
Tide generator 198, 222
Tide, harmonic components 3, 4, 149, 169, 261
Tide, meteorological effect on 5
Tide, neap 2, 3, 7, 8, 17, 19, 70, 113, 142, 143, 218, 239, 240
Tide, ocean 2–5, 149, 150, 151
Tide, semi-diurnal 3, 5, 115, 261
Tide, small amplitude 3, 5, 149–155
Tide, spring 2, 3, 7, 8, 16–19, 26, 43, 100, 115, 143, 218, 239, 240, 250, 258, 265, 267
Tracer tests 63, 74, 138–140
  fluorescent 78, 140
    for dispersion coefficients 75
    in Mersey estuary model 258
  heavy mineral 273
  radioactive 78
    for dispersion coefficients 75
    for sediment movements 140, 144, 263, 273
Training
  tidal-average circulation patterns 255, 256
  tidal discharges 267
  Liverpool Bay 250–257
  Lune estuary 264–269
  Ribble estuary 230–231
  works 117, 225, 239, 250, 252, 265, 276, 277
Transport scale 213–221
Turbulence 14, 15, 34, 45, 111, 118, 125, 136, 193, 203, 204
  modifications to 50–52, 99
  intensity 50–52, 100, 109
  time scale 34, 49, 62
Turbulent, diffusion 20, 54–58
  flow rate 50
    rough 42, 210
    smooth 210
  kinetic energy 51, 66, 67, 100, 187
  mixing 14, 16, 50, 70
  shear stress 34, 56
  similarity 203, 204
  structure of a flow 50, 52, 65, 107
Two-dimensional long crested waves 90
  mathematical models 148, 149, 158, 171, 172, 177, 178, 182, 183, 185–187

Upper Chesapeake Bay 85, 107, 108
Upper East river 181, 182
Upstream storage volume 63, 188–192

Velocity, critical erosion 87–89, 111, 112
Velocity, depth distribution, in salt wedge 69
Velocity, entrainment 67, 68
Velocity erosion correlation 105
Velocity field 49
Velocity, freshwater 63, 64, 178, 180–182, 184
Velocity, grain-size correlation 87, 88
Velocity head 128
Velocity, logarithmic depth-distribution 65, 76, 98, 99
Velocity, lower layer 72
Velocity measurement 133–135, 137–139
Velocity meter 49, *see also* Current meter
Velocity, parabolic depth distribution 56, 57

Velocity profile 41, 42, 76, 138, 139
reversed-flow type 73
Vertical mixing 70, 73, 100
stratification, degree of 65, 66, 73, 178
Viscometer 104
Viscosity 54, 99, 104, 106
dynamic 56, 57
kinematic 42, 67, 89
Volume concentration 52, 54
freshwater 188
inter-tidal 226
Von Karman's constant 65
vertical density gradients 100
Vortices 49, 56

Wash load 18, 226
Water quality 126, 147
problems 63, 177, 178, 187, 188
Wave action 2, 20, 55, 60, 68, 84, 85, 90, 105, 111, 126, 144, 147, 198, 221, 232, 250, 256, 273
Wave breaking 56, 111
Wave equation 149
Wave harmonic 149, 154
Wave height 111, 112, 167, 250
Wave length 3, 111, 167
Wave period 111, 112, 250
Wave phase 167
Wave/sediment movement correlation 111
Wave speed 157
Waves
bed forms due to 112
critical erosion velocity of 111
internal 66, 68, 72, 77
mass transporting action 67, 111, 112, 254
ocean wind 94, 180
orbital bed velocities 111, 112
progressive 114, 116, 152
damped 152
reflected 150, 157
standing 63, 150, 152
Wave, swell 111
Wear estuary 76, 187
Weser estuary 95
Westerschelde 114
Wind 36, 116, 125, 144, 198
action of 49, 55, 60, 65, 67, 68, 147, 149, 180
Wind erosion of dunes/sand-banks 84
Wind induced currents 67
residual flows 254
shear forces 55, 171, 187, 254
Wyre estuary 220

Zero bed slope 68, 72, 149, 152
Zero freshwater flow 13, 68
Zero net movement 13, 18, 19, 38, 115, 116, 256
Zero tidal-average flow 72
movement 258
Zircon 86, 272, 274
Zones
accretion 117, 118
coastal 60, 67, 149, 154, 178
freshwater flow 178
flood-dominant 256
homogeneous
estuarine 76, 77, 116, 181
tidal 17, 73, 75, 76, 168, 177, 178, 180, 181, 183, 184
water quality models for, 180, 182–184
near-bed 100
near-shore 22, 67, 178, 254
nodal shoaling 148
non-homogeneous estuarine 17, 73, 76, 177, 178, 180, 186
offshore 254
saline intrusion 71, 117
seaward estuarine 77
upper freshwater 69
Zooplankton 84, 85